# Thermodynamic Cycles for Renewable Energy Technologies

# IOP Series in Renewable and Sustainable Power

The IOP Series in Renewable and Sustainable Power aims to bring together topics relating to renewable energy, from generation to transmission, storage, integration, and use patterns, with a particular focus on systems-level and interdisciplinary discussions. It is intended to provide a state-of-the-art resource for all researchers involved in the power conversation.

**Series Editor**
**Professor David Elliott**
*Open University, UK*

**About the Editor**
David Elliott is emeritus Professor of Technology Policy at the Open University, where he developed courses and research on technological innovation, focusing on renewable energy policy. Since retirement, he has continued to write extensively on that topic, including a series of books for IOP Publishing and a weekly blog post for *Physics World* (physicsworld.com/author/david-elliott)

**About the Series**
Renewable and sustainable energy systems offer the potential for long-term solutions to the world's growing energy needs, operating at a broad array of scales and technology levels. The IOP Series in Renewable and Sustainable Power aims to bring together topics relating to renewable energy, from generation to transmission, storage, integration, and use patterns, with a particular focus on systems-level and interdisciplinary discussions. It is intended to provide a state-of-the-art resource for all researchers involved in the power conversation.

We welcome proposals in all areas of renewable energy including (but not limited to): wind power, wave power, tidal power, hydroelectric power, PV/solar power, geothermal power, bioenergy, heating, grid balancing and integration, energy storage, energy efficiency, carbon capture, fuel cells, power to gas, electric/green transport, and energy saving and efficiency.

Authors are encouraged to take advantage of electronic publication through the use of colour, animations, video, data files, and interactive elements, all of which provide opportunities to enhance the reader experience.

**Do you have an idea for a book you'd like to explore?**
We are currently commissioning for the series; if you are interested in writing or editing a book please contact Caroline Mitchell at caroline.mitchell@ioppublishing.org.

# Thermodynamic Cycles for Renewable Energy Technologies

**Edited by**
**K R V Subramanian and Raji George**
*Department of Mechanical Engineering,*
*Ramaiah Institute of Technology,*
*Bengaluru, India*

**IOP** Publishing, Bristol, UK

ISBN    978-0-7503-3711-3 (ebook)
ISBN    978-0-7503-3709-0 (print)
ISBN    978-0-7503-3712-0 (myPrint)
ISBN    978-0-7503-3710-6 (mobi)

DOI    10.1088/978-0-7503-3711-3

Version: 20220101

IOP ebooks

British Library Cataloguing-in-Publication Data: A catalogue record for this book is available from the British Library.

Published by IOP Publishing, wholly owned by The Institute of Physics, London

IOP Publishing, Temple Circus, Temple Way, Bristol, BS1 6HG, UK

US Office: IOP Publishing, Inc., 190 North Independence Mall West, Suite 601, Philadelphia, PA 19106, USA

Cover caption: Solar energy: Solar One, Barstow, California, which consists of a circular arrangement of 1,818 mirrors, each measuring 23x23 feet (7x7 metres). These mirrors focus the sunlight onto a hugh central receiver, which sits atop a 300 foot (91 metre) tower. The mirrors are computer controlled to track the path of the sun. Water is pumped through the receiver and heated to a temperature of 960 degrees fahrenheit. The resultant steam runs a turbine, producing 10 megawatts of power for eight hours a day.

Cover image credit: PETER MENZEL/SCIENCE PHOTO LIBRARY

# Contents

## 3 Storage of electricity generated from the renewable sources using electrochemical energy conversion devices 3-1

*Roushan Nigam Ramnath Shaw, Ravi Sankannavar, G M Madhu, A Sarkar, K R V Subramanian and Raji George*

## 5   Waste heat recovery      5-1
*Prakriti Gupta and G M Madhu*

## 6   OTEC Rankine and Stirling engines      6-1
*B V Raghuvamshi Krishna*

# Preface

The role of specialized thermodynamic cycles in renewable energy technologies such as solar thermal, ocean energy, wind energy, geothermal energy looks at effective and innovative implementation of the cycle with maximum payback.

In thermodynamics, the Carnot cycle has been described as being the most efficient thermal cycle possible, wherein there are no heat losses and consisting of four reversible processes, two isothermal and two adiabatic. It has also been described as a cycle of expansion and compression of a reversible heat engine that does work with no loss of heat. Moreover, there are vast amounts of renewable energy sources such as solar, thermal, tidal waves, wind power, geothermal, biomass and industrial waste heat. The moderate temperature heat from these sources cannot be converted efficiently to electrical power by conventional power generation methods. Therefore, how to convert these low-grade temperature heat sources into electrical power is of great significance. Thus, comparison of different specialized thermodynamic cycles helps to combat and find an appropriate innovation in the cycle for conversion.

This book will deal with innovations in technology of the Rankine, Stirling, Brayton, Kalina, OTEC Rankine, Goswami cycles and their application to renewable energy area.

This book exposes the reader to the nuances and innovations in thermodynamic cycles related to renewable energies and reveals the parameters for maximum payback.

We, the editors, Subramanian and George have drawn upon our research works and interests to collate and critically present the chapters and provide an insightful reading experience to readers at the juncture of thermodynamic cycles and renewable energy technologies. We thank the management of Ramaiah Institute of Technology (Shri M.R. Seetharam, Vice Chairman GEF and Director; Dr. NVR Naidu, Principal) for their constant encouragement and support. We also thank our family members for their support in undertaking this endeavor.

**Professor K R V Subramanian**
**Professor Raji George**

# Editor biographies

## K R V Subramanian

Dr K.R.V. Subramanian is a professor in the Mechanical Engineering department and a research coordinator at Ramaiah Institute of Technology, Bangalore, India. He earned his PhD from Cambridge University (UK) in 2006, specialising in nanotechnology. He has over 25 years of academic and industrial experience and has published over 100 journal and conference papers. He has been a co-principal investigator for many government-funded research projects.

## Raji George

Dr. Raji George is a professor and Head of the Mechanical Engineering Department at Ramaiah Institute of Technology, Bangalore, India. He earned his PhD from Visweswaraya Technological University (VTU) in 2008 and has more than 33 years' experience in academia and teaching. He was awarded best engineering teacher Gold medal award constituted by Sir M.V. foundation in 2009. He was awarded the Scientific Award of Excellence in 2011 by the American Biographical Institute. He has more than 40 journal and conference papers to his credit. He is the principal investigator on a major research project for Boeing Inc. USA.

# List of contributors

**Entesar H Betelmal**
Department of Mechanical and Industrial Engineering, University of Tripoli, PO Box 13275, University Road, Sidy Almasry Tripoli, Tripoli, Libya

**Gokmen Demirkaya**
TOBB Economy and Technology University, Mechanical Engineering Department, Engineering Faculty, Söğütözü Cad. No:43 06560 Söğütözü, Ankara, Türkiye

**Raji George**
Department of Mechanical Engineering, Ramaiah Institute of Technology, MSR Nagar, MSRIT Post, Bengaluru 560 054, India

**D Yogi Goswami**
Clean Energy Research Center, University of South Florida, ENB 260, Tampa, FL 33647, USA

**Prakriti Gupta**
Department of Chemical Engineering, M S Ramaiah Institute of Technology, MSRIT Post, MSR Nagar, Banglore 560 054, India

**B V Raghuvamshi Krishna**
Department of Mechanical Engineering, GITAM School of Technology, Nagadenahalli, Dodballapur Taluk, Bengaluru 562 163, India

**Naveen Krishnan**
Centre for Energy Studies, Indian Institute of Technology Delhi, India

**K Ravi Kumar**
Centre for Energy Studies, Indian Institute of Technology Delhi, India

**Martina Leveni**
The Ohio State University, College of Engineering Dept. of Civil, Environmental, and Geodetic Engineering, 483b Hitchcock Hall, 2070 Neil Avenue, Columbus, OH 43210, USA

**Lokesha**
Department of Mechanical Engineering, M S Ramaiah Institute of Technology, Bangalore, India

**G M Madhu**
Department of Chemical Engineering, Ramaiah Institute of Technology, MSR Nagar, MSRIT Post, Bengaluru 560 054, India
and
Centre for Advanced Materials Technology, Ramaiah Institute of Technology, MSR Nagar, MSRIT Post, Bengaluru 5607 054, India

**P B Nagaraj**
Department of Mechanical Engineering, M S Ramaiah Institute of Technology, Bangalore, India

**Ricardo Vasquez Padilla**
Faculty of Science and Engineering, Military Road, Lismore NSW 2480, Australia

**K S Reddy**
Department of Mechanical Engineering, Indian Institute of Technology Madras, India
and
University of Exeter, Stocker Road, Exeter EX4 4PY, UK

**Ravi Sankannavar**
Department of Chemical Engineering, Ramaiah Institute of Technology, MSR Nagar, MSRIT Post, Bengaluru 560 054, India
and
Centre for Advanced Materials Technology, Ramaiah Institute of Technology, MSR Nagar, MSRIT Post, Bengaluru 5607 054, India

**A Sarkar**
Department of Chemical Engineering, Indian Institute of Technology Bombay, Powai, Mumbai 400 076, India

**Roushan Nigam Ramnath Shaw**
Department of Chemical Engineering, Ramaiah Institute of Technology, MSR Nagar, MSRIT Post, Bengaluru 560 054, India

**K R V Subramanian**
Department of Mechanical Engineering, Ramaiah Institute of Technology, MSR Nagar, MSRIT Post, Bengaluru 560 054, India

**IOP** Publishing

Thermodynamic Cycles for Renewable Energy Technologies

**K R V Subramanian and Raji George**

# Chapter 1

## Innovations in vapour and gas power cycles

**P B Nagaraj and Lokesha**

Energy is one of the primary needs for human beings required in every walk of life. Sources of energy are mainly classified into fossil fuels, solar radiation, wind energy, tidal and geothermal. The increasing standard of living and growing world population increases the demand for energy throughout the world and this leads to the development of innovative concepts in improving the conversion efficiency of energy devices. Environmental problems and climate changes in the world become more serious issues due to the over-exploitation of fossil fuels. Low-grade heat sources such as solar, geothermal and industrial waste heat are the clean, renewable and low-cost energy sources and the utilization of these energy sources is of great significance to solve these global issues. The study of thermodynamic cycles gives an idea about energy conversion from one form to another, and these cycles are mainly categorized into power cycles and refrigeration cycles. These cycles are further classified into gas power cycles in which the working fluid is in gaseous form throughout the cycle and vapour power cycles, which use working fluid in the liquid phase, and are being alternatively vaporized and condensed.

In this chapter, an attempt is made to briefly discuss the vapour and gas power cycles, their merits and demerits and a comparative study of technologies innovated is made to improve efficiency of these cycles. A review of research work is carried out on modifications of the Rankine cycle to convert low-grade heat energy into electric power, investigations on cogeneration cycle and combined power and cooling cycles, as well as analysis with different working fluids. The thermodynamic analysis of the supercritical Rankine cycle has been reviewed, and cycle energy efficiency and exergy efficiency have been studied. A thermodynamic comparison of a supercritical carbon dioxide cycle with helium Brayton, superheated steam and supercritical steam cycles is done. The recovery of exhaust heat from the gas turbine to improve part load thermal efficiency, the investigation of combined cycle coupled with super critical carbon dioxide recompression and a regenerative cycle is discussed. The hybrid solar gas turbine technology and a system consisting of solar thermal-based combined power and cooling cycle, which is operated from low-grade energy source and its application to domestic and industrial needs, has been reviewed. The improvement in efficiency with high and intermediate temperature finite thermal sources by using organic flash cycle and performance of different configurations of the gas turbine engine operating as a part of the combined cycle power plant has been analysed.

The application of an alternate energy source for air conditioning in rural areas and its effect in the reduction of global warming and ozone depletion is discussed. The review of the application of solar thermal system drive vapour jet refrigeration cycles and the development of innovative thermodynamic cycles for effective utilization of low-temperature heat sources, such as solar, geothermal and waste heat sources, have been made. This chapter helps in the understanding of vapour and gas power cycle concepts and the technologies that have evolved to overcome the demerits and improve the efficiency of the cycle.

## 1.1 Introduction

A traditional way of converting heat into electric energy using water as a working fluid is adopted in the Rankine cycle. The arrangements of components and the corresponding $P$–$V$ and $T$–$S$ diagrams are shown in figures 1.1 and 1.2. The main components are the boiler, steam turbine, condenser and feed water pump. Ideally, the water in a saturated condition is pumped to the operating pressure of the boiler. The water enters the boiler and heats up to saturation or evaporation temperature (process 4-5) due to sensible heating. Then the water is completely evaporated into steam (process 5-1) at constant pressure $P_1$ and temperature $T_1$, and the heat supplied is known as latent heat of evaporation. The steam condition at the boiler outlet may be dry (condition 1), wet (condition 1′) or superheated (condition 1″). The vapour with one of these conditions enters the turbine, expands isentropically and delivers power. The exhaust steam from the turbine enters the condenser, rejects heat to cooling water and leaves the condenser as saturated liquid.

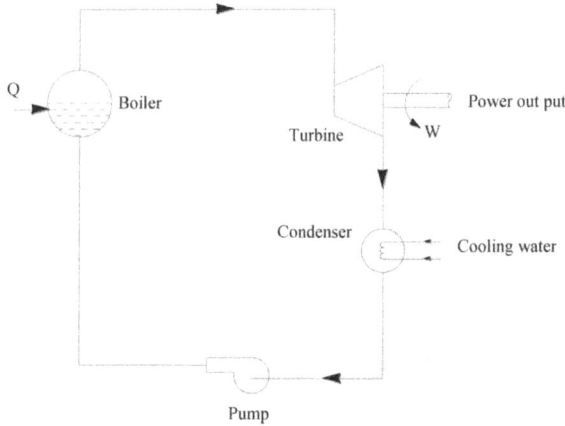

**Figure 1.1.** Components of the steam power plant working on the Rankine cycle.

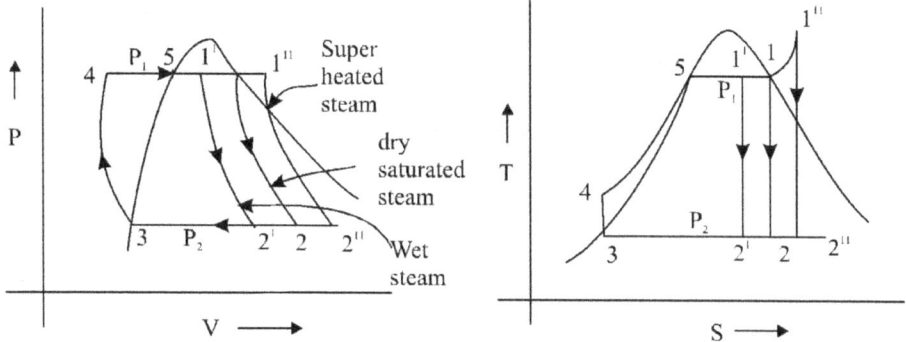

**Figure 1.2.** *P–V* and *T–S* diagrams for the Rankine cycle.

The Rankine cycle has some disadvantages, for example, the steam pressure in the condenser should be less than atmospheric pressure, which requires the desecrators unit to avoid air leaks in the last stage of the turbine. The molecularly heavier working fluid results in isentropic efficiency of up to 60%. The formation of droplets in the last stage of the turbine requires further superheating or reinforcing blade material to avoid erosion. The Rankine cycle with water is shown to be a poor working fluid for the conversion system, thus other organic fluids are recommended for use in the cycle.

The gas turbines are used in aircraft applications, turboshaft applications in electricity generation, the oil and gas industry, chemical and process industry applications and to recover the exhaust heat for the combined cycles. The efficiency of the gas turbine cycle can be improved by reducing excess air compressor load and recovering the energy that might otherwise be wasted. The use of a combined cycle power plant by adding a steam turbine bottoming cycle improves plant efficiency. The gas turbine can be used as a topping cycle. The combined cycle plant uses more than one cycle to convert energy into electricity and the first cycle uses working fluid with the highest temperature, second with intermediate temperature fluid and the third cycle with lower temperature working fluid. In general, combined cycle plants with more than two cycles are not preferred.

The non-conventional power cycles are developed to convert low-grade heat into power to utilise large amounts of waste heat generated in modern industries and the moderate temperature of the renewable energy source. Some of the important non-conventional power cycles are the zeotropic vapour cycle, organic Rankine cycle, trilateral flash cycle and supercritical Rankine cycle. Both thermoelectric devices and heat engines are used to convert thermal energy into power. Thermoelectric devices are small, costly, and the efficiency is low and limited to some special applications like vehicle heat recovery. In heat engines, the energy conversion is driven by a temperature gradient between thermal reservoirs. Advanced power vapour cycles are introduced to achieve higher cycle efficiency and energy sources may be geothermal, solar thermal, biomass and industrial waste heat. A combined cycle may also be adopted to obtain higher thermal efficiency.

## 1.2 Organic Rankine cycle

A study of different organic fluids, such as water, ethanol and toluene, was made and physical properties like latent heat, thermal stability, freezing temperature and environmental aspects are compared. The working principle is same as the simple Rankine cycle, but uses organic working fluids with low boiling points to recover heat from low-temperature heat source. The study recommended that the water is suitable for high-temperature waste heat recovery and organic fluids are suitable for the recovery of low-grade waste heat recovery. Different researchers have also examined pure organic fluids like HCFC123, HFC-254fa, isobutene, n-pentane and aromatic hydrocarbons. The advantages of the organic Rankine cycle are: (i) it requires less heating to evaporate the working fluid; (ii) the organic working fluid evaporates at lower pressure and temperature; and (iii) at the end of expansion, the working fluid is in a vapour state, hence superheating is not required to prevent the erosion of turbine blades. The sub cooling and superheating of working fluid can be done with organic Rankine cycles and this showed a negative ORC efficiency. Some attempts have been made to use Rankine cycles for geothermal heat recovery and it is a promising solution for a low-temperature liquid geothermal source. The efficiency of the organic Rankine cycle with dry fluids can be improved by using a regenerator. The heat recovery potential of basic ORC can be increased by superheating, heat recuperation, two-pressure level and supercritical cycle methods [1].

## 1.3 Organic flash cycle (OFC)

The organic flash cycle is another type of vapour power cycle used to improve the efficiency and temperature matching, and reduce exergy losses during heat addition. The liquid and vapour components of the mixture are separated to avoid the use of two phase expanders. In a trilateral flash cycle, the working fluid absorbs heat in a single phase liquid, hence no isothermal phase change exists (figure 1.3).

The cycle is similar to a flash steam cycle that uses high pressure and temperature geo-fluid in the liquid state which is extracted from the geothermal well. Then it is throttled to lower pressure or flashed to produce a mixture of liquid and vapour. The drawback of the steam flash cycle is the formation of moisture in the turbine resulting in the creation of a two-phase mixture with liquid droplet formation due to the isentropic expansion of wet fluid. This results in the erosion of turbine blades and requires reinforcing materials for blades [1].

The organic flash cycle (OFC) can be used as an alternative cycle to the vapour power cycle to improve the efficiency when high and intermediate temperature heat sources are used. The temperature matching can be greatly improved by using OFC. It also reduces exergy losses during heat addition in the cycle. The work involves comparison of different aromatic hydrocarbons and siloxanes as potential working fluids. The OFC, optimized basic organic Rankine cycle (ORC), a zeotropic Rankine cycle with binary ammonia-water mixture and transcritical carbon dioxide cycle are compared and showed that the use of aromatic hydrocarbons are better

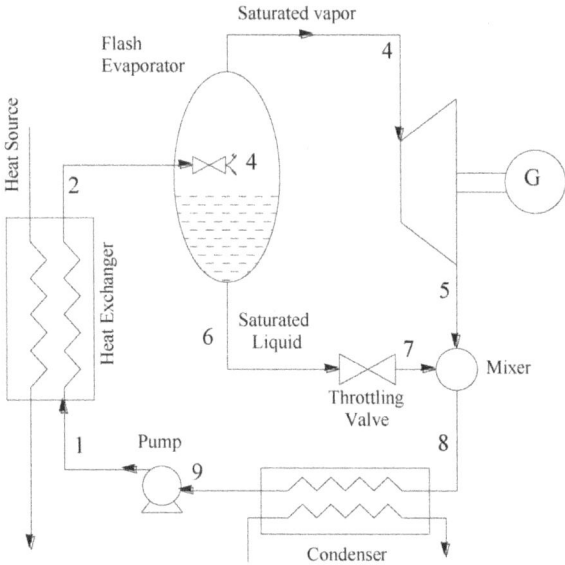

**Figure 1.3.** Schematic diagram of the organic flash cycle.

working fluids for ORC and OFC and resulted in higher power output with simple turbine design [2].

## 1.4 Zeotropic vapour cycle

A zeotropic mixture is a mixture with components that have different boiling points. The individual components of the mixture do not evaporate or condense at the same temperature as one substance. The mixture of nitrogen, methane, ethane propane and isobutene is a zeotropic mixture. These mixtures are characterized by non-isothermal phase transition at a constant pressure. The mixture possesses a property called temperature glide during the isobaric evaporation and condensation process. The main advantage of using a zeotropic mixture is that the power cycle can be operated at low pressure but the mixtures will have a lower convective heat transfer coefficient compared to pure substances. In the next section, a discussion on three different zeotropic vapour cycles is made [1].

## 1.5 The Kalina cycle

Kalina is a Russian scientist who invented the Kalina cycle. It is a power generation cycle and is the modified form of the Rankine cycle that uses ammonia-water mixture as the working fluid and converts thermal energy into mechanical power using two different working fluids. The cycle is more complex and requires more maintenance and is suitable to recover waste heat from power plants and solar collecting systems. A low-temperature Kalina power plant works with a source temperature of up to 150 °C and the high-temperature heat recovery system works in the range of 250 °C–600 °C of the source. The intermediate concentration of ammonia is pumped to a recuperator and is heated by using a brine solution in the evaporator. The liquid and vapour gets

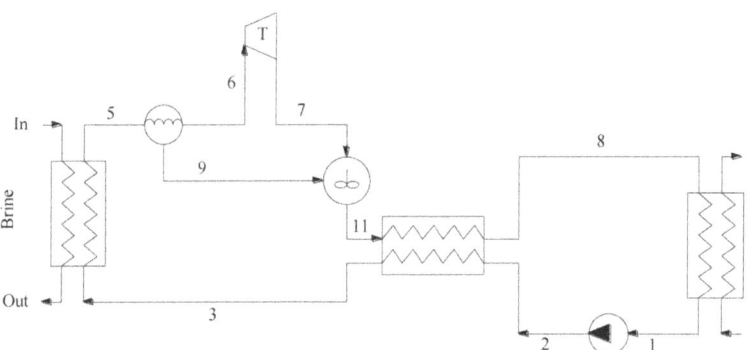

**Figure 1.4.** Arrangement of components in the Kalina cycle.

separated in state 5 and the vapour phase in state 6 enters into the turbine to generate power. The exhausts of the turbine and liquid extracted from state 5 are further mixed and get cooled in the recuperator and condenser. The efficiency of the Kalina cycle is around 45% and upto 52% with a gas turbine combined cycle plant. The application of the Kalina cycle is limited to medium-low temperature heat sources (300 °C–400 °C in heat recovery and 100 °C–200 °C in in binary geothermal plants) and small power conversion systems [1] (figure 1.4).

## 1.6 Uehara cycle

This cycle is an improvement over the Kalina cycle with the use of an additional turbine, a heater and after condenser. The working fluid used in the cycle is an ammonia-water mixture. The warm sea water passes through the evaporator to generate a vapour-liquid mixture. The vapour extracted from the liquid enters into turbine 1, after heating enters turbine 2. The mixed vapour from turbine 2 enters into the absorber and gets absorbed with ammonia-water. The remaining part is condensed to liquid by cold sea water. By simulation it was found that, for a 100 MW OTEC plant with 26 °C hot water and 4 °C cold water, R717 was recommended as the most appropriate working fluid in the cycle [1].

## 1.7 The Maloney–Robertson cycle

This cycle uses ammonia-water as the working fluid. This cycle is simple when compared to the Kalina cycle. In the Kalina cycle, during condensation the working fluid exchanges heat with the environment and this limits the temperature of the working fluid in the turbine. The Maloney–Robertson cycle uses the absorption condensation process to overcome this problem. The heater supplies rich ammonia to the superheater, and from there it is expanded in the turbine. Then it is mixed with a weak solution from the distillation unit and used to absorb rich vapour in ammonia to regenerate the base solution [1] (figures 1.5 and 1.6).

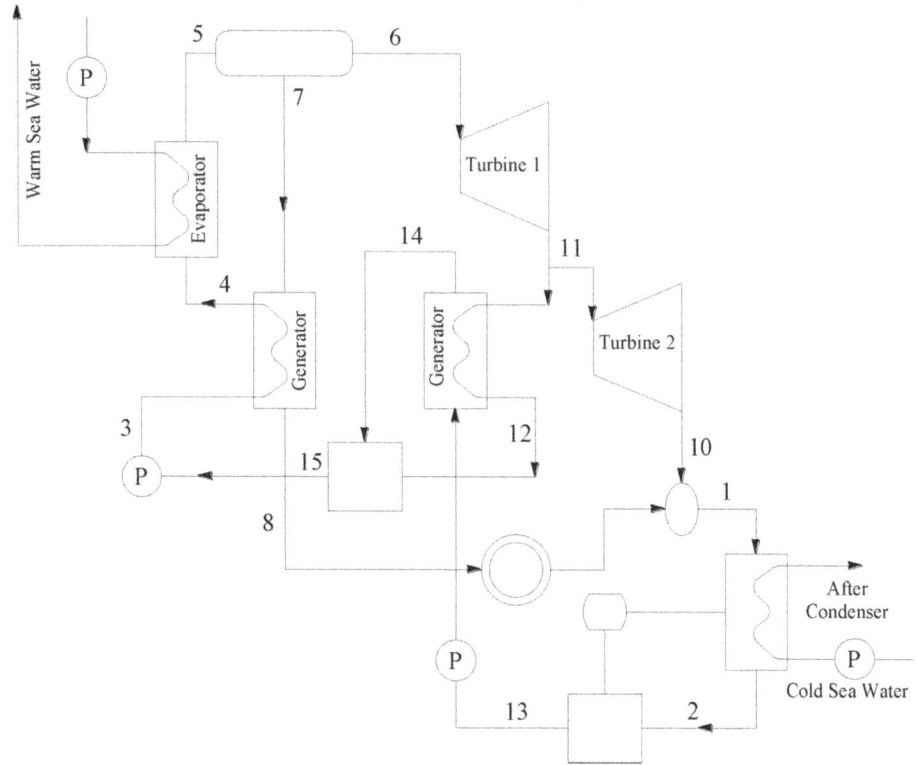

**Figure 1.5.** Uehara cycle.

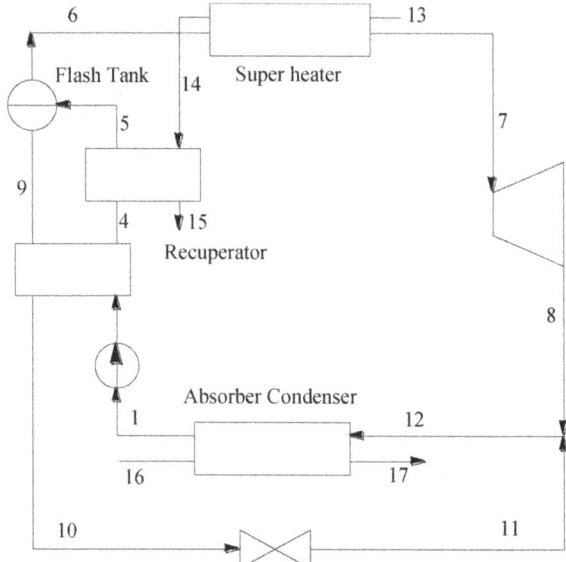

**Figure 1.6.** Schematic diagram of the Maloney–Robertson cycle.

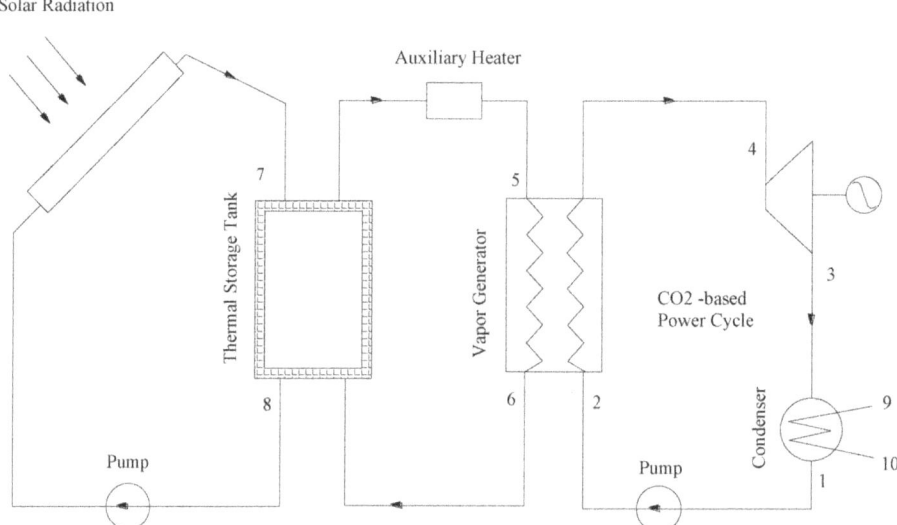

**Figure 1.7.** Solar type transcritical $CO_2$ power plant.

## 1.8 Transcritical and supercritical cycles

In this method, the heat is transferred to the working fluid of the cycle at a pressure above the critical pressure [1]. The transcritical cycle is one which operates partly under supercritical conditions. The cycle that operates completely in the supercritical region is known as the supercritical cycle. The working fluids such as carbon dioxide, helium, refrigerants and alkanes are used in these type of cycles. The thermal efficiency can be increased by optimising temperature or pressure in the condenser. Some research has been carried out to compare sub-critical and supercritical processes for different refrigerants and proved that propane or R143a is having a specific net power output greater than 40% at 150 °C of the geothermal source.

## 1.9 Carbon dioxide transcritical cycle

Carbon dioxide is a natural working fluid extensively used in supercritical and transcritical Rankine cycles. Carbon dioxide is a non-flammable, non-toxic working fluid with a low critical temperature of 31.18 °C and a high critical pressure of 7.38 MPa (figure 1.7).

It consists of a solar collector, expansion unit or turbine, condenser and a pump. Extensive research has been carried out by considering parameters such as thermal efficiency, specific power output, exergetic efficiency and the cost of the system. The effect of rise in turbine inlet temperature on exergy efficiency is also studied. An optimum value of turbine inlet pressure is determined for maximum values of system efficiency and net power output [1].

## 1.10 Combined power and cooling cycles

The combined power and cooling thermodynamic cycle improves efficiency as it generates power and cooling effect or refrigeration in the same cycle. A combination of Rankine cycle which uses ammonia as the working fluid and ammonia-water absorption refrigeration cycle is used. The figure shows the schematic diagram of the arrangement of components in the combined cycle. The simulation results showed that the working fluid enters the turbine at 193 °C and 2.76 MPa and leaves the turbine at 0.14 MPa. The system generates 2 MW electric power and 50 tons of refrigeration. This is a power cycle for low-cost solar thermal collectors, geothermal resources and waste heat from other power plants [1] (figure 1.8).

## 1.11 Combined cycle to recover exhaust heat from marine gas turbine

A combined cycle consisting of supercritical $CO_2$ recompression and regenerative cycle is developed to recover the exhaust heat from the marine gas turbine to improve part load thermal efficiency [3]. The parameters considered for the analysis are output power, exergy efficiency, heat exchanger area per unit power output and levelized energy cost. The genetic algorithm was used to obtain the optimum system parameters. The combined cycle supplies 80% of the propulsion power in case of gas turbine plant failure and is used as a backup generator.

The arrangement of the components is as shown in figure 1.9. It consists of a propeller system, marine power system, two shaft aero derivative marine gas turbines and a waste heat recovery system. The LM2500+ marine gas turbine is used. The waste heat recovery system consists of a supercritical carbon dioxide recompression cycle that acts as a topping cycle due to its higher efficiency and the supercritical carbon dioxide regenerative cycle which acts as bottoming cycle. The coupling cycles reduce exhaust outlet temperature and maximum waste heat

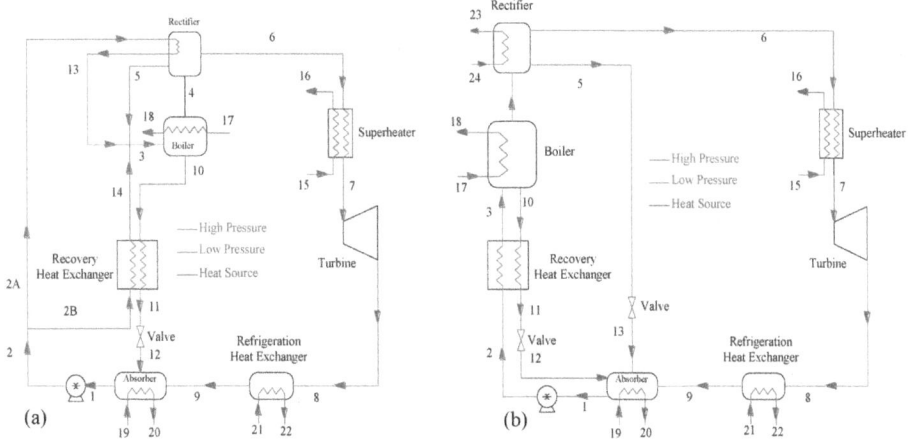

**Figure 1.8.** Combined cooling and power cycle (reference 1) with (a) internal cooling source and (b) external cooling source.

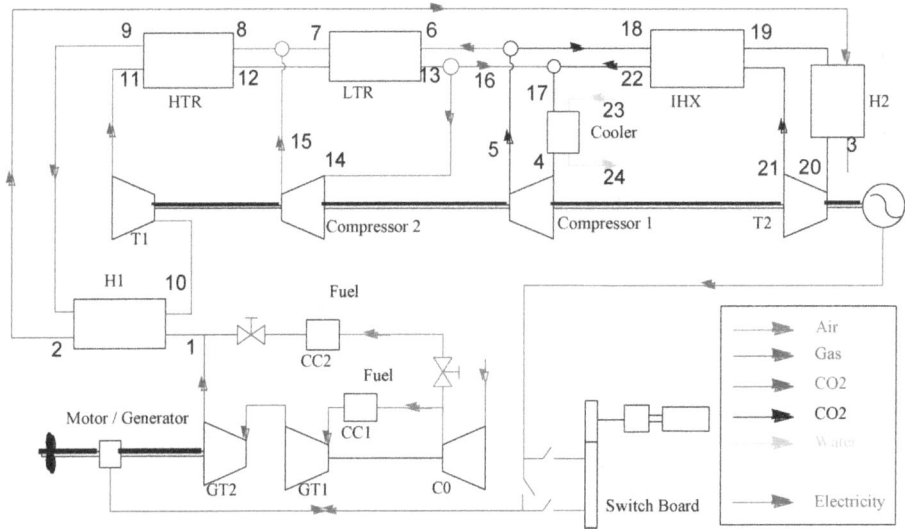

**Figure 1.9.** Schematic diagram of the combined cycle designed by Shengya Hou *et al.*

recovery is achieved. Maximum exhaust heat is absorbed by the topping cycle which results in lowering the outlet temperature of that cycle. A small amount of residual waste heat is available in the bottoming cycle and hence uses a regeneration cycle. For both the recompression cycle and regeneration cycle, the inlet and outlet pressures of the compressor are same. First the air is compressed in compressor C and sent to combustion chamber CC1 where the fuel burns to generate heat. The exhaust of CC1 is supplied to gas turbine GT1 which in turn drives compressor C and GT2 which in turn drives the generator. The two-shaft gas turbine design used here ensures that the propeller shaft speed is not affected by compressor speed. The exhaust from GT2 is the heat source for the combined cycle.

When the gas turbine fails, the control valve opens the high pressure air flow to CC2 and the exhaust from the CC2 acts as a hot source for the combined supercritical $CO_2$ system which generates power. The exhaust from the GT2 is used to heat $CO_2$ in heat exchanger H1. The power required to run compressors and generator is obtained by allowing the expansion of high temperature working fluid in turbine T1. The steam at condition 8 and 6 is preheated by passing it through a high-temperature recuperator and a low-temperature recuperator and is divided [4] into two paths: 16 and 14. The steam at 16 is cooled by seawater and then compressed by C1, and the steam at 14 is compressed by the C2 without a cooling process. The C1 supplies compressed steam to the LTR to absorb heat of steam coming from HTR [5] and then is mixed with compressed steam 15. Then it flows to the HTR to absorb heat of steam at condition 11 and the preheated steam 9 enters into H1 to complete the supercritical carbon dioxide recompression Brayton cycle. The temperature of the steam at the exit of H1 is still high and can be used as the hot source for the supercritical carbon dioxide regenerative cycle. The steam from the IHX is heated in H2 by using high-temperature steam from H1. The steam at condition 20 expands in

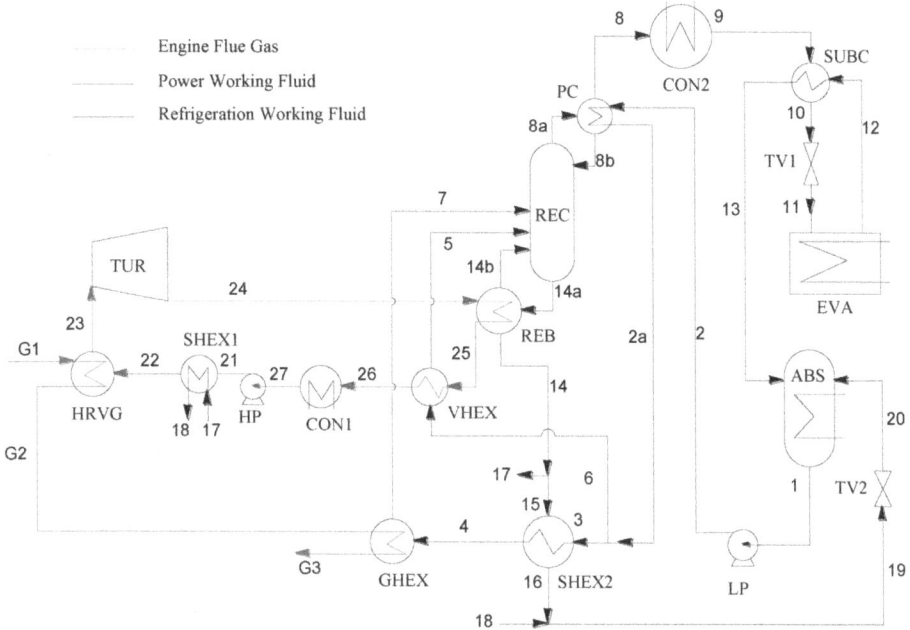

**Figure 1.10.** Schematic diagram of the combined power and cooling cogeneration system designed by Liuli Sun *et al.*

turbine T2 to run the generator. The steam expanded in T2 enters into the IHX to recover the heat to preheat steam 18. The steam from the IHX is mixed with steam 16 and the mixed steam 17 is cooled and then compressed by using C1. After compression, the steam is again divided in to two streams and one stream enters the IHX to recover waste heat. After preheating, the steam enters into H2 to complete a supercritical carbon dioxide regenerative cycle [3].

## 1.12 Power and cooling cogeneration system with a mid/low-temperature heat source

The ammonia-water mixture is one of the important working fluid mixtures extensively used in absorption refrigeration systems to provide comfort cooling conditions to the buildings, for process refrigeration in industries such as cold storage, food freezing and preservation [6]. A power and cooling cogeneration system using an ammonia-water mixture using mid/low temperature heat source is developed. Basically the cogeneration system is made in two parts: one consists of a Rankine cycle and other an absorption refrigeration cycle. A portion of waste heat with high temperature is utilized to generate power and remaining low temperature portion of waste heat is used for refrigeration. The refrigeration subsystem recovers the exhaust heat of the power subsystem.

The schematic diagram of power and cooling cogeneration system is as shown in figure 1.10. It consists of a Rankine subsystem which produces power and an absorption refrigeration subsystem. The ammonia-water mixture is pumped to high pressure and then preheated in a heat exchanger with the use of weak solution supplied by a rectifier. Then the solution is supplied to a generator to absorb the high-temperature portion of the heat source and gets converted into superheated vapour. This superheated vapour is allowed to expand in a turbine to produce power. The exhaust from the turbine is supplied to the re-boiler in the refrigeration cycle and heat required for the rectification process is extracted from this re-boiler. A partial condensation of exhaust vapour occurs and the mixture of liquid and vapour flows into VHEX for preheating a part of the strong solution. The refrigeration subsystem recovers the condensation heat and the solution with a small fraction of vapour is fully condensed to a liquid condition by using condenser 1.

In the cooling cycle, the absorber discharges a strong solution that is pumped by a low pressure pump to a partial condenser of the rectifier where the heat is recovered by a strong solution. The high-temperature, strong solution is again divided into two paths. One path is preheated in Solution HE2 (condition 4). The solution in condition 4 is heated in Gas HE by using low temperature waste heat and then sent to the rectifier. The second path of the strong solution with condition 6 enters VHE and after preheating is sent to the rectifier. The ammonia rich vapour [7] and weak solution [8] are obtained from paths 7 and 5 in the rectifier. The ammonia rich vapour with condition 8 gets condensed to a liquid refrigerant with condition 9 and is subcooled in a heat exchanger. After throttling, it enters into the evaporator to provide a refrigeration effect.

An another combined power/cooling cogeneration system based on an ammonia-water power cycle was introduced by Jiqiang Yin *et al* [9]. The schematic diagram is as shown in figure 1.11. The addition of the absorption-ejector refrigeration cycle results in the higher cooling capacity of a conventional combined system. The schematic diagram of the arrangement of components is as shown in figure 1.11. At the exit of the absorber, the basic saturated solution is pumped to the heat exchanger to recover the heat from the ammonia weak solution which is coming from the boiler. The boiler ammonia-water solution is separated into ammonia rich vapour and ammonia weak solution. After transferring heat to basic working fluid, the ammonia weak solution is throttled to low pressure. The vapour which is rich in ammonia is made to pass through a super heater for superheating and is allowed to expand in a turbine to generate power. After expansion to 1.9 MPa, a part of ammonia rich vapour is extracted and is supplied to the rectifier, and the remaining working fluid expands in the turbine. In the ejector, a very high vacuum is created due to the high velocity of the primary flow and this entrains the secondary flow from evaporator. Both primary and secondary flow are mixed in the mixing chamber and then enters into a condenser where the vapour rejects heat to surroundings and condensed to liquid. The ammonia liquid is then throttled and enters into the evaporator to produce refrigeration. The ejector entrains some of the evaporated vapour and remaining is absorbed by the weak solution in the absorber to complete the cycle. The thermal efficiency of 21.34% and exergy efficiency of 38.95% can be achieved.

The following assumptions are made in the analysis of the cycle:

1. A steady state operation is assumed by neglecting heat losses and pressure drops.

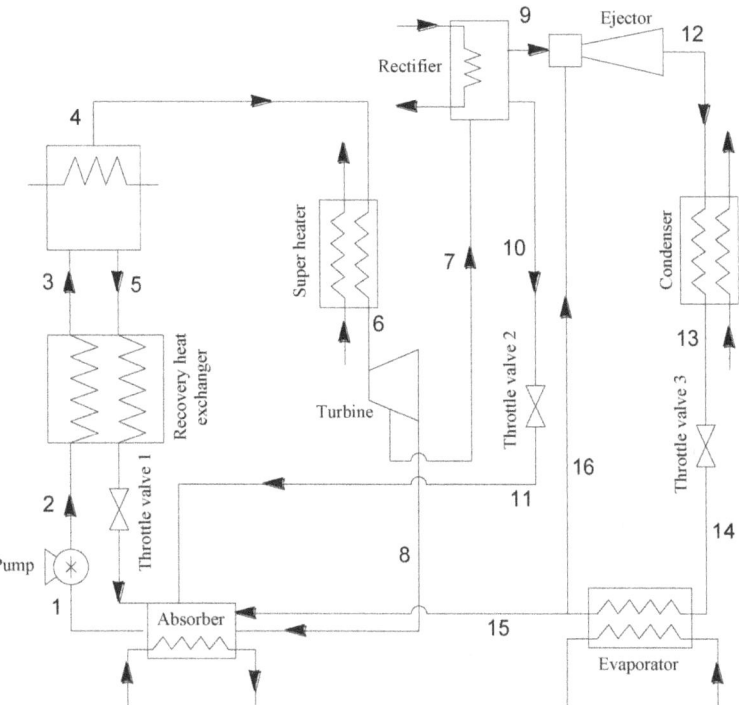

**Figure 1.11.** Schematic diagram of the power/cooling cogeneration system by Jiqiang Yin *et al.*

2. The working fluid is considered as the saturated liquid at the outlet of the absorber and condenser.
3. A saturated vapour is assumed at the rectifier outlet and weak solution from the rectifier is considered as saturated liquid.
4. A saturated vapour is assumed at the evaporator outlet.

A power and cooling cogeneration cycle is investigated by Zeting Yu *et al* [10] in which a modified Kalina cycle (subcycle 1) and ammonia absorption refrigeration cycle with pre-cooler (subcycle 2) are interconnected by various components such as mixers, splitters, absorbers and heat exchangers [10]. The cycle uses ammonia-water mixture as the working fluid. A separate mode is used to adjust cooling to power ratios in the system without splitting or mixing processes in the subcycles. The engine flue gases, high temperature fuel cells and industrial waste heat are the sources that drive the system to generate power and to produce a cooling effect.

The schematic diagram for the arrangement of components is as shown in figure 1.12. In subcycle 1, the basic concentration solution ammonia-water mixture is pumped to rectification pressure (process 1-2) and splitter SP1 divides it into two paths 3A and 3B. The solution in condition 3B passes through HE1 and HE2 and heated by turbine exhaust [4] before being sent to REC1. In rectifier REC1, pure ammonia and weak solution gets separated. The REC1 sends ammonia vapour to condenser1 and reboiler REB, where separate heat is added. After condensation in

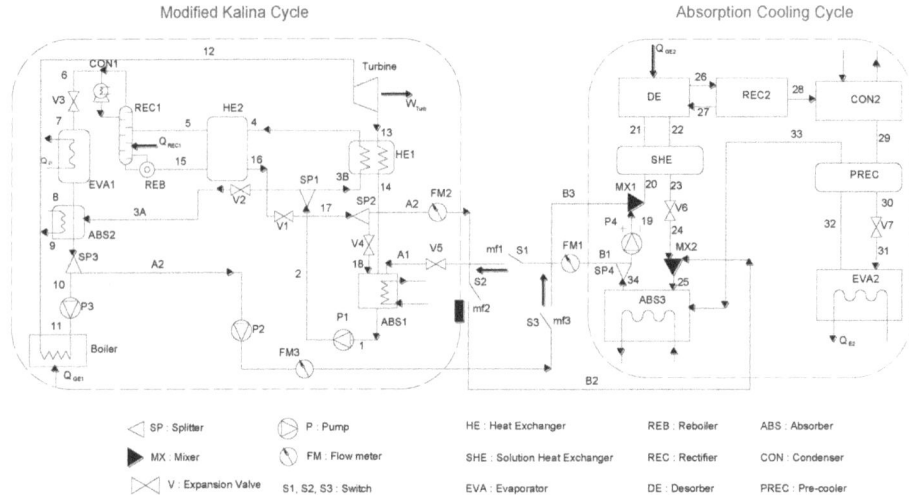

**Figure 1.12.** Schematic diagram of the combined power and cooling cycle designed by Zeting Yu *et al* (reference 10).

CON1 and throttling, the strong concentration saturated liquid enters into the evaporator where it absorbs heat to produce the refrigeration effect. Then it is absorbed by the fluid from path 3A in absorber ABS2 and this results in concentration solution [2]. Splitter 3 divides stream 9 into two paths, a portion of which is supplied to the absorption cooling system and the remainder is pumped to a HP turbine through the super heater of the boiler. After superheating, the vapour is expanded in the turbine to produce power.

In the absorption cooling cycle, the absorber ABS3 sends a portion of solution to ABS1 of subcycle 1. The remaining solution with condition 35 is pumped to heat exchanger SHE [11] where the solution gets heated after mixing with high concentration solution from subcycle 1. Then the solution is supplied to the desorber or generator. The refrigerant is generated in desorber DE by adding heat to the liquid solution. The hot vapour at high pressure enters into rectifier REC2 and condenser CON2. The SHE exchanges heat between the weak liquid solution from DE and cold strong solution and mixes with the solution from subcycle 1. The refrigerant from CON2 is subcooled by PREC and the preheated vapour is supplied to the absorber.

The performance of the cycle is affected by parameters such as mass fraction of ammonia, operating pressure and temperature of the system and outlet pressure and temperature of the evaporator and absorber.

## 1.13 Advanced hybrid solar tower combined-cycle power plants

Most of the currently installed solar thermal power plants are based on parabolic trough technology and these plants uses Rankine power cycles with low-temperature steam turbines (less than 400 °C) [12]. The hybrid solar gas turbine technology is one of the most promising non-conventional type power plants. The thermal energy

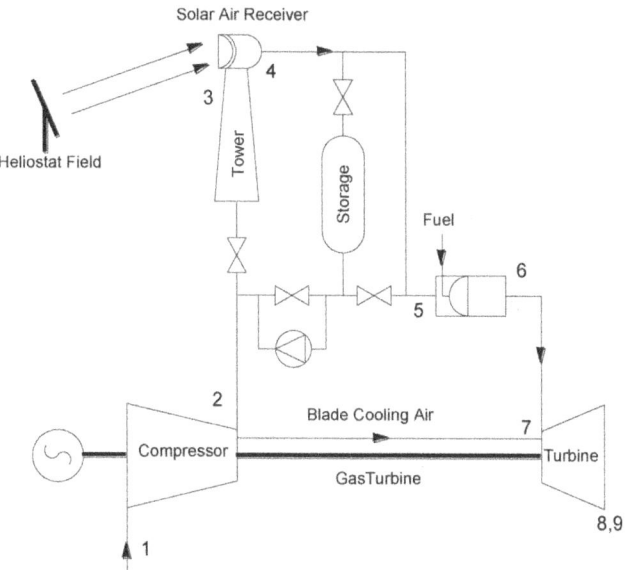

**Figure 1.13.** Hybrid solar gas turbine integrated with TES designed by Spelling J *et al.*

storage and combined power cycles are integrated to enhance the economic viability of technology. Due to advancement in the technology of solar air receivers, higher temperatures can be achieved and increase in conversion efficiency of solar energy is also possible. This decreases size of the collector for a given power output. More than 50% reduction of electricity costs can be achieved with the use of an advanced solar hybrid combined cycle when compared to parabolic trough power plants.

The schematic diagram for the arrangement of components for the combined cycle hybrid solar gas turbine (HSGT) plant is as shown in figure 1.13. A heliostat solar collector is used and the power generation is done at the base of the tower. The concentric piping arrangement is done for ducting of compressor air up and down the tower. The heat exchange takes place between hot air from the receiver and compressed air so that hot air gets cooled. The fuel consumption in the combustion chamber decreases due to the preheating of the compressed air. The preheated air is then entering into the combustion chamber of the gas turbine where the fuel is injected. A desired temperature is maintained at the outlet of the combustion chamber and can be maintained as stable by adjusting the fuel flow to the combustor. A high-temperature thermal energy storage (TES) is arranged parallel to sthe olar receiver to increase the solar heat supplied to the power cycle. The excess energy stored during the daytime can be used during the night or during the time when solar energy is not available. The TES consists of a matrix of solid with high specific heat to store thermal energy and pressurized air is used as heat transfer fluid to charge or discharge the system. The exhaust from the turbine is made to pass through a heat recovery steam generator and steam cycle.

## 1.14 Combined power and cooling cycle with two turbines

A common working fluid such as ammonia is needed when combined power and cooling cycles are used for power generation [13]. The proposed plant works on both individual and combined mode.

The assumptions made in the thermodynamic analysis are:

- The pump and turbine isentropic efficiency is 75%.
- For both high-pressure and low-pressure turbines, the degree of superheat is assumed to be 10 °C.
- At the exit of evaporator, the refrigerant temperature is considered as 10 °C.
- The beam radiation is assumed as 650 W m$^{-2}$ and global radiation is considered as 900 W m$^{-2}$.

The arrangement of the components in the cycle is as shown in figure 1.14. A common generator that runs with solar parabolic concentrating collector is used to integrate both power and cooling cycles. The output of the generator is converted to useful energy with two turbines. The vapour produced by the super heater and reheater is used for running the turbines. The ammonia still exists in the saturated vapour state at the outlet of HPT hence less energy is required to convert it into dry vapour. At the outlet of HPT, depending on the requirement, the refrigerant is divided into two paths one for the power cycle and the other for the cooling cycle. A portion of saturated vapour is circulated in the components of the cooling cycle and the condenser load is reduced by HPT. The remaining refrigerant enters into components of the power cycle, gets reheated before it enters into LPT. The refrigerant makes its exit from the power cycle as saturated vapour. The flow rate of ammonia vapour can be regulated at the outlet of HPT according to the requirement of power and cooling.

## 1.15 Supercritical Rankine cycle for a modern steam power plant

The thermodynamic analysis of the supercritical Rankine cycle for a modern steam power plant with capacity 1200 MW has been done [7]. The properties of steam are estimated by a software code using Matlab. The cycle uses both single reheat and double reheat methods. Figures 1.15 and 1.16 show arrangements of components for a supercritical Rankine cycle with a single and double reheater. The steam with supercritical temperature enters the turbine and is allowed to expand to state 2. The steam at state 2 is tapped and reheated again by sending it to the boiler and the remaining expansion takes place to low pressure from 3 to 4. After condensation, the feed water is pumped back to the boiler to complete the cycle. In the supercritical cycle with a double reheater, steam is expanded until state 2, tapped and reheated by using the boiler. After expansion to state 4, further steam is reheated to increase its temperature and the remaining expansion takes place in the turbine i.e., process 5-6 (figures 1.15 and 1.16).

Parabolic concentrating Collector

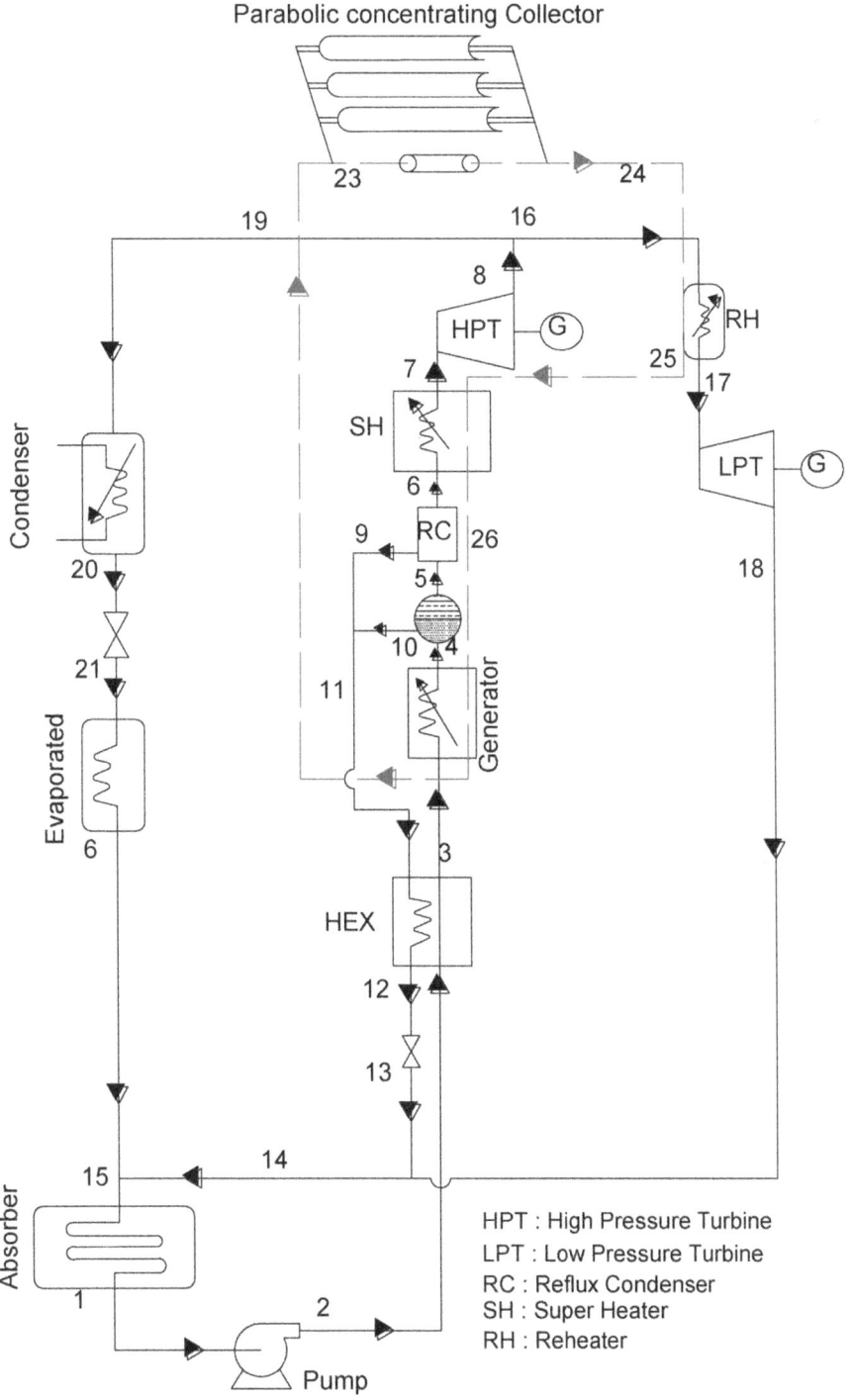

**Figure 1.14.** Schematic diagram of combined power and cooling cycle with two turbines (reference 13).

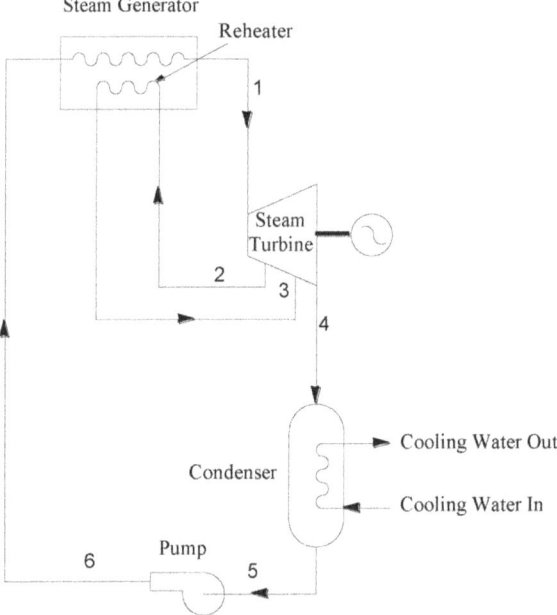

**Figure 1.15.** Schematic diagram of the supercritical cycle with a single reheater.

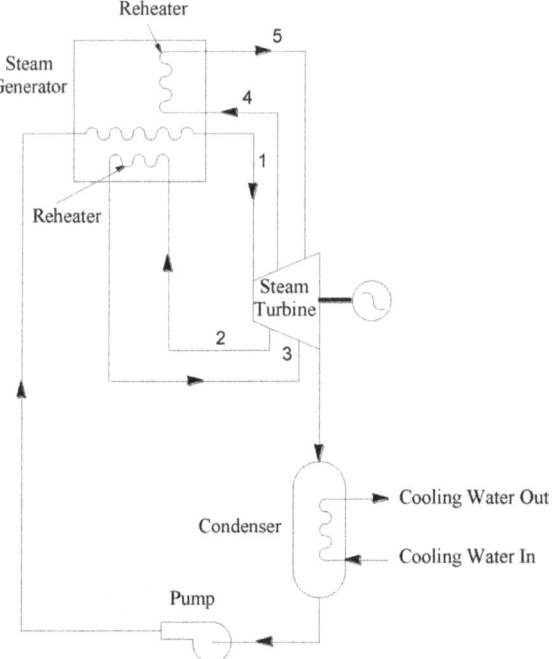

**Figure 1.16.** Schematic diagram of the supercritical cycle with a double reheater.

## 1.16 Combined gas-steam power plant with a waste heat recovery steam generator

The Rankine cycle plant efficiency ranges between 30% and 35% and combined cycle plants have efficiency in the range of 50% to 60%. The use of newer gas turbine technology has improved the combined cycle efficiency up to 65%. Approximately 3% of the capital expenditure could be saved with a 1% increase in efficiency for the same size of the plant. The steam power plant requires 4 to 5 years of time to design, construct and commission the plant, whereas the natural gas combined cycle plant requires 2 to 3 years.

The combined cycle power plant has the following advantages:
- These plant displace the coal of the steam power plant.
- It has highest efficiency up to 60% compared to steam power plant which has efficiency up to 35%.
- The plant erection time is less compared to the steam plant.
- The combined cycle plant is environmentally friendly, produces less emissions of unburnt hydrocarbons, CO and oxides of nitrogen.

In the concept of combined cycle power plant, the gas turbine cycle and steam turbine cycle are combined and gas turbine engine is used as a part of the combined cycle to improve the thermal efficiency [14]. The gas turbine plant generates electricity and high pressure steam required for the steam turbine plant is produced by exhaust heat rejected from the gas turbine plant. This produces more electricity. An after burner may be used to raise the gas turbine exhaust temperature. The coupling between two different systems is complicated and hence the design of the combined system is more complex. For the combined power plant, the maximum efficiency would be obtained at which the gas turbine cycle would have neither its maximum efficiency nor its maximum specific work output. The efficiency of the combined cycle decreases with gas turbine reheating or supplementary heating and the intercooler of the gas turbine cycle has less of an effect on the performance of the combined cycle. A comparison of simple combined cycle (combined gas turbine/ steam turbine plant) with various configurations like combined reheat gas turbine/ simple steam turbine power plant, combined intercooled gas turbine/steam turbine plant and combined steam-cooled turbine blades gas turbine/steam turbine plant is done. The cost of electric power generation is cheap in the simple combined gas turbine/steam turbine plant and efficiency and power output can be improved by adopting an intercooler (figure 1.17).

## 1.17 Cogeneration plants

Cogeneration means the production of more than one form of energy in a single plant. Cogeneration plants are used for electric power generation and steam production for the use in process industries, to generate power and exhaust gases from turbines are used for preheating air in furnaces or in absorption cooling systems or for heating fluids in various process applications. A typical cogeneration

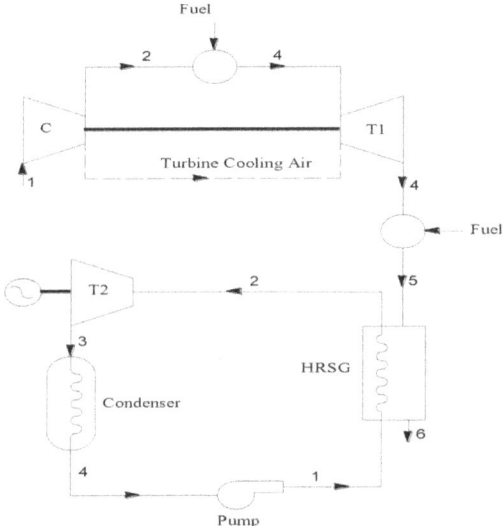

**Figure 1.17.** Schematic diagram of a combined gas-steam power plant with a waste heat recovery steam generator.

plant consists of a heat recovery steam generator (HRSG) or waste heat boiler which recovers waste heat of exhaust gases from the gas turbine to generate steam. The steam could be used to drive a cooling system, to produce more power or directly used in an absorption chiller to produce refrigerated air.

In a gas turbine, the power output decreases with the increase in atmospheric temperature. The turbine output decreases by 4% with an ambient temperature raise of 5.6 °C. The methods used to cool the compressor inlet in a gas turbine plant are:

Evaporative system — Conventional evaporative coolers or direct water fogging. The air temperature at the compressor inlet is reduced below ambient temperature by allowing air flow through media blocks. As air flows, it absorbs latent heat from water which is sprayed over media blocks and water evaporates by absorbing latent heat. This decreases compressor work and increases turbine output. In another method, the fog is produced from water by using high-pressure nozzles. This fog evaporates by absorbing heat at the compressor inlet, thus providing the cooling effect.

Refrigerated inlet cooling systems—Absorption/mechanical refrigeration:

This method is a more effective way of producing a cooling effect when compared to the evaporative cooling system as the temperature at the compressor inlet is reduced by about 25 °C to 30 °C. The absorption system uses lithium bromide as the absorber and water as the refrigerant. It can cool inlet air temperature up to 10 °C. The heat for the absorption chiller can be provided by the gas, steam or exhaust of the gas turbine. In a mechanical or vapour compression refrigeration system, the refrigerant vapour is compressed by a centrifugal compressor. After compression, the refrigerant vapour is sent to a condenser and then the condensed vapour is expanded in an expansion valve to produce the cooling effect.

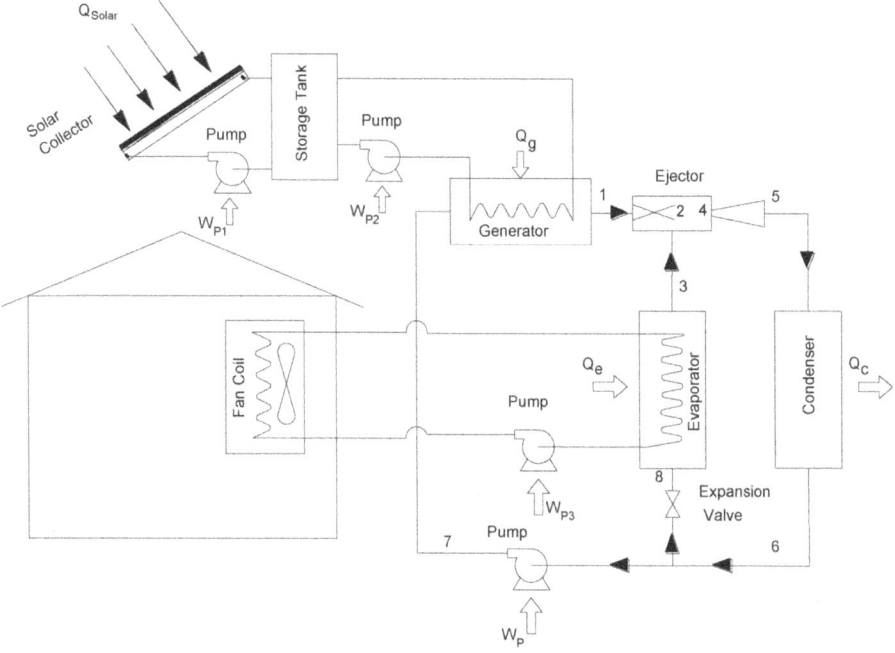

**Figure 1.18.** Schematic diagram of VJRC (reference 16).

Combined evaporative and refrigerated inlet system — he inlet air temperature is decreased by using chiller system which is assisted by the evaporative cooler.

Thermal energy storage system—These intermittent systems produce cold during off-peak period and chill the inlet air during hot hours of the day.

The gas turbine cycle plants possess the advantages of high cycle efficiency, less installation time, low carbon dioxide emissions and $NO_x$ emissions. In recent years, innovative modifications have done with the Brayton cycle like the cascaded humidified advanced turbine cycle, advanced integrated gasification combined cycle, hybrid cell combining gas turbine, a high-pressure solid oxide fuel cell, humid air turbine cycle etc., and cycle efficiency has been improved to a great extent [15].

## 1.18 Vapour jet refrigeration cycle

The use of a vapour jet refrigeration cycle (VJRC) driven by a solar-thermal system is an alternate system for a vapour compression refrigeration cycle in rural areas where there is no electricity [16]. This decreases global warming and ozone depletion. VJRC uses water as the refrigerant, simple in construction, robust, less maintenance and has no mechanical moving parts (figure 1.18).

It consists of an ejector, generator, condenser and pump in the power cycle and ejector, evaporator, expansion valve and the condenser in the cooling cycle. In the generator, heat is added to the primary fluid at high pressure. The high pressure heated primary fluid is then sent to the supersonic ejector where the velocity of fluid increases above the Mach number through the supersonic nozzle. The pressure of the working fluid decreases to evaporator pressure. This pressure drop in the nozzle entrains the secondary fluid vapour from the evaporator. The two fluids completely mix in the ejector mixing chamber and the velocity of the mixed fluid decreases below the Mach number by passing it through a diffuser. In the diffuser the fluid pressure increases to condenser pressure. Then the fluid is condensed to liquid by dissipating heat to ambient air. The circulating pump is used to pump the liquid from the condenser to the generator and this completes the power cycle. The remaining portion of the fluid expands in an expansion valve and in the form of the mixture, it enters into evaporator of the refrigeration cycle. As the mixture passes through evaporator, it absorbs heat and gets converted into vapour, thus produces the cooling effect. This vapour is again entrained by the primary fluid of power cycle and enters into the ejector to complete the refrigeration cycle.

## 1.19 Organic Rankine cycle/vapour compression cycle for producing cooling effect by utilising solar energy

In recent years, the energy required to produce the cooling effect has increased enormously due to the effect of global warming [5]. The burden on the use of fossil fuels and environmental issues can be greatly resolved by using solar energy for the cooling of food, beverage and for seafood preservation. The development of a cooling or refrigeration device driven by solar energy is very useful for food and vaccine preservation. Many researchers worked in the field of solar powered refrigerators in which energy conversion occurs by absorption/adsorption refrigeration cycle and organic Rankine cycle/vapour compression cycle (ORC/VCC).

A typical organic Rankine cycle/vapour compression cycle for producing the cooling effect by utilising solar energy is as shown in figure 1.19. It consists of a solar collector (generator), expander, compressor, condenser (one for cooling system and the other for the power system), throttle valve, evaporator and pump. The working fluid with low boiling point absorbs heat and gets evaporated in a solar collector. The vapour fluid expands in an evaporator, condensed and pumped back to the solar heater to complete the cycle. The expander drives the compressor of the refrigeration cycle and the desired cooling effect is obtained in the evaporator of the refrigeration cycle. Both ORC and VCC uses the same working fluid to avoid gas separation due to leakage. R600, R245fa and R600A can be used as working fluids.

## 1.20 Ejector-absorption combined refrigeration cycle

In this system, the COP of the cycle is improved by using an ejector to increase refrigerant pressure in the vapour absorption cycle. $H_2O$/LiBr combination is used as the working fluid in the cycle. The water vapour at the exit of the generator is divided into two paths. One enters into low pressure generator and gets condensed into saturated liquid. The liquid water is throttled in TV4 to reduce its pressure to condenser pressure before it enters into the condenser. The second path of water vapour at state 2 is injected into an ejector through a valve. The remaining vapour from the LP generator at state 6 also enters into the ejector. The mixture [2] formed due to the mixing of vapour at 7 from the ejector and stream 8 from TV4 is

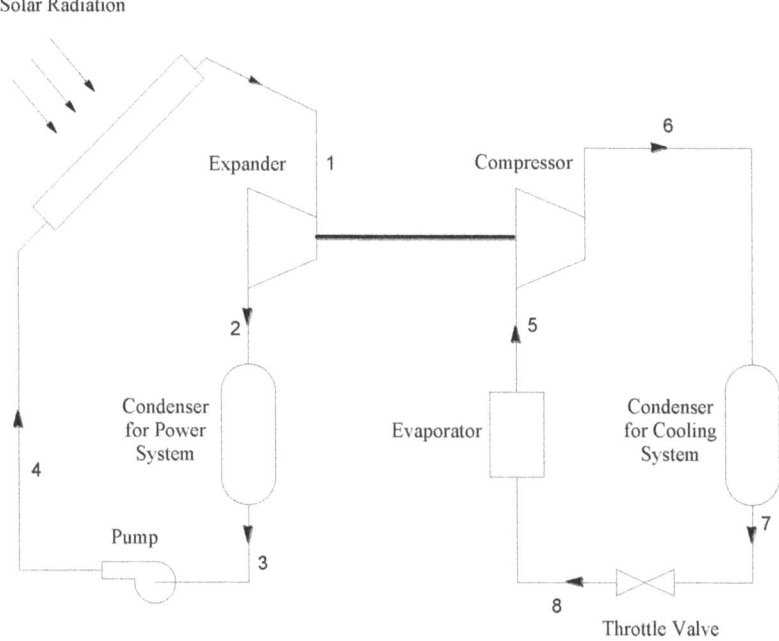

**Figure 1.19.** Schematic diagram of the system.

condensed. The pressure of the condensate decreases to evaporator pressure by TV5 before it enters into the evaporator. Further evaporation take place fully in the evaporator. The mixture formed due to the mixing of the stream from TV3 [17] and vapour from evaporator [5], enters into the absorber. The function of the absorber is similar to that of the standard vapour absorption cycle. The high pressure liquid at state 15 from the pump is divided into two paths. One path passes to SHX1 and the other to the LP generator through SHX2. The LP generator sends the ammonia solution to TV3 through SHX2 and thus completes the cycle. When the source temperature is low, the stop valve will be fully opened and the throttle valve regulates the flow rate of stream at state 4. When the source temperature is high, the stop valve will be fully closed and the cycle works as a double-effect cycle and results in higher COP (figure 1.20).

## 1.21 Absorption cycle integrated with a booster compressor

The arrangement of components for an absorption cycle with a booster compressor is as shown in figure 1.20. The use of the compressor booster improves the performance of the vapour absorption refrigeration cycle. The high concentration refrigerant solution is pumped to the generator through the solution heat exchanger. The weak solution coming from the generator exchanges heat in SHE and thus heats up the strong solution of ammonia. In the generator, ammonia is separated from strong solution and the weak solution is returned to the absorber after passing through SHE to transfer heat to a strong solution and is allowed to pass through the

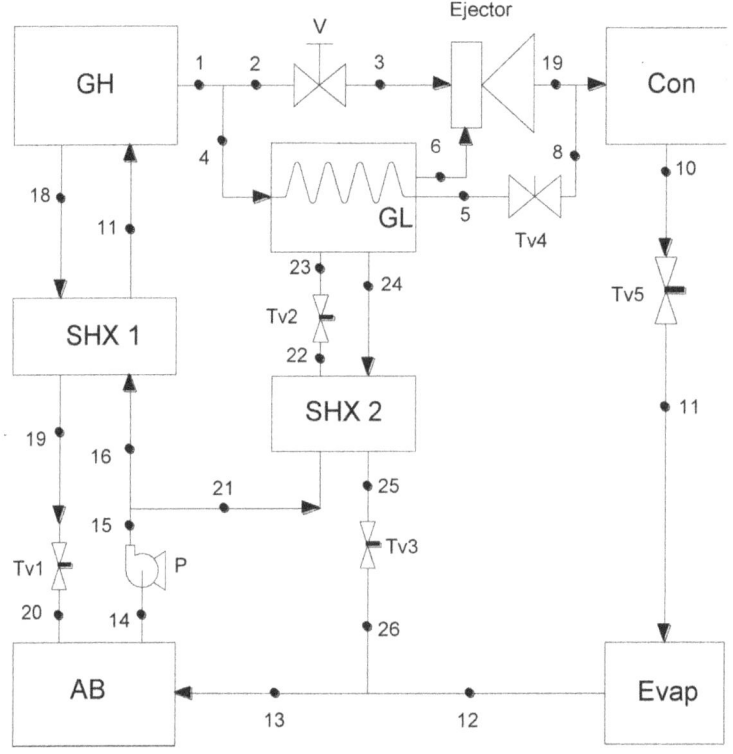

**Figure 1.20.** Block diagram of ejector-absorption combined refrigeration cycle (reference 8).

expansion valve. The ammonia vapour produced from the generator gets condensed in a condenser relatively at high pressure and decreases its pressure after passing through an expansion valve (process 2-3). In the evaporator, liquid ammonia absorbs heat and thus produces the refrigeration effect. The resulting vapour is sent to the compressor which is placed between the absorber and evaporator, and then to the absorber (figure 1.21).

## 1.22 Generator–absorber–heat exchanger (GAX) absorption refrigeration cycle

This system reduces losses due to expansion and increases COP. The saturated strong ammonia solution is pumped through the absorber to the GAX desorber. The refrigerant with high concentration of ammonia vapour is produced in the generator and weak solution goes back to the absorber through the expansion valve and GAX absorber. The liberated heat increases temperature of the solution passing through GAXD. The ammonia vapour produced in the generator passes through the GAXD, rectifier and then the condenser where it dissipates heat to surroundings and produces liquid ammonia. In the pre cooler, the liquid ammonia exchanges heat with ammonia vapour leaving the evaporator. The expansion valve decreases

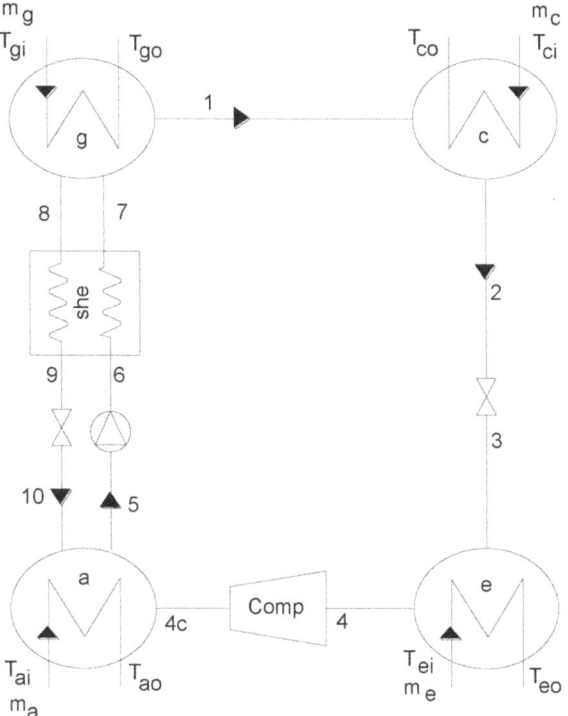

**Figure 1.21.** Block diagram of the absorption cycle integrated with a booster compressor between the evaporator and absorber (reference 8).

pressure of liquid ammonia to that of evaporator pressure. In the evaporator, it absorbs heat and thus produces the cooling effect.

The GAXA sends weak ammonia solution to the absorber where it absorbs ammonia vapour coming from evaporator to complete the cycle. The COP of the cycle varies from 0.7 to 1.1.

## 1.23 Hybrid generator–absorber–heat exchanger (HGAX) absorption refrigeration system

The arrangement of components for the HGAX cycle is as shown in figure. This cycle is similar to the GAX cycle. A compressor sends the high pressure and temperature ammonia vapour from the evaporator to the absorber. Hence the cycle recovers more heat compared to the GAX cycle. The absorber is at a lower temperature and requires a cooling medium different from atmospheric air. The COP of this cycle varies in the range of 1 to 1.88 (figures 1.22 and 1.23).

## 1.24 Triple-effect absorption refrigeration system (TEAR)

The arrangement of components for TEAR is as shown in figure 1.24. This cycle is used to improve the performance of the vapour absorption refrigeration cycle.

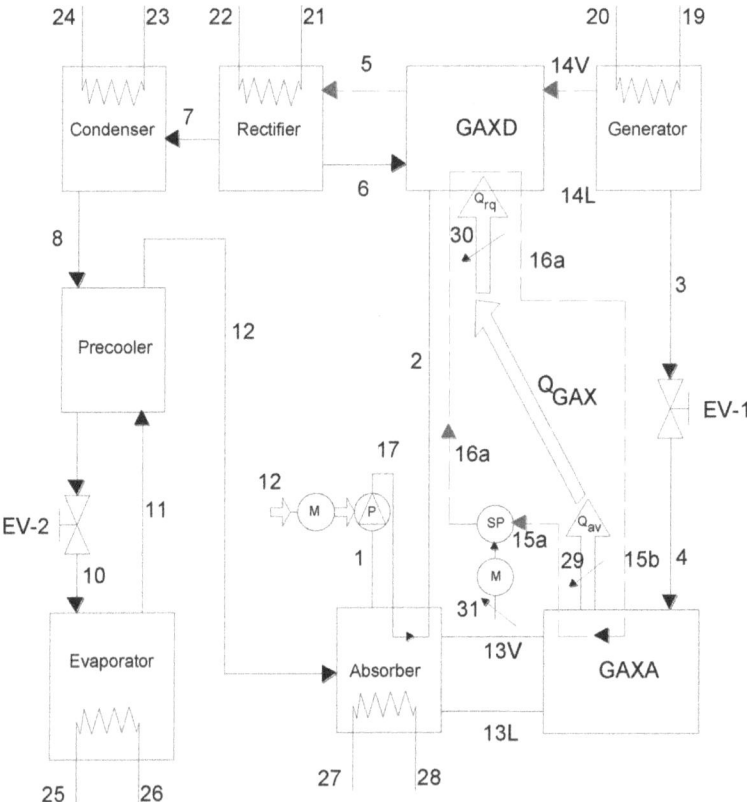

**Figure 1.22.** Block diagram for GAX absorption refrigeration cycles (reference 8).

The system consists of three sets of generators and condensers. The weak ammonia solution is pumped through three preheaters before it enters into main generator G. The ammonia vapour is generated by heating the strong ammonia solution in generator G. The ammonia vapour gets condensed in condenser C4 and heat dissipated during this condensation is utilised by generator G3. The strong solution from the main generator at state 8 is supplied to G3 after rejecting heat to strong solution in preheater 3. The generator G3 produces more refrigerant vapour. Then the refrigerant vapour enters into condenser C3 and during its condensation exchanges heat with generator G2. In the second preheater heat exchange take place between strong solution from G3 and weak solution from the first preheater and strong solution enters to G2. In G2, still more vapour is boiled off to enter into main condenser, thus the total refrigerant entering into the main condenser is the sum of refrigerants supplied from all generators. After condensation in the main condenser, the liquid refrigerant is throttled and entered into the evaporator where it absorbs heat to produce the cooling effect (figure 1.24).

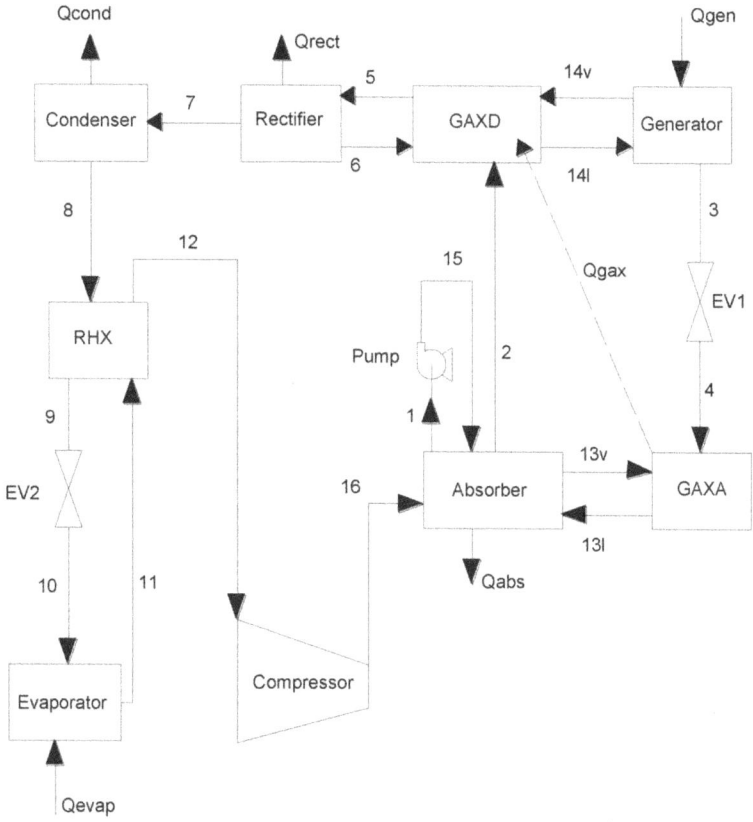

**Figure 1.23.** Block diagram of a hybrid GAX absorption refrigeration cycle (reference 8).

## 1.25 Thermodynamic optimization of combined gas/steam power plants

The energy and exergy analyses is done for the different configuration lime steam bottoming cycle with the heat recovery steam generator, heat exchanger and secondary bottoming cycle [18]. The arrangement of components for a combined gas/steam power cycle with a steam generator for heat recovery is as shown in figures 1.25, 1.26 and 1.27. A heat exchanger is used to recover the heat of exhaust gases from the gas turbine and this heat is used to generate steam, which can be used as working fluid to run Rankine cycle. The power output of this cycle is limited to 1000 kW. The temperature range exhaust gases in the heat exchanger ranges from 727 K to 1359 K.

In another configuration, an additional heat exchanger is used between the compressor and combustion chamber of the topping cycle to extract the energy of exhaust gases. The energy extracted is used to preheat the air before it enters into the combustion chamber, thus reducing the fuel required for combustion and increases thermal efficiency.

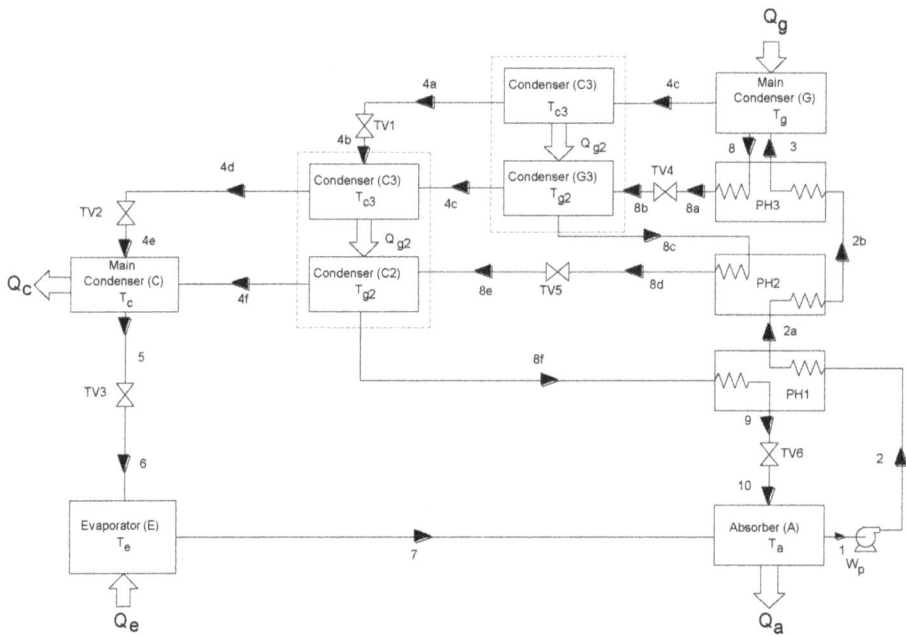

**Figure 1.24.** Block diagram of the triple-effect absorption refrigeration system (reference 8).

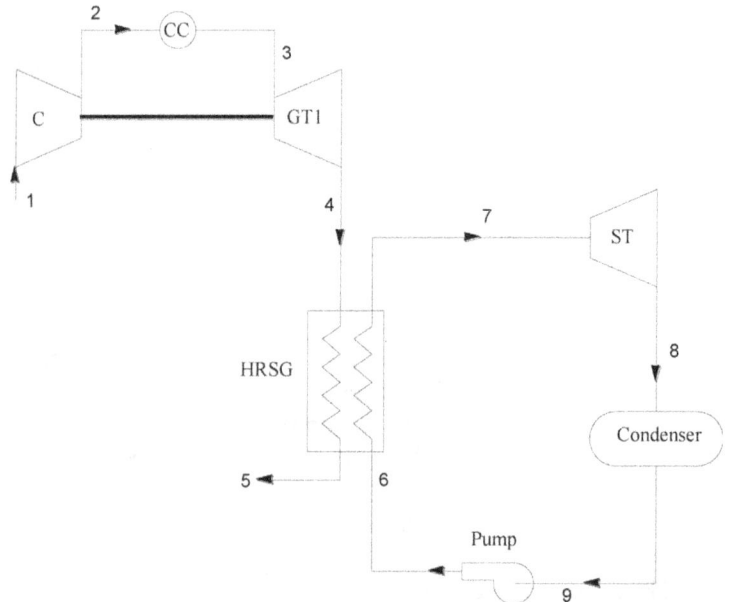

**Figure 1.25.** Schematic diagram of the combined gas/steam cycle with HSRG.

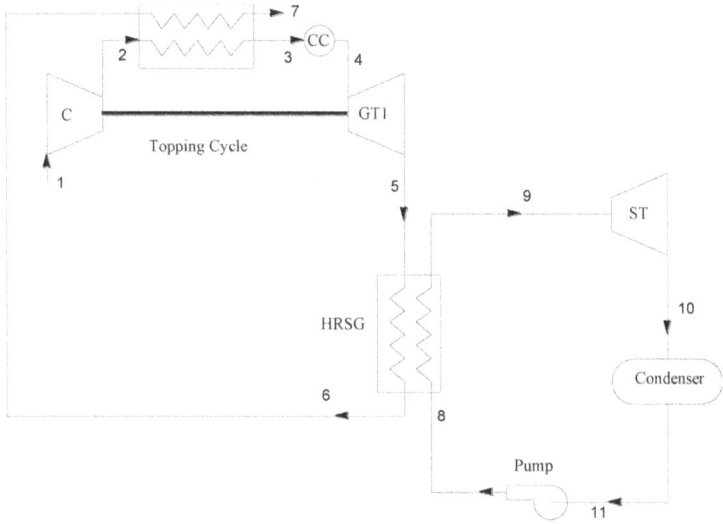

**Figure 1.26.** Schematic diagram of the combined gas/steam cycle with the HSRG and heat exchanger.

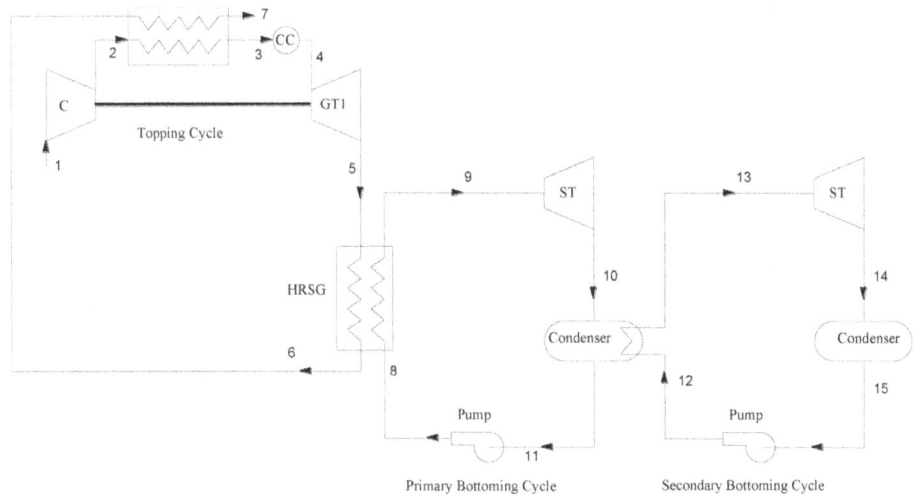

**Figure 1.27.** Schematic diagram of the combined gas/steam cycle with the HSRG, heat exchanger and secondary bottoming cycle.

A secondary bottoming cycle may be added to utilize the heat loss from the condenser of the Rankine or bottoming cycle. The refrigerant R134 can be used as working fluid for the secondary bottoming cycle. The steam flow rate can be adjusted in such a way that the temperature of gas leaving the heat exchanger used in the primary bottoming cycle should be greater than the temperature of the gas leaving the compressor. The mass flow rate of the refrigerant should be such that the power output of the secondary cycle should be less than that of the primary bottoming cycle.

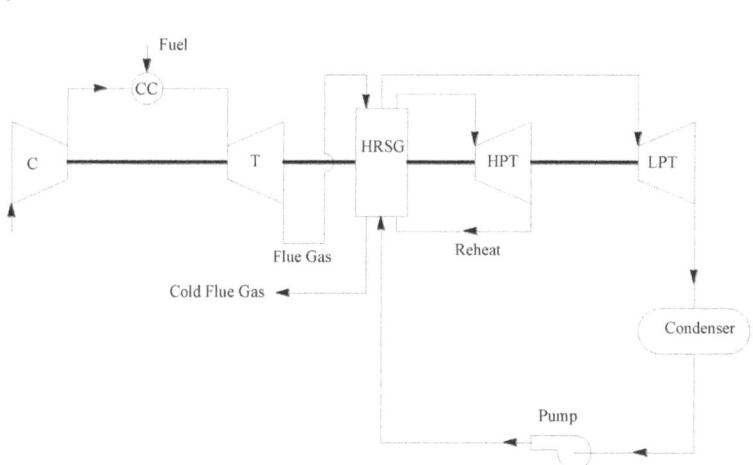

**Figure 1.28.** Block diagram of a combined cycle power plant.

One of the major challenges for mankind is to convert the present energy source into a sustainable one [19]. In the last few decades, innovative technologies have been developed for efficient energy conversion. The fossil fuels such as oil, gas and coal are non-renewable in nature and consist of hydrocarbons. After burning, these fossil fuels release an enormous amount heat energy, carbon dioxide and water vapour. Finally the heat energy is used in turbines for power generation through generators.

The combined cycle power plant is the combination of gas turbine and steam turbine in which the heat of the exhaust from the gas turbine is used to generate steam in a boiler, and the generated steam is used to run a steam turbine to produce power [19]. The combined plants can work with higher efficiency in the range of 50% to 60% with low pollutant emissions. These plants may consist of a single shaft, in which case both the gas turbine and steam turbine are connected to one generator in a tandem arrangement or may consist of multi shafts in which a separate generator is used for both the gas turbine and steam turbine.

The high-pressure, high-temperature gas stream can be generated by several methods [20]. The high pressure and temperature gas stream can be generated by use of the pressurized fluidized bed combustion plant, high-pressure molten carbonate fuel cell and high-temperature solar power plants. This stream of gas is used to drive the gas turbine and exhaust of the turbine may be used to produce steam to run a steam turbine or can provide fuel for fuel cells. In general, a combined cycle with a gas turbine as the topping cycle and steam turbine as the bottoming cycle is used. A third cycle may also be included to recover the low-grade heat remaining after steam generation. This can be accomplished by using a closed cycle turbine such as an organic Rankine cycle. These cycles convert low-grade heat energy into electricity and are used in geothermal plants where the geothermal reservoir temperature is low.

The arrangement of components for a typical combined cycle is as shown in figure 1.28. It consists of a compressor, gas turbine, heat recovery steam generator and a steam turbine. The compressor, combustor and gas turbine forms the topping cycle and generates its own electric power. Then the steam bottoming cycle is added. The exhaust of the gas turbine is at a high temperature of around 400 °C to 500 °C and this waste heat is recovered in HRSG to generate steam. Most of the heat available in the exhaust of the gas turbine is used to generate steam. The steam is available at different pressure levels in HRSG. The steam at highest pressure is sent to first HP stage of 2 stage steam turbine, then it is reheated before it is supplied to second and intermediate stages. The last pressure level is used for steam cooling. Then the steam is expanded in a steam turbine to produce more electricity.

The use of newer materials and innovative designs resulted in the improvement of efficiency of gas turbine cycles. A maximum inlet temperature of around 1600 °C is achieved with new design concepts and efficiency of more than 60% is obtained. In a standalone gas turbine, the highest efficiency is possible with the lowest exhaust gas temperature. But for the combined cycle, it is required to allow the exhaust gases at significantly high temperature from the turbine. So a balance of energy recovery in each cycle is required.

## 1.26 Integrated solar combined cycle power plant (ISCC)

The combined cycle power plant uses various energy sources to reduce the cost of power generation and to improve the efficiency. The integrated solar combined cycle plant is one in which the solar heat energy is collected and added to the obtained by burning the fuel in a conventional plant [20]. An array of parabolic trough solar collectors is used to extract the solar thermal energy. The tube arrangements are made along the length of each trough at the focus of the parabola in such a way that the energy collected in the solar collector is focussed to transfer its heat energy to fluid flowing through the trough. The temperature of the working fluid in the pipe can be increased up to 550 °C by pumping it through a number of parabolic trough collectors. The heated fluid is then sent to a heat exchanger which is the part of HRSG of the combined plant, where steam is generated. This improves efficiency of the combined plant. Ideally the solar input should be 10% or even less than the total energy input. Thermal energy storage systems may be included to account for fluctuation in the availability of solar energy and the day–night cycle. The start-up time for power generation from the steam turbine is less (figure 1.29).

## 1.27 Supercritical-$CO_2$ closed Brayton cycle ($sCO_2$-CBC) control in a concentrating solar thermal (CST) power plant

The role of small- and medium-scale concentrating solar thermal power plants is significant in power generation and can replace the expensive and polluting diesel generator-based power plants. Furthermore, the cost of electric power generation can be greatly reduced by using a supercritical carbon dioxide closed Brayton cycle power plant. The thermal efficiency of these plants is as high as 50%, the working fluid used is cheap, non-toxic, non-flammable, stable at high temperature, has low critical pressure and critical temperature. The power plant is more compact, simple in design, avoids two-phase flow complications and eliminates the heat transfer losses in intermediate heat exchangers.

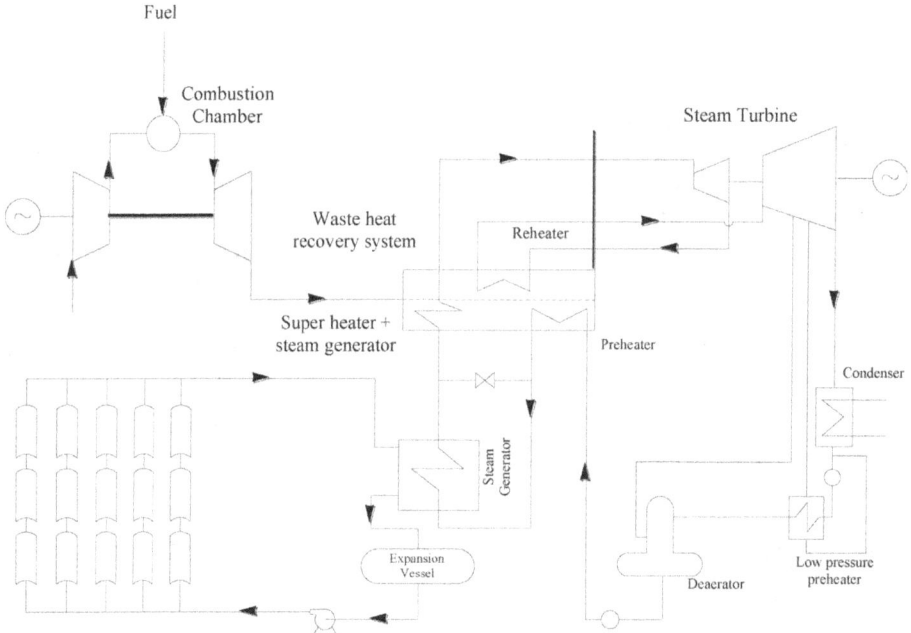

**Figure 1.29.** Schematic diagram of the integrated solar combined cycle power plant (reference 20).

The overall solar to electric conversion efficiency can be greatly improved by using a direct heated concept in which the working fluid carbon dioxide is used as a heat collection medium. This reduces the thermal losses in the intermediate heat exchanger. Figure 1.1 shows the block diagram of a simple recuperated $SCO_2$ CBC in a direct heated and air cooled CST power plant. It consists of a compressor to increase pressure of the working fluid $CO_2$, a heater to replace solar collector receiver, a recuperator to recover the available/unused energy of the turbine, and a cooler to absorb heat rejected from the cycle. The working fluid $CO_2$ expands in the turbine and produces useful work at high pressure and temperature and the generator coupled to the turbine produces electricity. The compressor and turbine are mounted on a common shaft rotating at a constant synchronous speed of 60 000 rpm through a gear box. The gear box is used to decrease the generator speed to grid frequency of 50 Hz. The heat exchangers used in the plant are the absorber tube of the solar collector receiver, recuperator and a cooling tower in which $SCO_2$ cooling takes place by exchanging heat with air (figure 1.30).

## 1.28 Cascaded humidified advanced turbine (CHAT)

An innovative modification to the Brayton cycle has been made in the cascaded humidified advanced turbine (CHAT) plant [11]. The gas turbine-based power generation plant uses intercooling reheat and humidification. The existing heavy duty gas turbine is integrated with an additional shaft consisting of industrial compressors and high pressure expander to improve efficiency of the gas turbine.

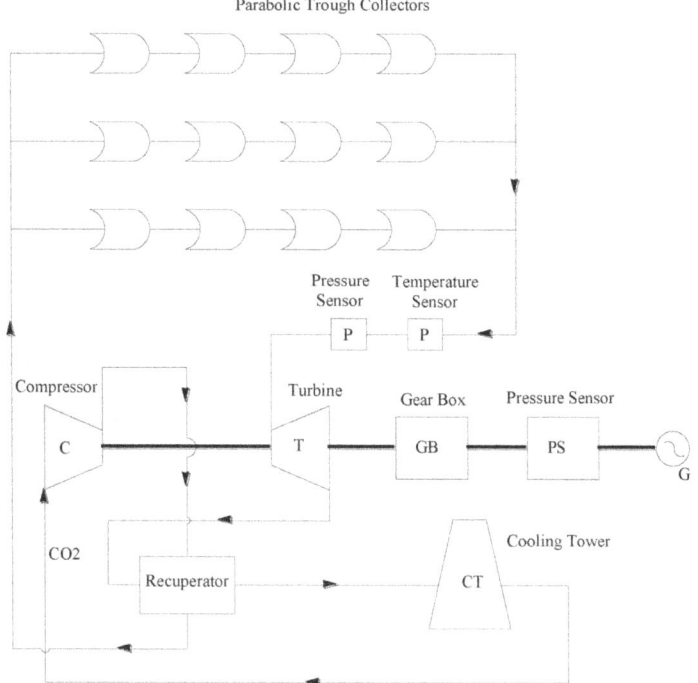

**Figure 1.30.** Schematic diagram of $SCO_2$ CBC power plant (reference 21).

The system offers lower emissions due to air humidification. The effective integration of CHAT is possible with coal gasification.

The arrangement of components in the CHAT plant is as shown in figure 1.31. The components of the CHAT cycle are as follows:

A power generation shaft with the LP turbine, combustor and LP compressor, generator, shaft auxiliaries and controls.

The power-balanced second shaft consisting of an intermediate compressor, high pressure compressor, high pressure combustors and turbine assembly. The IP and HP compressors are driven by power obtained from HP turbine.

Different types of heat exchangers which includes compressor intercoolers, recuperators, water heaters/economizers and fuel heaters.

Air saturators/humidifiers which is a packed column for direct heat and mass transfer from hot water to moving air stream.

The LP compressor draws atmospheric air on the power generation shaft and discharge it to the IP compressor on the second shaft after passing through the water-cooled intercooler. Then the air delivered to a HP compressor through the second intercooler. In the HP compressor, the air is compressed to system pressure and is sent to the air saturator or humidifier where the humidification and preheating of air takes place by the heat and mass exchange. The air stream is then passed to the recuperator for further preheating by using turbine exhaust heat. Next the air stream is sent to tbe HP combustor/turbine assembly where the fuel addition takes place and

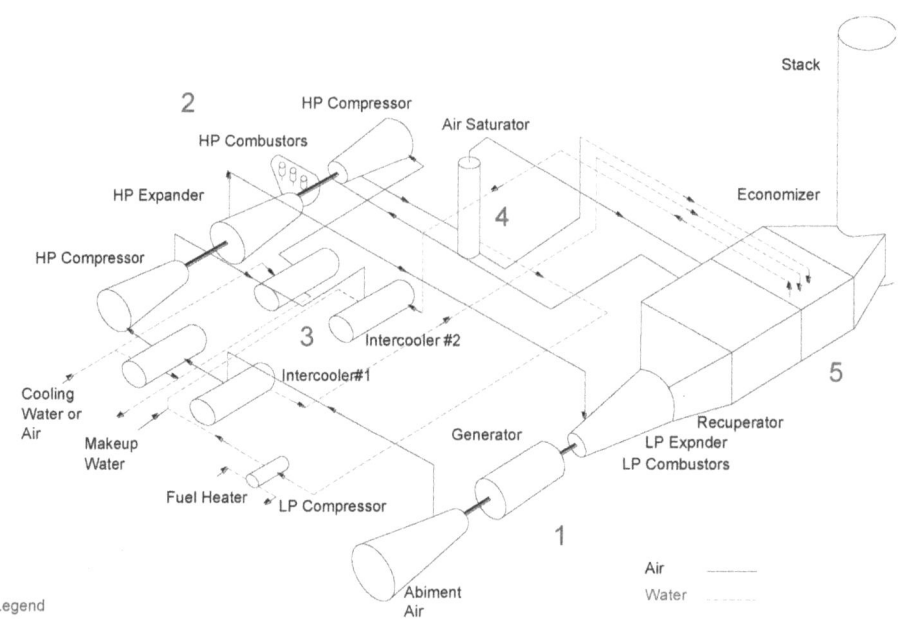

**Figure 1.31.** Details of arrangement of components in the CHAT plant.

Legend
1 : Modified Wresting house 501 F combustion turbine
2 :Trubo-expander and compressors package supplied by dresser- rand
3 : Inter-cooler - Typical shell and tube heat Exchange
4 : Air Saturator - Typical industrial column
5 : Heat recovery unit - Consisting of a Recuperate and water heater i similar to an HRSG

the hot gas is expanded in the HP turbine before it is supplied to the LP combustor/turbine. The temperature of the flow is increased to the desired value in LP combustors. The exhaust of the turbine is sent to the recuperator and water heating sections, and is exhausted through the stack.

The main features of CHAT power plant are:

The discharge pressure of LP compressor is greatly reduced by using the power-balanced shaft. For the same overall cycle pressure ratio, the LP compressor pressure ratio is significantly less and this decreases its power consumption. For the same turbine power generation, the LP compressor consumes less power and that increases net power obtained by the cycle.

The power required by the LP compressor is further reduced by using motive air humidification as it decreases air flow. The cycle heat is effectively utilised due to the presence of moisture content in the air.

The use of the saturator and recuperator increases the temperature of the motive air, resulting in the decrease of fuel consumption by HP combustors.

## 1.29 Advanced integrated coal gasification combined cycle

The thermal efficiency of the conventional coal fired power plants can be greatly improved by using the integrated coal gasification combined cycle (IGCC) and the integrated coal gasification fuel cell combined cycle (IGFC) [22].

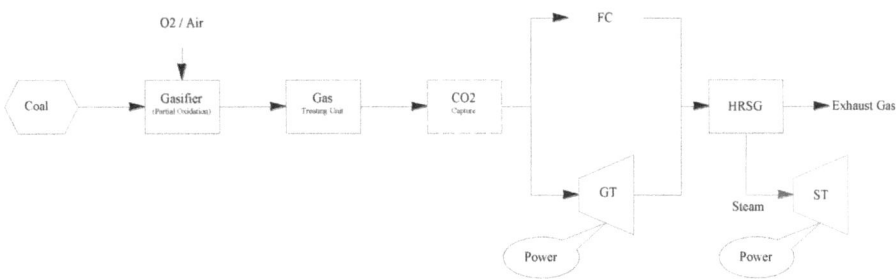

USC-Ultra super critical
FC- Fuel cell
GT- Gas turbine
ST- Steam turbine
HRSG- Heat recovery steam generator

**Figure 1.32.** Details of the arrangement of components in the IGCC/IGFC plant.

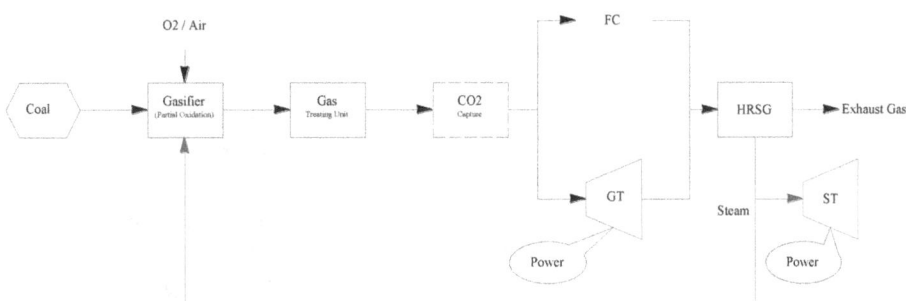

**Figure 1.33.** Details of the arrangement of the components in the advanced IGCC/IGFC plant.

Figure 1.32 shows the arrangement of components for the IGCC system. In this system, coal particles are pulverised and injected into an entrained flow gasifier where endothermic reaction occurs and some coal burns by using oxygen or air to supply heat and this results in the formation of $H_2$ and CO. Then the cleaned flue gas enters into the gas turbine for power generation. A heat recovery steam generator is used to recover waste heat from the exhaust of the gas turbine. The resulting steam from the steam generator is supplied to the steam turbine for further power generation.

The thermal efficiency of the plant ranges from 45% to 50%. The carbon dioxide emissions are greatly reduced by using solid oxide fuel cells (figure 1.33).

## 1.30 Advanced integrated coal gasification combined cycle

The thermal efficiency of coal fired plants is further increased by using advanced IGCC/IGFC. In this system, the heat required for the endothermic gasification

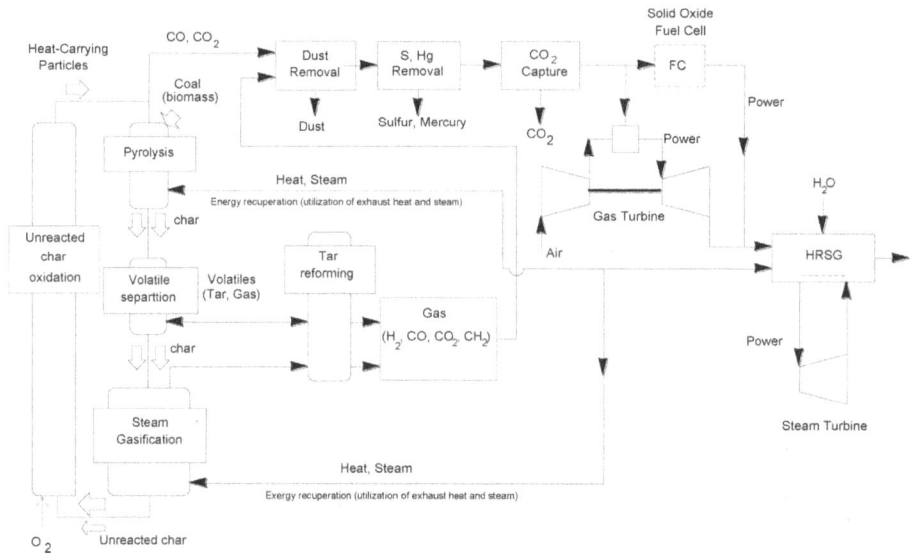

**Figure 1.34.** Schematic diagram of the A-IGCC/IGFC system with the TBCFB gasifier.

reaction is obtained by the heat of exhaust gases from the gas turbine and SOFC. The resulting exergy of H2 and CO gases (chemical energy) from the gasification process is greater than the thermal energy of the exhaust and is recuperated by gasification. This decreases the oxygen required in the gasifier and hence increases thermal efficiency of the IGCC system.

A triple-bed combined circulating fluidized bed (TBCFB) gasifier is used for the A-IGCC/IGFC system. The schematic diagram of the arrangement of components is as shown in figure 1.34. The TBCFD consists of a downer pyrolyzer for the pyrolysis of coal to produce volatiles (gas and tar) and solid char, a bubbling fluidized bed char gasifier and a riser unreacted char combustor. In the downer pyrolyzer, immediately after the coal particles heat up, tar decomposes and volatile and char are gets separated in a gas-solid separator before the char enters into the gasifier.

The fluidized bed gasifier uses steam from the HRSG for gasification of char, which is separated and the unreacted char enters into the riser combustor. The heat required for endothermic pyrolysis and gasification reactions is obtained combustion. The heat of the steam from HRSG is used for pyrolysis. The additional tar-reformer transfers tar in the separated volatile to the gas turbine for power generation after passing through the gas cleaner.

## 1.31 The solid oxide fuel cell (SOFC) and gas turbine (GT) hybrid system

The efficiency of power generation can be greatly improved by using fuel cells with conventional turbine power plants. The higher operating temperature of some fuel cells, like solid oxide fuel cells (SOFCs) and molten carbonate fuel cells (MCFCs) etc., makes them suitable for use in hybrid or combined systems [23]. The use of SOFC power system produces less harmful chemical and acoustic emissions than conventional systems and are the most promising potential alternative power generation systems.

## 1.32 Solid oxide fuel cell (SOFC)–gas turbine hybrid system

The Brayton cycle is the reference cycle used for the working of the gas turbine plant. The gas turbine plant consists of a compressor, combustor and a turbine. The cycle may consists of several compressors and turbines. The plant efficiency ranges between 30% and 40% and it goes up to 60% by adding topping cycle. A gas turbine may be directly or indirectly connected to the SOFC. In case of indirect connection, a heat exchanger replaces the combustor of the gas turbine and the compressed air is heated by using the fuel cell exhaust. The SOFC operates at atmospheric conditions.

Figure 1.35 shows block diagrams for the direct connection of SOFC with the gas turbine. The SOFC and after burner are used instead of the combustor of the gas turbine. The compressor delivers high pressure air to SOFC. The exhaust of SOFC enters into afterburner and high temperature and pressure exhaust enters into turbine. The stream entering into the SOFC is preheated by recovering waste heat from the turbine exhaust in a heat exchanger.

## 1.33 Helium Brayton cycles with solar central receivers

The helium Brayton cycles are used to achieve higher thermal efficiency [24]. Helium possesses good heat transfer characteristics and a closed helium cycle is suitable for cogeneration applications as the rejected waste heat is available relatively at high temperature. The higher cycle efficiency can also be obtained by using the super-critical Brayton cycle, which uses carbon dioxide as the working fluid. The fluids show heat transfer characteristics similar to liquids in the supercritical region and thus reaches the highest temperatures. This decreases compressor work and increases the efficiency of the regenerative heat exchanger.

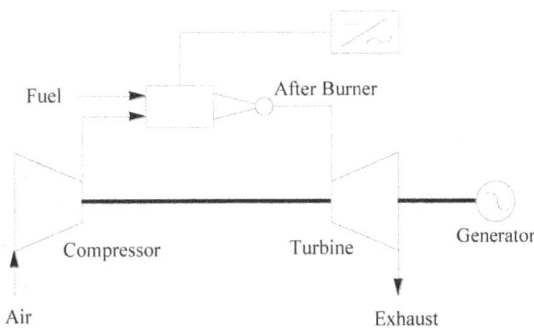

**Figure 1.35.** Gas turbine engine as a bottoming cycle in a SOFC-GT hybrid system.

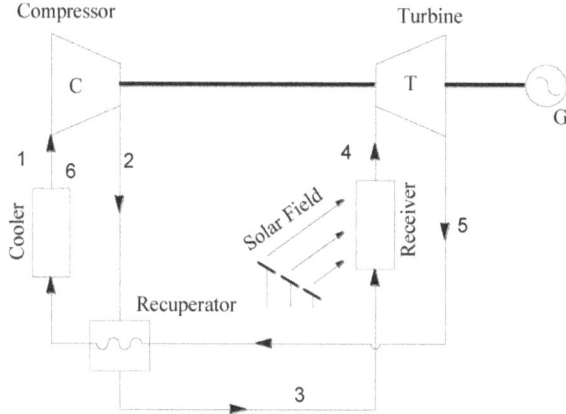

**Figure 1.36.** Block diagram of a typical helium Brayton cycle with single compression.

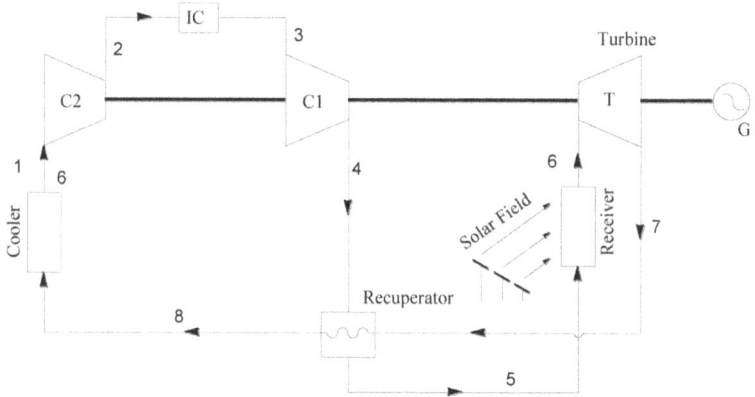

**Figure 1.37.** Block diagram of a typical helium Brayton cycle with two-stage compression.

The helium Brayton cycle shown in figure 1.36 consists of a compressor and expander or a turbine [25]. The exhaust of the turbine passes to a recuperator where the heat of exhaust gases is recovered to preheat the compressor outlet fluid before it enters into the solar receiver. A cooler is placed between recuperator and compressor and cools the helium before compression. The temperature at the inlet of the compressor is 303.15 K and pressure ratio is 2.53.

The efficiency of the cycle can be further improved by using two compressors with an intercooler between the stages as shown in figure 1.37. The overall pressure ratio is maintained 2.53. After the first stage of compression, the helium is cooled in an intercooler to its initial temperature of 303.15 K and then enters into the second compressor. The use of multi-stage compression improves the thermal efficiency by 3% to 4% compared to single-stage compression.

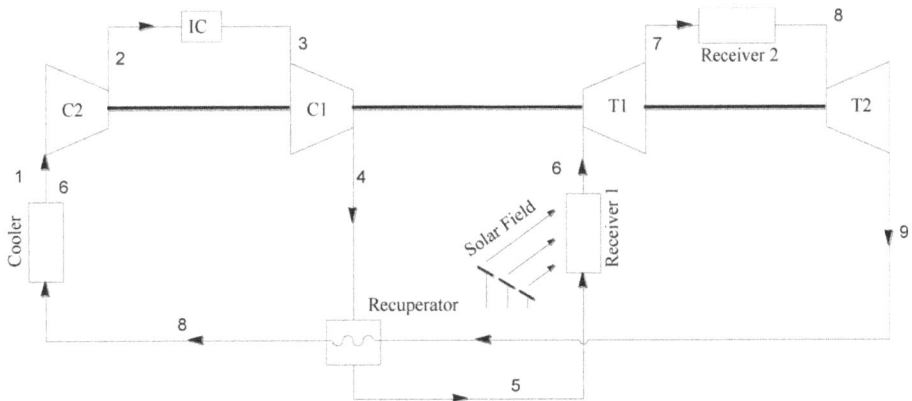

**Figure 1.38.** Block diagram of a typical helium Brayton cycle with two-stage compression and two-stage expansion.

The thermal efficiency of the helium Brayton cycle is further improved by using two-stage expansion and including a sequential heating as shown in figure 1.38. The cycle consists of two compressors with an intercooler and expansion is also divided among two turbines. There are some additional pressure losses in the second receiver and hence the sequential heating is not recommended. However, the thermal efficiency is increased by 2.5% when compared to the previous cycle.

## 1.34 Advanced power cycles for concentrated solar power

The advanced power cycles for concentrated solar power (CSP) have been introduced [17]. The gas turbine combined cycle with CSP is one of the innovative concept available with the highest efficiency. The increase in turbine inlet temperature increases cycle efficiency, thus CSP provides higher temperatures to match the most efficient cycles. A number of configurations are available to convert concentrated solar energy into electricity, one is Brayton cycles based on the closed loop supercritical carbon dioxide and the other is the air turbine combined cycle.

## 1.35 Solar gas turbine systems

The use of solar gas turbine systems results in a significant decrease of size of solar field and this reduces the cost required for the CSP system. In this system, the pressurized air is partially or fully heated by solar energy and the conventional fuel combustor is either supported or fully replaced by a solar receiver. The figure shows one such solar gas turbine system with a bottoming steam cycle.

The temperature at the inlet of the turbine is around 1500 °C in modern gas turbines. The CSP systems will not provide such high temperatures. The parabolic dishes and solar towers can generate temperature up to 1000 °C. Two methods are adopted to overcome this problem: they are (i) solar system can be used as a

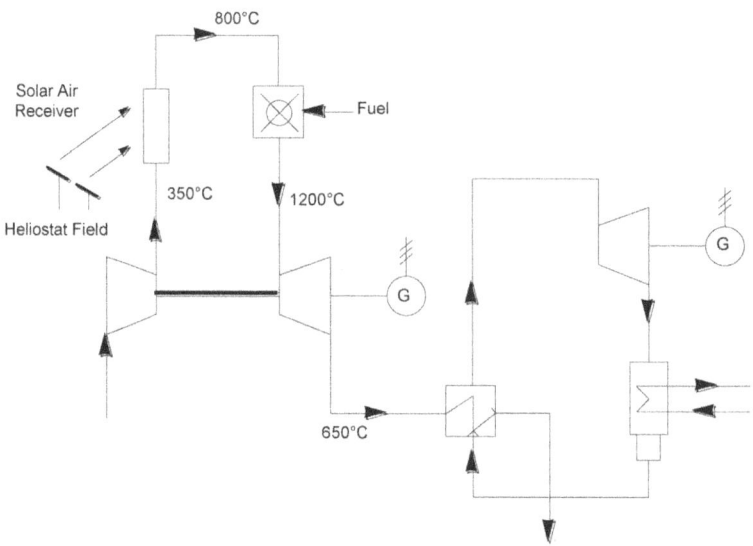

**Figure 1.39.** Block diagram of a solar gas turbine system with the bottoming steam cycle.

preheater to obtain required turbine inlet temperature with fuel combustion, (ii) develop solar specific gas turbine models to obtain turbine gas inlet temperature around 1000 °C. The following modifications are required in the conventional gas turbine to adopt solar heating:

Air path—Extraction of compressed air and reintroduction of preheated air

Solarized combustor in parallel or series connection

Optimization for solar operation

Figure 1.39 shows the arrangement of components for a solar gas turbine with the combined cycle configuration. The combined cycle configuration is similar to conventional power plants.

## 1.36 Integrated solar combined cycle systems (ISCCS) and the bottoming cycle storage systems (BCSS)

The system is based on the use of decoupled Brayton and Rankine combined cycles. In the integrated solar combined cycle system, only conventional fuel drives the topping gas turbine cycle and the bottoming steam cycle is operated by solar energy by adding solar generated steam. In the bottoming cycle storage system, solar mode or solar-hybrid mode drives the gas turbine. The exhaust of the turbine is used to run the bottoming cycle and to charge the medium temperature storage system. The bottoming cycle directly uses the gas turbine exhaust or storage system or can operate in a mixed mode (figure 1.40).

A higher power output is possible during sunshine hours as power is delivered from both the gas turbine and bottoming cycle and the bottoming cycle works for the remaining time to generate lower power when sunshine is not available [26].

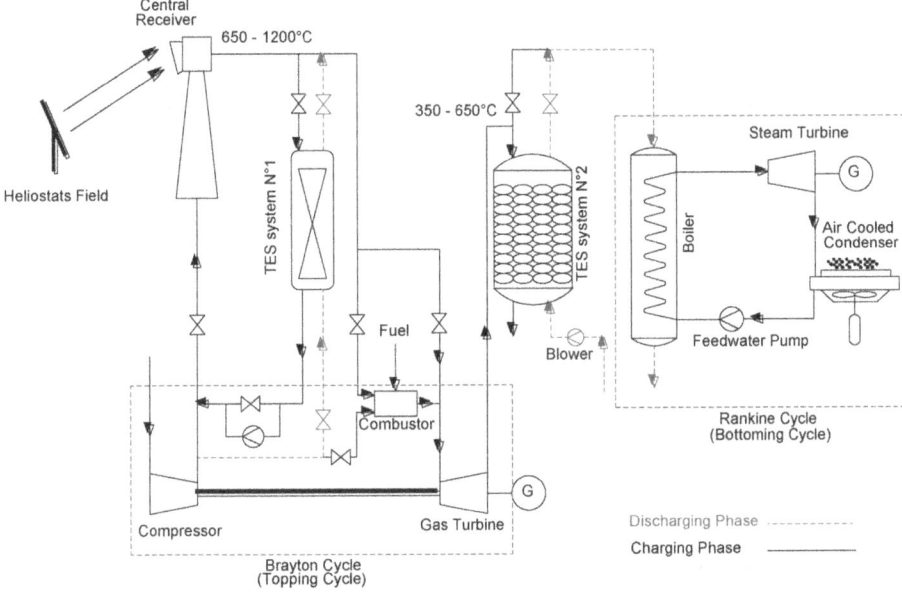

**Figure 1.40.** Gas turbine with a bottoming cycle storage system.

# References

[1] Saadatfar B, Reza F and Fransso T 2014 Thermodynamic vapour cycles for converting low-to medium-grade heat to power: a state-of-the-art review and future research pathways *J. Macro Trends Energy Sustain. JMES* **2**

[2] Hou S, Wu Y, Zhou Y and Yu L 2017 Performance analysis of the combined supercritical CO2 recompression and regenerative cycle used in waste heat recovery of marine gas turbine *Energy Convers. Manag.* **151** 3–85

[3] Sun L, Han W, Jing X, Zheng D and Jin H 2013 A power and cooling cogeneration system using mid/low-temperature heat source *Appl. Energy* **112** 886–97

[4] Yin J, Yu Z, Zhang C, Tian M and Han J 2018 Thermodynamic analysis and multi-objective optimization of a novel power/cooling cogeneration system for low-grade heat sources *Energy Convers. Manag.* **166** 64–73

[5] Yu Z, Han J, Liu H and Zhao H 2014 Theoretical study on a novel ammonia–water cogeneration system with adjustable cooling to power ratios *Appl. Energy* **122** 53–61

[6] Spelling J, Laumert B and Fransson T 2014 Advanced hybrid solar tower combined-cycle power plants *Energy Procedia* **49** 1207–17

[7] Shankar R and Srinivas T 2014 Parametric optimization of vapor power and cooling cycle *Energy Procedia* **54** 135–41

[8] Peyyala N R and Govindarajulu K Thermodynamic analysis of 1200 MW coal based supercritical thermal power plant with single and double reheating *IJESRT Int. J. Eng. Sci. Res. Technol.* http://www.ijesrt.com 677–84

[9] Ho T, Mao S S and Greif R 2012 Comparison of the Organic Flash Cycle (OFC) to other advanced vapor cycles for intermediate and high temperature waste heat reclamation and solar thermal energy *Energy* **42** 213–23

[10] Ebaid S Y M and Al-hamdan Q Z 2015 Thermodynamic analysis of different configurations of combined cycle power plants *Mech. Eng. Res.* **5** 89–113

[11] Sider I and Sider K 2020 Evaluation of vapor jet refrigeration cycle driven by solar thermal energy for air conditioning applications: case study *Adv. Sci. Technol. Eng. Syst. J. ASTESJ* **5** 646–52

[12] Hu B, Bu X and Ma W 2014 Thermodynamic analysis of a rankine cycle powered vapor compression ice maker using solar energy *Sci. World J.* **2014** 742606

[13] Kishkin A, Delkov A V and Melkozerov M G 2017 Theoretical and experimental research of organic Rankine cycle steam turbine plants *IOP Conf. Series: Materials Science and Engineering* **255** 012012

[14] Dubey A M 2020 Modified vapour absorption refrigeration cycles *Int. J. Ambient Energy* (London: Taylor and Francis)

[15] Khan M N and Tlili W A K 2017 Thermodynamic optimization of new combined gas/steam power cycles with HRSG and heat exchanger *Arab. J. Sci. Eng.* 4547–58 (Berlin: Springer)

[16] Breeze P 2016 Combined cycle power plants *Gas-Turbine Power Generation* (Amsterdam: Elsevier) pp 65–75

[17] Singh R, Manzie C and Kearney M P 2014 Improving performance of extremum-seeking control applied to a supercritical-Co2 closed Brayton cycle in a solar thermal power plant *The 4th Int. Symp. - Supercritical CO2 Power Cycles* (*Pittsburgh, PA, 9–10 September*)

[18] Islam M M, Hasanuzzaman M, Pandey A K and Rahim N A 2020 *Energy for Sustainable Development* (Amsterdam: Elsevier) pp 19–39

[19] Boyce M P 2012 *Combined Cycle Systems for Near-Zero Emission Power Generation* (Woodhead Publishing Limited) pp 01–43

[20] Bhargava R K, Bianchi M, De Pascale A, Negri di Montenegro G and Peretto A 2007 Gas turbine based power cycles - a state-of-the-art review *Int. Conf. on Power Engineering-2007* (*Hangzhou, China, 23–27 October*) pp 309–19

[21] Nakhamkin M, Swensen E C, Wilson J M, Gaul G and Polsky M 1996 The cascaded humidified advanced turbine (CHAT) *J. Eng. Gas Turbines Power* **118** 565–71

[22] Fushimi C and Guan G 2017 Advanced integrated coal gasification combined cycle: current status of development *Encyclopedia of Sustainable Technologies* (Amsterdam: Elsevier) Vol 3 pp 423–39

[23] Saisirirat P 2015 The Solid Oxide Fuel Cell (SOFC) and Gas Turbine (GT) hybrid system numerical model *Energy Procedia 79 Int. Conf. on Alternative Energy in Developing Countries and Emerging Economies* pp 845–50

[24] Kusterer K, Braun R, Moritz N, Lin G and Bohn D 2012 Helium brayton cycles with solar central receivers: thermodynamic and design considerations *Proc. of ASME Turbo Expo 2012*

[25] Dunham M T and Iverson B D 2014 High-efficiency thermodynamic power cycles for concentrated solar power systems *Renew. Sustain. Energy Rev.* **30** 758–70

[26] Stein W H and Buck R 2017 Advanced power cycles for concentrated solar power *Solar Energy* (Amsterdam: Elsevier)

**IOP** Publishing

Thermodynamic Cycles for Renewable Energy Technologies

**K R V Subramanian and Raji George**

# Chapter 2

# Vapour cycles for concentrating solar power generation using novel working fluids

**K Ravi Kumar, Naveen Krishnan and K S Reddy**

Concentrating solar power (CSP) is one of the most promising techniques to convert solar radiation into electricity with high dispatchability. Based on the heat extraction from the CSP technologies, it can be classified into direct and indirect steam generation. On the basis of the number of working fluids, the Rankine cycle can be classified as the primary, secondary and tertiary cycles. In this chapter, novel working fluids such as liquid metals and organic fluids along with water are discussed for the above-mentioned Rankine cycles. The cycle efficiency for various working fluids with and without nanofluids in the organic Rankine cycle are discussed. 4-E (Energy-Exergy-Environment-Economic) analysis of power generation from various CSP technologies are discussed. Also, the recent advancements in the solar concentrator, falling particle receiver, integration of thermal energy storage, and the deployment of supercritical $CO_2$ as the heat transfer fluid are discussed.

## 2.1 Introduction

The energy demand is increasing continuously due to rapid industrialization and improved lifestyles. Solar, wind, hydro, biomass, geothermal, etc., are the natural resources that can be used to satisfy the energy demand in several countries, since the renewable energy sources are sustainable and mitigating the emission of $CO_2$ into the atmosphere. Solar energy has the greatest potential among renewable energy technologies. Concentrating solar power (CSP) plants have attracted wide interest in recent years since the electricity generation method is clean by harnessing solar energy. The process of generating electricity requires a high temperature and the CSP technologies can generate high-temperature heat. It can be achieved using a concentrator to reflect solar radiation on the absorber surface. The heat can be directly used to generate steam or by circulating primary fluid to collect the heat and transfer it to working fluid using the heat exchanger to generate steam. It drives the steam turbine to generate electrical energy [1]. During the daytime, the thermal energy collected by the CSP systems is directly used to generate steam and in the course of cloudy periods and off-sunshine hours, the energy stored in the thermal energy storage (TES) can be drawn to generate steam [2]. In recent years, explicitly the growth of CSP plants is substantial as shown in figure 2.1 [3].

The CSP plants have the advantage of using tracking that aids in harnessing more solar radiation during the daytime [4]. The CSP technologies are classified on the basis of the concentrator design as line- and point-focussing systems and it is shown in figure 2.2.

## 2.2 Concentrating solar power technologies

The nuclear fusion reaction persistently takes place at the sun, leading to an enormous amount of energy getting released and reaching the Earth's surface in the form of radiation. According to an International Energy Agency (IEA) report in 2014, energy from the Sun is estimated as 1575 EJ to 49 837 EJ. The estimated energy is much higher than annual energy (573 EJ) consumed across the world in 2014. In the same report, the IEA vividly states that solar energy is a standalone energy source to meet the energy demands of the world. In 2014, along with solar energy, other renewable energy resources, such as wind and geothermal, together constitute 1.4% of the world's energy consumption. It clearly shows that solar energy is underutilized. The reports estimated that the Sun has provided solar energy continuously for 4 billion years. The solar radiation reaching the Earth's surface is approximately 1000 W m$^{-2}$ during clear sky conditions. It is sufficient to satisfy the low-temperature needs such as hot water supply for residential and industrial purposes.

In the need for high temperature application, especially for electricity generation, the Sunlight needs to concentrate on the receiver using the reflector. The concentrating solar power plant uses the reflectors to concentrate the solar radiation on the receiver surface and convert it to heat. It can be used for various applications such as generating electricity by driving the conventional steam turbine, and processing heat for industrial applications such as water desalination, oil recovery, chemical production, food processing, mineral processing, etc. In the process of generating

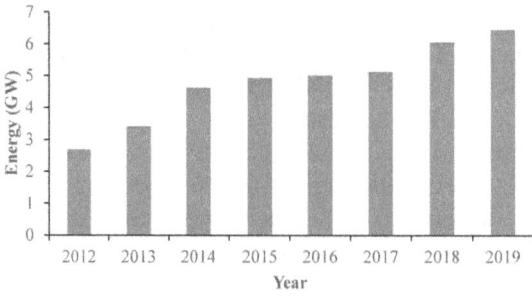

**Figure 2.1.** Development of the CSP plant in recent years.

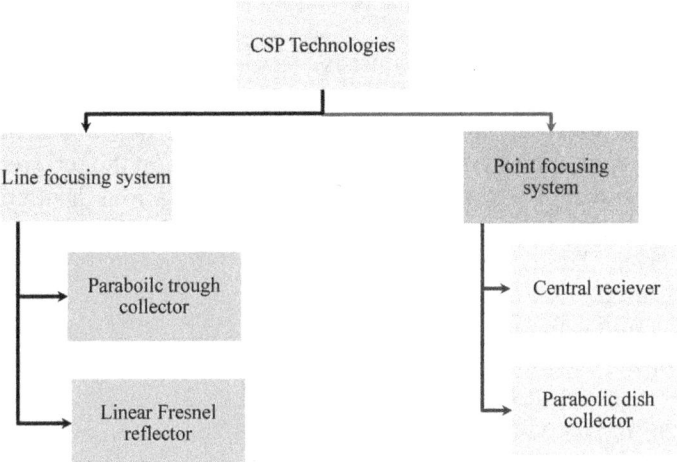

**Figure 2.2.** Classification of CSP technologies.

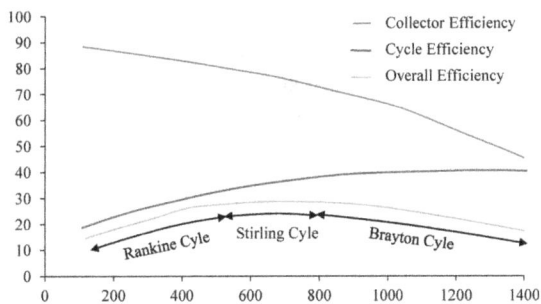

**Figure 2.3.** Efficiencies of the CSP systems, power generation cycle and overall power plant.

electricity, the CSP plant plays a pivotal role and in recent years efforts have been taken to reduce the production cost. The production cost of electricity can be reduced by 50%–60% by reducing the number of components enlarging the setup, technical improvements and volume of production. Nonetheless the mismatch occurs between the supply and demand from most of the renewable energy technologies. Hence, it is desirable to integrate the thermal energy storage to supply heat to the power generation cycle during cloudy periods, off-sunshine hours and to improve the overall efficiency of the plant. The thermal energy storage system integrated over the CSP system acts as an intervening step for generating electricity. In other cases, the CSP systems can be also integrated with a conventional power plant for providing supplementary heat to the feed water heaters during the daytime [5]. The solar collector efficiency, engine efficiency and overall or combined plant efficiency of concentrating power plants are represented in figure 2.3. It is evident that the collector efficiencies gradually decrease with an increase in operating temperature and the cycle efficiency increases with an increase in temperature. However, the overall efficiency of the CSP plants increases with an increase in temperature up to a certain point and then gradually decreases [6–8].

### 2.2.1 Parabolic trough collector

A parabolic trough collector (PTC) consists of a series of parabolic-shaped reflectors to reflect the Sunlight on the absorber surface. The parabolic trough collectors are oriented in the north–south horizontal axis and track the position of the Sun in the sky from east to west. To ensure the continuous focusing on the receiver/absorber, a single tracking system is usually employed. In PTC, various heat transfer fluids (HTF), such as thermal oil, molten salt, water, etc., can be used to extract the heat from the receiver. The commercial PTC uses a diphenyl oxide/biphenyl eutectic mixture and is used as the HTF and the maximum temperature that can be obtained in the collector and is limited to around 400 °C. If the temperature level exceeds the specified limit, the decomposition of the HTF will occur. The working principle of the PTC is also shown in figure 2.4.

Nonetheless the silicon oil mixtures were tested and the results show that they can go up to the temperature of 450 °C. The efficiency of PTC-based plants is found to be increased with a wet cooling tower instead of a dry cooling tower. The annual solar to the electrical conversion efficiency of the PTC plant is reported as 15%. It was reported that increasing the size of power blocks would increase efficiency by 1% [9]. The PTC plants integrated with thermal energy storage (TES) have attracted wide interest because of providing a constant supply of heat throughout the designed operating hours. Figure 2.5 shows the integration of PTC with a two-tank TES system. The TES can be charged during excess heat generated from the solar field. The heat is extracted from the TES during the fluctuation in the solar radiation and also during off-sunshine hours [10–12].

### 2.2.2 Linear Fresnel reflector

The series of flat reflectors are either arranged horizontally or aligned in such a way so that it concentrates the Sunlight on the absorber surface. A linear Fresnel reflector (LFR) uses the flat mirror as opposed to PTC. In the LFR system, unlike PTC the

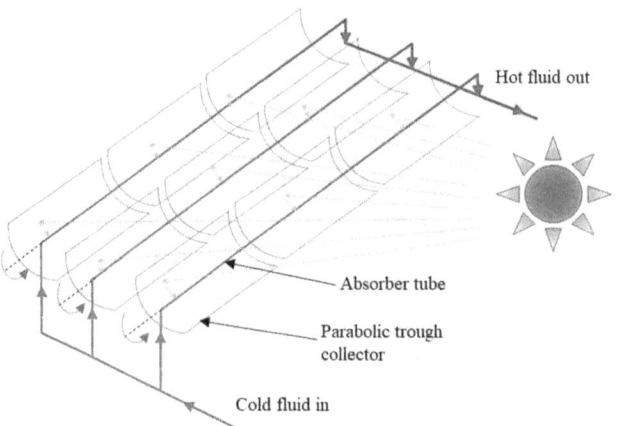

**Figure 2.4.** Schematic of the parabolic trough solar collector.

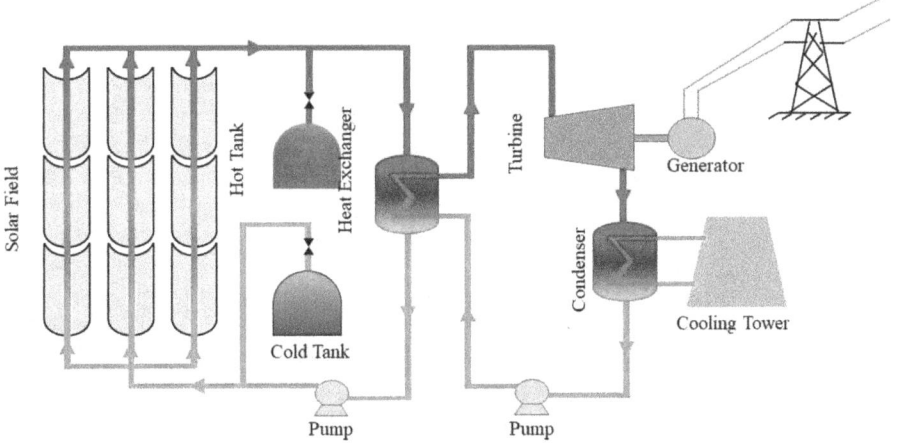

**Figure 2.5.** Parabolic trough collector integrated with TES.

absorber is fixed and mirrors are tracking the Sunlight and reflect them on the absorber surface. The aperture size is large and it aims at low cost as it is not affected by wind loads, unlike PTC. LFR also uses the secondary mirror over the absorber surface to reflect the Sunlight on the absorber surface. The primary mirrors are mounted in such a way that it collects the maximum amount of Sunlight in a day. The HTF, such as water, is passed into the absorber tubes and it gets converted directly into steam at 285 °C [13]. This steam that can be used to drive the turbine without any additional heat exchanger requires a high temperature from other working fluids. This type of system is simple in construction and less efficient compared with other CSP technologies. It is a standalone system and well suited for remote areas.

In earlier days, the LFR system is coupled with conventional fossil fuels for power production during cloudy days and night time. The energy acquired per unit area is lesser than PTC and but it is simpler in construction. LFR does not require high pressure flexible joints as that of PTC and a parabolic dish collector [1]. The receiver is held stationary and it allows a dense packing of mirrors close to each other. Nevertheless, the flat positioning of the reflecting mirror inherently leads to optical losses and it could account for 20%–30% of annual energy output as compared with PTC. LFR is more often used for DSG and 3% of CSP plants in the world are based on LFR. The first commercial LFR plant was constructed in Spain with a reflector surface area of 302 000 m$^2$ and a capacity of 30 MW. It provides clean energy to 12 000 homes and it mitigates 16 000 metric tonnes of $CO_2$ emission into the atmosphere. The limitations of using LFR is the fixed receiver and it results in higher optical losses for large-scale power plants. The compact linear Fresnel collector (CLFR) comprises small Fresnel reflectors are arranged and increases the intensity by 30% as compared to normal LFR. An additional set of mirrors are placed across the absorber surface to refocus the concentrated solar radiation on the absorber surface and it occupies less space as compared with LFR. CLFR is the most effective

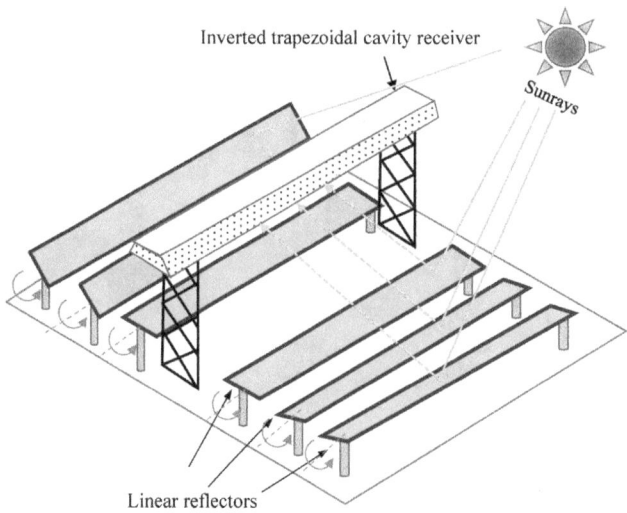

**Figure 2.6.** Schematic of the linear Fresnel reflector.

one as compared with LFR and can be scaled as per user requirements. The schematic of LFR is shown in figure 2.6.

### 2.2.3 Central receiver

The central receiver is also known as the solar tower. The central receiver consists of a large number of heliostats and the surface area of heliostats ranges from 20 to 200 m². The receiver is placed at the uppermost part of the tower and it transfers the heat into the working fluid. The distance between the receiver and heliostats may be up to a km. The receiver is designed based on the types of HTF used, mode of energy transfer and materials used. Two types of receivers may be used in the central receiver, (a) external receiver and (b) cavity receiver. The external receiver consists of a cylinder composed of many boiler tubes carrying the HTF. The limitations of using an external receiver are more convective and radiative losses as compared with a cavity receiver since the external receiver is largely exposed to the atmosphere [14]. The cause for the cavity in the receiver is to reduce the heat losses through the aperture of the receiver. In the cavity receiver, the air is used as the HTF and makes it difficult because of poor heat transfer characteristics. They opted for the porous structure to increase the residence time of the particle. The various components and schematic of the central receiver are represented in figure 2.7.

The central receiver can achieve a very concentrated ratio. The central receiver has been the suitable option for utility scale power generation at a lower cost compared to PTC and LFR since it has a high temperature range up to 1000 °C. The central receiver has attracted wide interest in recent years and several heat transfer fluids are exploited to extract the heat from the receiver. About 20% of CSP plants are central receivers in 2018 and liquid or gas to be used as a heat transfer medium in the central receiver. Molten salt is commercially available and used as a heat transfer

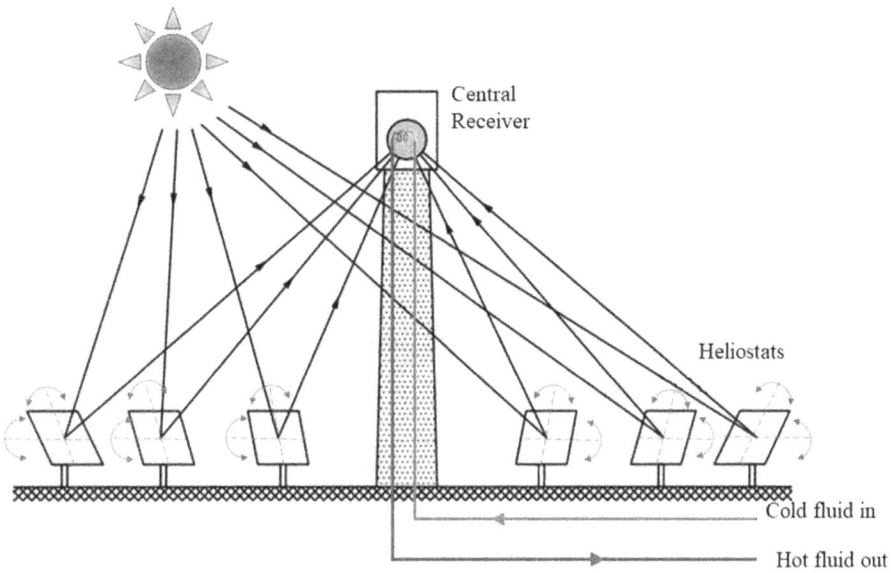

**Figure 2.7.** Schematic of the central receiver system.

medium in the central receiver. One of the central receivers was erected in Spain with a power output of 19.9 MW of electrical energy. The molten salt from the cold storage tank (290 °C) is pumped to the receiver and heated up to 565 °C and again it passes to the heat exchanger for producing steam at 540 °C and it goes to the cold storage tank. It consists of 2650 heliostats with each heliostat area of 115 m$^2$ [15]. The storage system is designed for 15 h and it can be operated throughout the day in summer. The central receiver was first commercially available in the market with water as the HTF and steam are generated directly in the receiver to generate the electricity. The central receiver with air as a heat transfer medium was erected in Germany. The air was sent to the receiver and it was heated up to 700 °C and this hot air may be directly used in the heat recovery steam generator (HRSG) to generate the steam. During the rest of the time, the hot air is stored by the regenerator type and the direction is reversed in the discharging state. The current research on the central receiver is about the usage of particles as the heat transfer medium to extract the heat at high temperature from the central receiver.

### 2.2.4 Parabolic dish collector

The parabolic dish collector (PDC) has a paraboloidal shape concentrator and receiver to transfigure the radiation into heat. The solar concentrator is a three-dimensional concentrator and it can track the Sun in both elevations as well as the azimuth angle. The size of the solar concentrator ranges from a few square meters to hundreds of square meters owing to constraints on the wind load. There are two methods to convert the concentrated solar radiation into electricity in the PDC. The heat collected by the receiver is transferred to the steam engine to produce electricity.

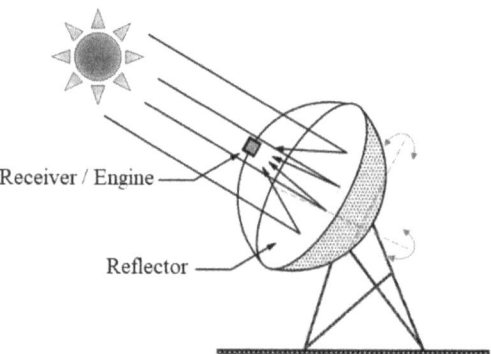

**Figure 2.8.** Schematic of the parabolic dish collector.

The Stirling engine is the most common type of engine used in PDC. The reflecting mirrors are used to concentrate the Sunlight at a focal point on the receiver. In the Stirling engine, the receiver consists series of tubes containing the heat transfer medium as hydrogen or helium. In another configuration, the HTF is circulated through the tubes to collect the heat from the receiver and then thermal energy is converted into linear motion using a piston cylinder arrangement. These linear motions are again converted into rotary motion and then are coupled with a generator to produce electricity.

The Stirling engine in the dish produces the maximum efficiency of 31.25% by converting the incident solar radiation into electricity including the parasitic losses. Hence, the Stirling dish engine is more efficient as compared to other CSP technologies. Several thermodynamic cycles have been tried with parabolic dish collectors such as the open and closed Brayton cycle, Rankine cycle, and Stirling cycle. The dish Stirling engine has experimented with different working fluids such as ammonia, methane and ethane. The thermal efficiency of all the working fluids such as water, ammonia and ethane was reported as 31%, 39% and 38%. The results show that organic working fluids such as ammonia and ethane have significantly higher thermal efficiency than water [16, 17]. PDC systems have been developed as independent systems in remote locations with a reasonable amount of solar radiation. The range for independent PDC is from 9 to 25 kW [9]. The independent system is mainly used for the irrigational purpose and generating electricity for regional communities. Solar PV generates electricity at a lower cost as compared to electricity generation from the PDC and makes a tough task for the PDC. A future research topic is the integration of TES with PDC. The schematic of the PDC is shown in figure 2.8.

### 2.2.5 Heat transfer fluids

Concentrating solar power plants have been recognized as one of the most commercial applications in electricity generation. Nonetheless, the continual dispatch of electricity generation leads to integration with the thermal energy storage (TES) system. The TES system not only provides continuous dispatchability, it also improves the capacity factor of the plant. The storage system can be classified into active and passive storage systems. In an active storage system, the heat transfer takes place, employing forced convection and in the passive storage system, the heat transfer occurs by natural convection or density gradients. More than 80% of the present CSP technologies are PTCs in the construction stage and operation.

(a) **Thermic fluids** Thermal oil is used as the HTF in these technologies to improve the performance of the system. The temperature of the thermal oil is limited to 390 °C and it heats the secondary HTF in a hot tank up to 385 °C. The advantage of using thermal oil as the HTF is its thermal stability and almost constant thermal conductivity at all temperatures up to 400 °C. Therminol VP-1 is a mixture of dinphenyl oxide and biphenyl is the most commonly used thermal oil. It has been tested to show that it is less prone to eye irritation and respiratory problems and it is the most suitable fluid in the temperature less than 400 °C. Regarding the economical aspect, the cost of therminol per kg is reported as $265 compared with other thermal oils. Other than therminol, mineral oil, synthetic oil, and silicone oil are the other thermal oils used as the HTF in the CSP plant. These three oils are having almost the same thermal conductivity at all temperatures and the temperature limit is less than 400 °C. The disadvantage of using these thermal oils cannot be applicable for high-temperature limits and expensive when compared with other HTFs. Dow corning 550 silicon oil was tested experimentally and the results were proven that it has good heat transfer characteristics and saturation rate is low in comparison with steam. The advantage of using synthetic oil as the HTF is constant thermal conductivity over the wide range of temperature, less prone to flammability and corrosiveness but the drawback is the rate of pumping is high owing to the low heat capacity. The seven different thermal oils such as Therminol D12, Marlotherm X, Santotherm LT, Santotherm 59, Syltherm XLT and Syltherm 800 were tested experimentally in PTCs under climatic conditions of Algeria. These thermal oils show thermal stability up to the temperature range of 300 °C–400 °C. In the economic aspect, the maximum cost was incurred by Santotherm LT and Syltherm 800 were reported as the best compared to other thermal oils. Some of the paraffinic oils such as Gulf security 53, Dowtherm HP, Exxon Cloria HT-43, Shell Thermia 33 and Texatherm $H_2O$ are used as HTFs and operate in the temperature range of 70 °C–550 °C.

(b) **Direct steam generation** The process of generating steam directly by circulating the water as the HTF into the absorber is known as direct steam generation (DSG). The water is initially preheated, evaporated and lastly superheated in the solar field at the pressure of 110 bar and 500 °C. The advantage of DSG using water as the HTF which eliminates the usage of the heat exchanger, usage of the primary HTF, the pumping effort and the handling of the primary and secondary HTF. The configuration of PTCs is different from the conventional steam generation process. Aside from electricity generation, the DSG can be used for industrial process heating such as enhanced oil recovery, pulp and paper, textile, food and beverage, pharmaceutical, mineral processing, etc [18–20].

(c) **Molten salt** Molten salts are generally used in the solar tower because of their high temperature applications. The molten salts have been tried in PTCs in later years with a combination of eutectic mixtures as liquid salts. The eutectic mixture is a combination of two or more solid substance (two or more types of molten salts) bringing out the liquid phase at certain thermodynamic conditions. These mixtures are typically known as liquid salts. The reasons for choosing salt over thermal oil is high storage capacity, high specific heat capacity and thermal conductivity and high density at low pressure. The potential salt substance is nitrate salts. The maximum allowable temperature for sodium and potassium combination is 550 °C and the crystallization incurs at the temperature of 238 °C. Adding additional materials such as nanoparticles into nitrate salts will lower the crystallization temperature. The combination of sodium nitrate and potassium nitrate is typically known as solar salts. It is one of the most widely used CSP plants since it has optimized cost and good thermal properties.

The Hitec salt (7% of sodium nitrate, 53% of potassium nitrate, 40% of sodium nitrite) and Hitec$^{TM}$ XL (7% of sodium nitrate, 45% of potassium nitrate, 48% of calcium nitrate) are the other two salts having lower freezing points and it does not provide the same thermal stability as offered by solar salt. The molten salts are inexpensive as compared with thermal oils [9]. The combination of molten salt with thermal oil leads to a 16.4% reduction of LCoE and a 20% reduction for the complete switchover to molten salts. The lowest value of LCoE was reported for solar salt as $5450 MWh$^{-1}$. The primary benefit of using molten salt as the HTF in CSP plants is storing the thermal energy above 500 °C. The solar salt was used as the HTF in the PTC plant and tested experimentally for more than 2500 h under various climatic conditions of Egypt [21]. Under several circumstances such as start-up, tracking of solar collector, draining, off sunshine hours, the solar salt performed better than the other molten salt. The power supply of 680 kW from the solar collectors is required for the start-up and draining completes less than 1 h with a capacity of 10 000 m$^2$ PTC solar field. Three salts such as solar salt, Hitec salt and Hypo-Hitec salt were simulated and investigated the physical properties. The limitation of using molten salt is higher melting temperature and freezing occurs at temperatures of 120 °C to 220 °C [4].

## 2.3 Thermodynamic analysis of vapour power cycles

A cycle persistently transforms heat into work in which the working fluid is circulated continuously and undergoes a phase change. The working fluid undergoes various operations such as compression in the pump, evaporation in the boiler, expansion in the turbine, condensation in the condenser and is known as the vapour power cycle. It is also known as the Rankine cycle. The following sub-section elucidates the basic vapour power cycles in detail.

### 2.3.1 Rankine cycle

A Scottish physicist named William J.M. Rankine developed the Rankine cycle in 1859. The conventional steam turbine generates the power using steam as a working fluid and follows the sequence of four processes such as compression, evaporation, expansion, and condensation is known as the Rankine cycle. The working fluid proceeds through the various components of the power cycle without the pressure drop due to friction and the irreversibility is known as the ideal Rankine cycle. The working fluid which persistently changes its phase from liquid to vapour and vice-versa is the basic cycle for the operation of power plants [22, 23]. The essential components involved in the Rankine cycle are shown in figure 2.9. The various processes that occur in the Rankine cycle are elucidated in figures 2.10 and 2.11.

**Isobaric heat addition (2–3):**
Conventional fossil fuels, such as coal, diesel and lignite, are liberating the heat during combustion and the heat is transferred to the water by indirectly using the boiler as a heat exchanger. Water in the form of compressed liquid enters into the boiler and the output from the boiler is saturated/superheated steam. The energy addition in the boiler is given by:

$$q_{in} = h_3 - h_2. \tag{2.1}$$

**Isentropic expansion (3–4):**
The saturated/superheated steam from the boiler enters into the turbine. The steam expands at constant entropy in the turbine where the kinetic energy of the steam gets converted to mechanical energy. The turbine is connected to the generator and it generates electricity. The work done in the turbine is given by:

$$W_{turbine,\ out} = h_3 - h_4. \tag{2.2}$$

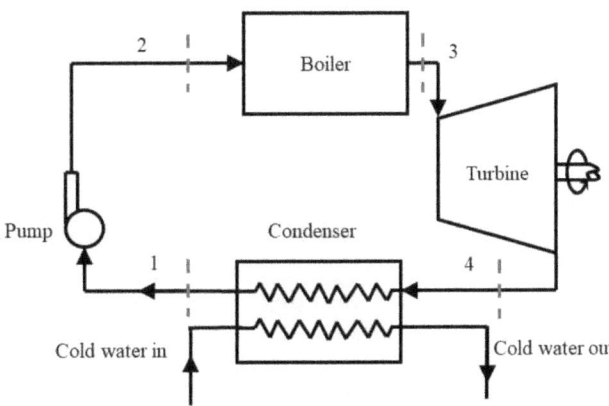

**Figure 2.9.** Schematic of the organic Rankine cycle.

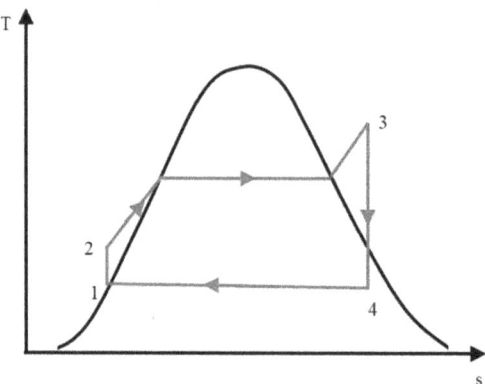

**Figure 2.10.** T–s diagram of the Rankine cycle.

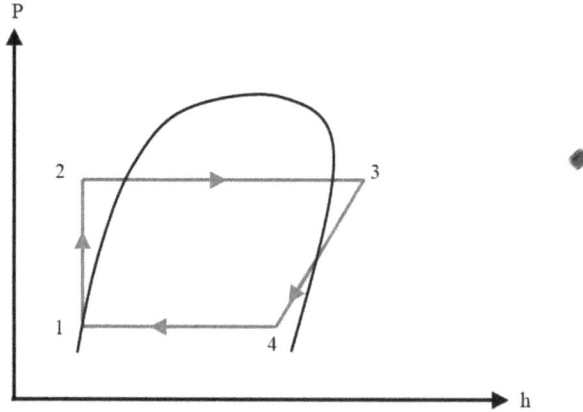

**Figure 2.11.** Pressure-enthalpy diagram.

### Isobaric heat rejection (4–1):

The expanded steam from the turbine enters into the condenser and changes its phase from vapour to liquid by losing the heat into the condenser. The heat can be transferred from steam to water circulated through the tubes in the condenser. The heat rejection in the condenser is given by:

$$q_{out} = h_4 - h_1. \tag{2.3}$$

### Isentropic compression (1–2):

The water from the condenser is fed to the pump. In the pump, the water is compressed by giving the work in the course of the process. Mostly, the work done by the pump is neglected because it is very small as compared to the power generated from the turbine. The work done by pump is given by:

$$W_{pump,\ in} = h_2 - h_1. \tag{2.4}$$

The thermal efficiency is given as

$$\text{Thermal efficiency } (\eta_{\text{th}}) = \frac{\text{Net work done}}{\text{Heat input}} \tag{2.5}$$

$$\eta_{\text{th}} = \frac{W_{\text{Turbine}} - W_{\text{Pump}}}{q_{\text{in}}}. \tag{2.6}$$

(a) **Non-ideal Rankine cycle** The non-ideal Rankine cycle is known as the real Rankine cycle. In this real Rankine cycle, the expansion in the turbine and work done by pump are irreversible and the entropy increases in these two processes. Owing to saturated steam, during expansion in the turbine, the water droplets formed rapidly while expanding. It causes erosion and corrosion in the turbines and had an impact on the efficiency [24]. This drawback can be overcome by heating the saturated steam into superheated steam. The actual scenarios can be represented in the *T–s* diagram as shown in figure 2.12.

(b) **Rankine cycle with reheat** The objective of using the reheat concept is to curtail the amount of wetness in the steam during the expansion process. The reheat concept consists of two turbines. The first turbine ingests the steam directly from the boiler at higher pressure. After the expansion of steam in the high-pressure turbine, the steam is again sent to the boiler for reheating. The pressure of steam coming from the reheater is typically one-fourth of the initial pressure of the steam from the boiler [23]. Nevertheless, the temperature of the steam in the reheater is equal to that of the steam from the boiler. It has the advantage of a reduction in condensing the steam during the expansion, blade corrosion and erosion, and also increases in

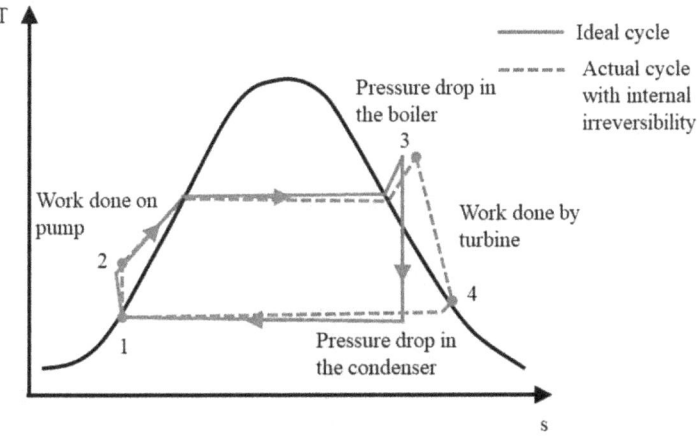

**Figure 2.12.** Steam Rankine cycle.

cycle efficiency. The schematic and *T–s* of reheating stages were represented and shown in figure 2.13(a) and 2.13(b) respectively.

(c) **Rankine cycle with regeneration** The name regeneration evolved with the Rankine cycle is owing to the sub cooled liquid from the condenser being heated by trapping the steam at various positions in the turbine. This type of heating may be referred to as direct contact heating [25]. The preheating of water before sending it to the boiler decreases the workload of the boiler and thus enhances the efficiency of the cycle. The steam which used for preheating the water is known as bleed steam. The schematic and variations are represented in the *T–s* diagram as shown in figure 2.14(a) and 2.14(b) respectively.

(d) **Rankine cycle with reheat and regeneration** The reheating of steam may be implemented in the Rankine cycle when the vaporization pressure is high. The effect of reheating alone on the thermal efficiency of Rankine cycle is very small. Regeneration or the feed water gets heated by extracting the bleed steam from turbine has a significant impact on the cycle efficiency. Terminal temperature difference (TTD) is one of the important parameters to measure the performance of the heat exchanger, condenser, and feed water heaters. In the case of the feed water heater, TTD is defined as the difference between the temperature of saturation steam extracted from

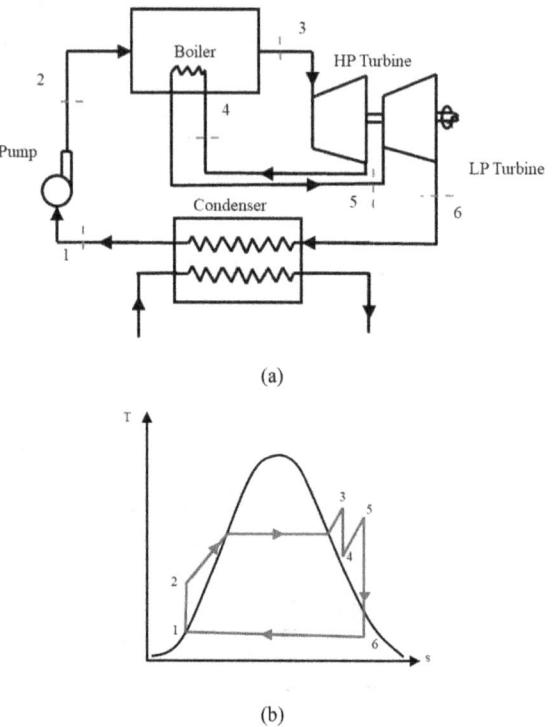

Figure 2.13. Rankine cycle with reheat. (a) Schematic. (b) T–s diagram.

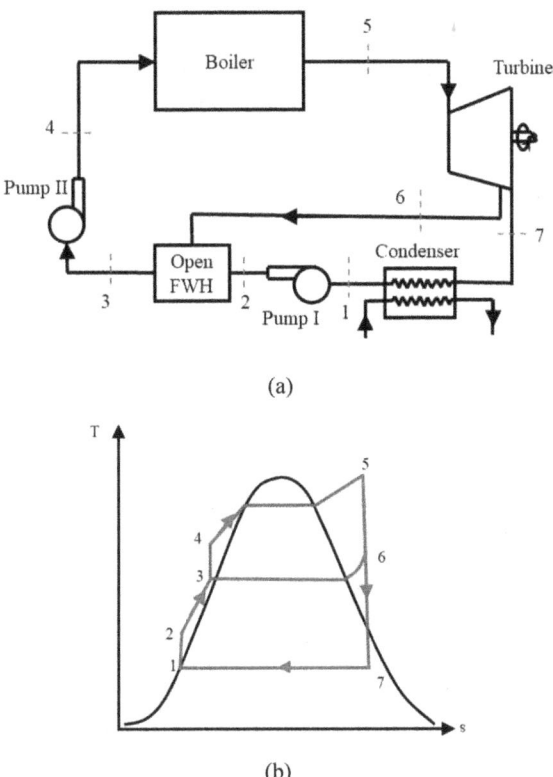

(a)

(b)

**Figure 2.14.** Rankine cycle with regeneration. (a) Schematic. (b) T–s diagram.

turbine and outlet temperature of the feed water. TTD is typically used to control the quantity of bleed steam extracted from the turbine. The modern-day steam power plants are equipped with reheat and regenerative concepts and it is shown in figures 2.15(a) and (b) respectively.

(e) **Organic Rankine cycle** Instead of using water and steam, organic fluids such as butane, hexane, pentane, etc., are used to produce electrical energy at low temperatures is known as the organic Rankine cycle. These are organic cycles deployed for low-temperature applications such as solar ponds, waste heat recovery in the range of 80 °C–350 °C. The thermal efficiency of the cycle is low as compared to the actual Rankine cycle because of the low operating temperature [26]. In the Rankine cycle, the boiling point of the working fluid higher than water can also be used as a working fluid. The working fluid has a significant effect on the cycle efficiency because of the properties of the working fluid.

(f) **Exploitation of nanoparticles in organic Rankine cycle** The demand of novel working fluids in the organic Rankine cycle (ORC) has been increased persistently to mitigate the adverse effects of existing working fluids. The typical working fluids used in the ORC are toluene, octamethyltrisiloxane (MDM), cyclohexane, n-pentane, R245, R113, R611 and n-hexane. The

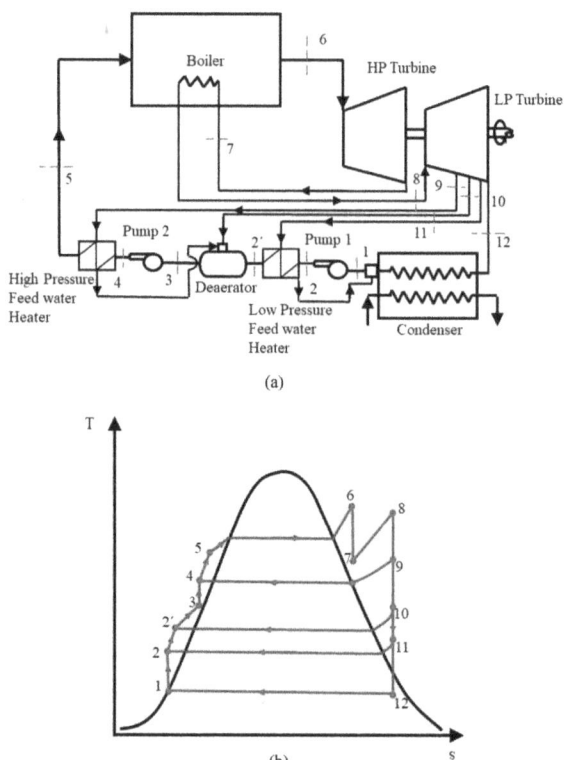

Figure 2.15. Rankine cycle with reheat and regeneration. (a) Schematic. (b) T–s diagram.

addition of nanoparticles into working fluids helps to improve its thermo-physical properties. The working fluid with the inclusion of nanoparticles in heat exchangers offers better heat transfer coefficients as compared to the base fluids. Despite this fact, the difficulty in using the nanofluids in ORC is their poor thermal stability, the clogging of nano particles in the base fluid may cause erosion in the parts of flow path, and the cost of the nano-particles. Table 2.1 represents the thermophysical properties of commonly used nanoparticles at ambient temperature.

On average, the usage of nanofluids in ORCs yields the result as 4% decrement in the heat exchanger area and 18% increment in pressure drop with a volume concentration ratio of 1% [26]. To date, only one study has been reported related to the addition of nanoparticles into organic working fluids used in ORCs. Most of the experimental studies and analyses were performed using standalone organic working fluid without direct inclusion of nanoparticles into it. However, the nanoparticles gets mixed up with base fluid and use them to collect/transfer heat from solar collector fields and the heat exchanger to the ORC. The novel working fluid named R245fa/MIL101 was tested experimentally in an ORC and the performance was compared with pure organic working fluids [27]. MIL101

**Table 2.1.** Thermo-physical properties of nanoparticles.

| Property/nanoparticle | $Al_2O_3$ | CuO | Cu | ZnO | Ag | $SiO_2$ | $TiO_2$ | SWCNT |
|---|---|---|---|---|---|---|---|---|
| Density, $\rho$ (kg m$^{-3}$) | 3950 | 6320 | 8933 | 5610 | 10 490 | 2650 | 4230 | 2170 |
| Thermal conductivity, $k$ (W m K$^{-1}$) | 25 | 33 | 401 | 13 | 407 | 1.4 | 8.4 | 3000 |
| Heat capacity, $C$ (kJ kg-K$^{-1}$) | 0.85 | 0.55 | 0.385 | 0.43 | 0.24 | 0.93 | 0.68 | 2.1 |

(Material Institut Lavoisier composed of trimeric chromium (III) octahedral clusters interconnected by 1,4 benzenedocarboxylates) was termed as a metal organic heat carrier and it is capable of undergoing endothermic and exothermic reaction. The sorption and desorption process provides additional heat extraction and rejection as compared with the pure liquid phase. The result shows that heat exchanger area was considerably reduced as compared with existing working fluids. The sensitivity analysis was performed and it was found that 6% incremental of enthalpy gain [28]. Optimization of different organic fluids with a solar-driven ORC was performed. It comprises a collector, storage tank, and an ORC cycle. The study involves evaluation of different nanofluids and organic fluids, and the optimal combination of nanofluids and organic fluids was also found. The PTC was used as a solar collector for all types of evaluation. The base fluid was syltherm 800 and $Al_2O_3$, CuO, $TiO_2$, and Cu are used as nanoparticles for the evaluation. The nanofluids are used to collect the heat from the collector to thermal energy storage tank.

Thermal oil was used as the HTF in thermal energy storage tank and the organic fluids such as toulene, MDM, cyclohexane and n-pentane in the ORC operates the power cycle. These organic fluids have high critical temperature and it can be easily operated with the temperature range of the PTC. The optimization parameter such as concentration ratio and pressure ratio are used since it is directly associated with turbine inlet pressure. Among these organic fluids, toluene was the best organic fluid followed by MDM, cyclo hexane, n-pentane and CuO offered better results followed by Cu, $Al_2O_3$, and $TiO_2$. The toluene and CuO combination yields a better result with the optimal concentration of 3.98% and the optimum ratio of 75.51%. The power produced from the system was also enhanced by 1.75% as compared with the base fluid [29]. Energy, exergy and economic analysis were performed with three different experimental setups. The experimental setup 1 was the PTC-driven simple Rankine cycle, experimental setup 2 consists of ORC integrated over a simple Rankine cycle in which the heat rejected from the condenser drives the ORC cycle and the experimental setup 3 comprises a combined SRC and ORC cycle in which both of them are driven by the PTC. Four different organic working fluids, such as R113, n-hexane, cyclohexane, and toluene, are tested with experimental setup 2 and experimental setup 3. Therminol VP-1 was used as the base fluid and $Al_2O_3$ was used as the nanoparticle in the experimental investigation. The results vividly show that experimental setup 3 yields the maximum power output while using toluene as compared with other two experimental setups. Despite this fact, the energy, exergy and economic performance of experimental setup 3 is low as compared with other two setups. The reason for low thermo-economic performance of experimental setup 3 was a high total cost along with low energy and exergy efficiencies [30].

The performance of the solar-driven ORC was performed in a different cavity of the parabolic dish collector. The nanofluid was prepared by the addition of $Al_2O_3$ in to thermal oil and R611 was used as working fluid in ORC. Three different cavities such as hemispherical, cuboidal and cylindrical shapes were tested with different concentration ratio of nanoparticles into the base fluid ranging from 1% to 5% and size of the nanoparticles from 5 nm to 50 nm. Better results were provided by the hemispherical cavity with the maximum thermal efficiency of 66.24%. The outlet temperature of the cavity was found to be increasing by decreasing the size of the nanoparticle and the turbine inlet temperature of 2 MPa was found to be optimum [31]. The solar-driven ORC was tested with multi-walled carbon nano tubes (MWCNT) with oil as the nanofluid and R113 were used as the HTF in the ORC. The parabolic dish collector with two types of cavity receivers, such as hemispherical and cylindrical, were also tested. The efficiency of the ORC was found to be 21.4% for the hemispherical cavity and 17.8% for the cylindrical cavity. The LCoE was found to be 0.092 for the hemispherical cavity and 0.094 for the cylindrical cavity. The hemispherical cavity was proven to be most efficient in terms of energy, exergy and economically [32]. From the above studies, it is clearly found that the inclusion of nanoparticles leads to heat transfer enhancement and better system efficiency. The key parameters such as concentration ratio, size of the nanoparticles, and the pressure ratio are used to determine the optimal nanofluids suitable for the organic Rankine cycle.

### 2.3.2 Binary vapour power cycle

In general, water is used as a working fluid in the primary vapour power cycle. The concept of a binary vapour power cycle arises to overcome the shortcomings of water used as working fluid and introduce the concept of two working fluids. The following are the important characteristics of a working fluid that should possess:
- High thermal conductivity
- Low triple point temperature
- High critical temperature
- Maximum pressure assured safe conditions
- High enthalpy of vaporization
- Condenser pressure should not be too low
- Easily available, inert, inexpensive and non-toxic

None of the working fluids alone possess all the worthwhile properties of a working fluid. Each working fluid possesses a unique property but not all desirable properties. Because of this, it may create a thought in the researcher's mind to make use of two different working fluids for power generation. The one working fluid is in the high-temperature region and the other at the low-temperature region and it is known

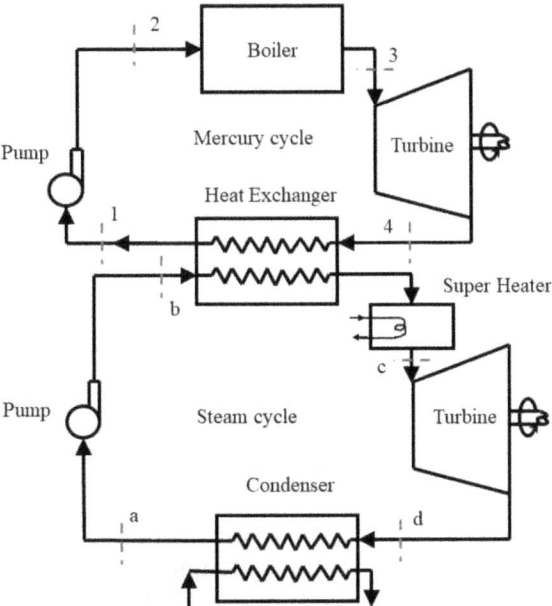

**Figure 2.16.** Schematic of the binary vapour cycle.

as a binary vapour power cycle. Mercury is the most common type of working fluid used in the binary vapour cycle along with the water as the working fluid. The critical pressure and temperature of mercury are 180 bar and 898 °C respectively which makes them act as a topping cycle. The water is used for the low-temperature range as compared with mercury. The mercury vapour is generated from the boiler and it has been expanded in the turbine. After the mercury vapor gets expanded in the turbine, it is sent to the condenser for heat rejection. The mercury condenser acts as an auxiliary boiler to the steam cycle and it converts the water into steam. The steam goes further back to the boiler and gets heated up again to generate the superheated steam. The schematic of the binary vapour cycle and $T$–$s$ diagram are shown in figures 2.16 and 2.17 respectively.

### 2.3.3 Tertiary vapour power cycle

The vapour power cycle in which three different working fluids are used to produce power by the Rankine cycle is called the tertiary vapour power cycle. The organic working fluids are having a low-temperature range to extract the heat from a low-temperature source. Along with the binary vapour power cycle, the addition of organic working fluid leads to power production at the low-temperature range as well. For instance, the sulphur-dioxide is used as the third working fluid in the mercury vapour power cycle. It is also known as a coupled cycle. The sulphur-dioxide utilizes the low temperature from the condensation of steam. In the coupled cycle, the mercury used as working fluid is the topping cycle and sulphur-dioxide is the bottoming cycle. The other working fluids, such as ammonia, freons, etc., can

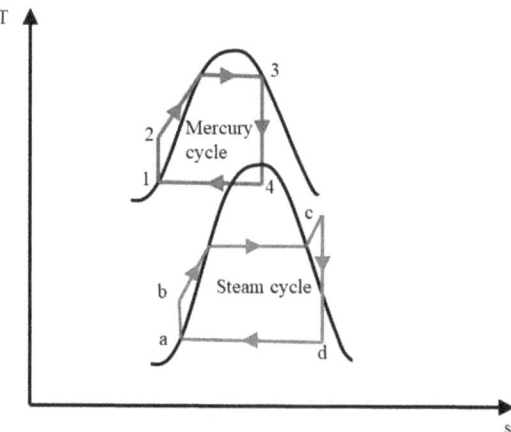

**Figure 2.17.** T–s diagram of the binary vapour cycle.

perform as a power cycle by utilizing the low-temperature source. These refrigerants may be used as the working fluid in the bottoming cycle. In other terms, sodium–mercury–steam cycle is known as a tertiary cycle. Sodium and mercury are used to utilize the high-temperature range and steam is used to utilize the medium-temperature range in the tertiary cycle. The schematic and *T–s* diagram of the tertiary vapour power cycle is shown in figures 2.18 and 2.19.

## 2.4 Novel working fluids for vapour cycles

In recent decades, power generation from low-temperature energy sources, such as solar, heat from biomass, and geothermal energy, turned out to be one of the major research domains. Generally, these energy sources have less energy density as compared with conventional energy sources. The power plants using water as the HTF operating on the Rankine cycle cannot exploit the heat sources completely. An alternative working fluid that accomplishes the power generation based on the Rankine cycle from the low-temperature sources is known as the organic Rankine cycle. The organic Rankine cycle (ORC) uses unconventional working fluids (organic fluids, inorganic fluids) to generate power from low-temperature sources. The installed capacity of ORC-based power plants using novel working fluids is shown in figure 2.18. In 2016, the installed capacity of the power plant is 2749 MW$_e$ from 563 plants, and the planned capacity is 523.6 MW$_e$ from 75 plants.

Figure 2.20 clearly shows that geothermal power plants mostly use the organic Rankine cycles compared to other low-temperature sources. The new working fluids in the Rankine cycle, which are more effective than conventional working fluid i.e. water, is known as the novel working fluid. The physical properties such as the quality of the vapour after expansion have a great impact on the cycle efficiency. Generally, the working fluids are classified based on the quality of vapour at the exit of the turbine. These fluids can be classified into three categories, namely dry, wet, and isentropic working fluids [33]. In the dry working fluid, the initial state of the working fluid is saturated vapour at the turbine inlet and remains vapor at the end of

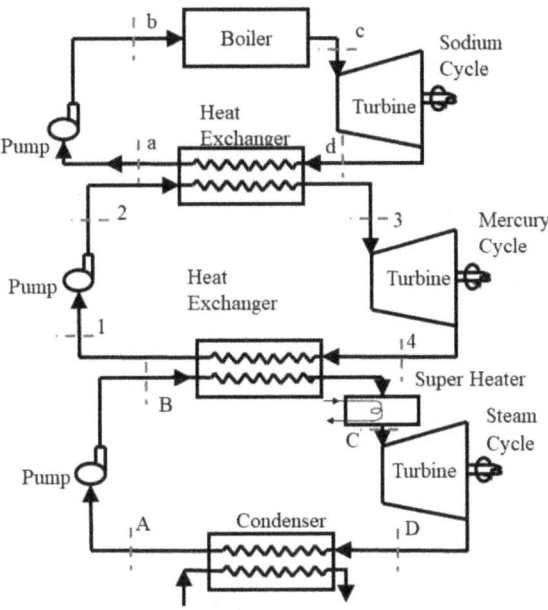

**Figure 2.18.** Schematic diagram of the Na-Hg-Steam tertiary cycle.

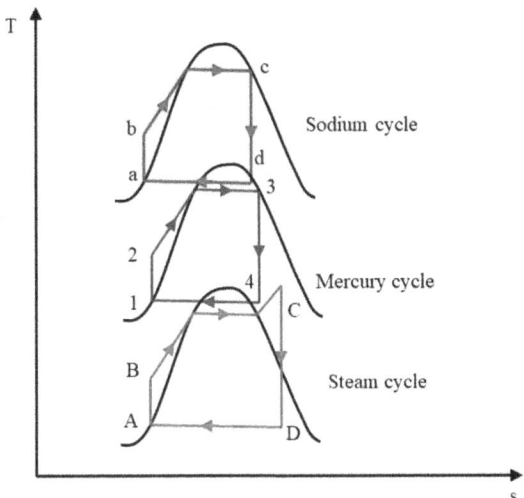

**Figure 2.19.** T–s diagram of the Na-Hg-Steam tertiary cycle.

the expansion process in the turbine with variation in entropy [34]. The dry working fluid ends in a single-phase region as the dry or superheated vapour region. The dry fluid acquires a slope, which is positive ($dT/ds > 0$) in the saturated curve. During higher temperatures, the saturation curve progresses through the maximum entropy and the slope has some negative portion in it. The superheated vapour at the starting stage of the expansion process and ends with two phases i.e. liquid and vapor is

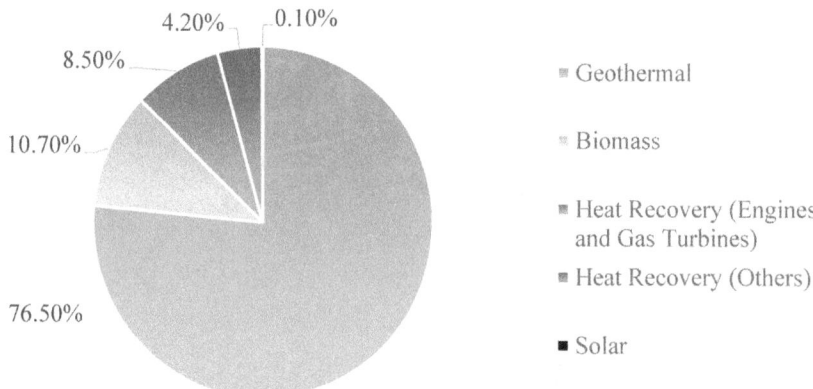

**Figure 2.20.** Installed capacity of ORC-based power plants around the world in 2016.

known as the wet working fluid. The wet working fluid acquires a negative slope ($dT/ds > 0$) in the saturated vapour curve of the temperature entropy diagram. It implicitly conveys the fact that entropy gradually rises with a fall in saturation temperature. The isentropic fluid is defined as the expansion process that takes place at constant entropy [35]. The slope of the isentropic fluid is infinite ($dT/ds > \infty$).

The wet working fluid requires a super heater to keep away from more droplet evolution at the shallow pressure stage of the turbine. The droplet formation should be within the range to avert the erosion of turbine blades. The concept of minimizing the droplet formation implicitly increasing the isentropic efficiency of the turbine. In the case of dry working fluids, the super heater is not required because of their shape of the saturation curve. The vapour at the exit of the turbine is still in the superheated region, although the superheated state at the exit is disadvantageous since the cooling load requirement is high and there are significant heat losses in the condenser. A heat exchanger named a recuperator is used to preheat the water. This recuperator also converts the superheated steam to saturated steam. This method of recovering the heat explicitly enhances the performance of the ORC. The super-heater or recuperator is not required in the case of the isentropic working fluid because the expansion line coexists with the saturation curve. From figure 2.21 superheating takes place before the expansion and in real cases, the expansion line runs slightly ahead of the saturation curve. In isentropic fluids, the expansion line does not penetrate deep into the superheated region. The novel way of classifying the working fluid is based on the low-temperature points on saturated liquid, critical temperature of the fluid, and local entropy points [36, 37]. Some of the novel working fluids and the characteristics are represented in table 2.2.

The novel concept of the combined ORC and direct vapour generation (DVG) is tested experimentally. The thermo-physical properties of dry and isentropic working fluids are presented in table 2.3. The investigation of the effect of fluid on the efficiencies of the ORC shows that, except for the collector efficiency, the efficiency of ORC and overall system efficiency increases with increases in critical temperature. The ORC, collector, and system efficiency of R236fa and benzene with conditions at

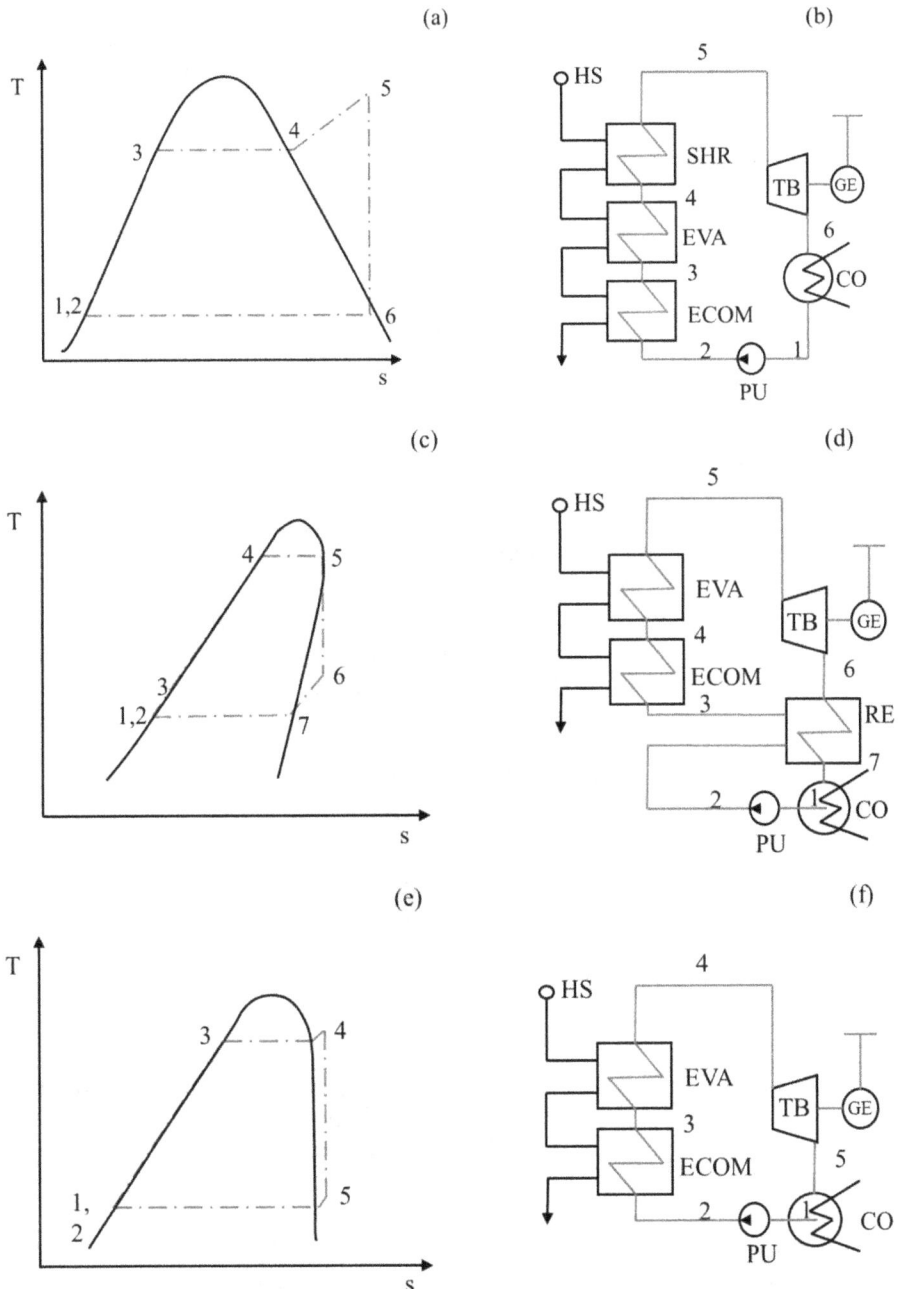

**Figure 2.21.** The different processes in the T–S diagram and their corresponding cycle layouts of (a), (b)—Wet working fluid (c), (d)—Dry working fluid (e), (f)—Isentropic working fluids. [CO—Condenser, ECOM—Economiser, EVA—Evaporator, GE—Generator, HS—A heat source, PU—Pump, RE—Recuperator, SHR—Superheater, TB—Turbine.]

**Table 2.2.** List of low-temperature working fluids with thermo-physical properties.

| Substance | Molecular mass (g mol$^{-1}$) | $T_b$ (°C) | $T_{cri}$ (°C) | $P_{cri}$ (MPa) |
|---|---|---|---|---|
| R245fa | 134.05 | 14.90 | 154.05 | 3.640 |
| R123 | 152.93 | 27.82 | 183.68 | 3.662 |
| R141b | 116.95 | 32.05 | 206.81 | 4.460 |
| RC318 | 200.05 | −6.0 | 115.2 | 2.78 |
| Isobutane | 58.12 | −11.7 | 134.7 | 3.63 |
| Isopentane | 72.15 | 27.8 | 187.2 | 3.38 |
| n-pentane | 72.15 | 36.1 | 196.6 | 3.37 |
| n-hexane | 86.175 | 68.71 | 234.67 | 3.034 |
| n-heptane | 100.2 | 98.38 | 266.38 | 2.736 |
| n-decane | 142.28 | 174.12 | 344.55 | 2.103 |
| Cyclohexane | 84.161 | 80.736 | 280.49 | 4.075 |

the critical temperature of 120 °C and average solar radiation level of 800 W m$^{-2}$ are reported as 10.59, 56.14, 5.08% and 12.5, 52.58, 6.57% respectively. The difference in collector efficiency between these two working fluids gets increased at the low level of solar radiation. Amidst all working fluids, R123 displays overall better performance and the most adapted working fluid ORC-based power generation [38].

### 2.4.1 Selection criteria for novel working fluids

The working fluid is one of the most important substances that play a major role in the conversion efficiency and economics of the cycle. Hence, special attention is needed before selecting a suitable working fluid for the Rankine cycle. The other properties of the working fluids, such as efficiency, availability, toxicity, flammability, and cost, are also considered while selecting the working fluid.

(a) **Critical parameters** The point at which the pressure and temperature of the working fluid are indistinguishable as vapour or liquid is known as the critical point. The same can be applied to properties of the fluid, such as pressure, temperature, density, and volume, known as critical pressure, critical temperature, critical density, and critical volume [39]. To improve the heat transfer characteristics, the critical point should be approximately ahead of the evaporating point.

(b) **Cycle pressure** The pressure of the working fluid from the evaporator section should not immoderate. The moderate pressure of the working fluid helps to avoid mechanical stress unless the design of the system requires high pressure. The safety and the cost have also been some of the factors to look out for if the evaporator section produces the steam at high pressures [40]. The minimum pressure in the cycle should be greater than the atmospheric pressure. The air from the atmosphere tried to get into the system, if the system has pressure lower than that of atmospheric pressure.

**Table 2.3.** Thermo-physical properties of working fluid for the organic Rankine cycle.

| Working fluid | Critical temperature (°C) | Critical pressure (MPa) | Thermal conductivity at 100 °C (W m K$^{-1}$) | Toxicity | Flammability |
|---|---|---|---|---|---|
| Benzene | 288.9 | 4.89 | NA | High | High |
| Cyclohexane | 280.49 | 4.07 | 106.73 | High | High |
| Acetone | 234.95 | 4.7 | NA | Low | High |
| R113 | 214.06 | 3.39 | 53.85 | Low | Non flammable |
| Isopentane | 187.2 | 3.39 | 83.02 | Low | Extremely |
| R123 | 183.68 | 3.66 | 58.19 | High | Non flammable |
| R245ca | 174.42 | 3.92 | 76.13 | NA | NA |
| Cis-butene | 162.6 | 4.22 | 82.89 | Non toxic | Extremely |
| Trans-butene | 155.46 | 4.03 | 81.36 | NA | Extremely |
| R245fa | 154.01 | 3.65 | 64.29 | Low | Non flammable |
| Butane | 151.98 | 3.79 | 78.23 | Low | Extremely |
| Isobutene | 144.94 | 4.00 | 76.94 | Non toxic | Extremely |
| R236ea | 139.29 | 3.50 | 55.15 | NA | NA |
| Isobutane | 134.66 | 3.63 | 66.57 | Low | Extremely |
| R236fa | 124.92 | 3.2 | 51.89 | Low | Non flammable |
| R-C318 | 115.23 | 2.78 | 44.88 | Low | Non flammable |
| R227ea | 101.75 | 2.9 | 48.25 | Low | Non flammable |

Also, the sealing requirements could be large enough to prevent the atmospheric air from entering into the system. By keeping the safety and economic aspects, the pressure in the heat exchanger must be kept in the range of 0.1–2.5 MPa.

(c) *T–s* **diagram** The low-power output delivered by utilizing the low temperature need does not require the wet working fluid, since the wet working fluids require superheating to avoid moisture formation in the turbine. The isentropic and dry working fluids do not show the moisture formation effect after the expansion process. If isentropic fluids are chosen, the isentropic fluids have the property of remains in the vapour form after the expansion process. If the slope of the saturation curve is more positive in the dry fluid, then it produces less efficiency.

(d) **The specific volume of the fluid** The working fluid must possess a low vapour and liquid specific volume. This is one of the major parameters which influence the rate of heat transfer in the heat exchanger. The specific volume of vapour is directly associated with the cost and size of the components. The high specific volume of liquid possesses the large volumetric flows needs a large expander in the system. Hence, the specific volume of fluid should be low at condenser pressure, the work required for pumping is also less.

(e) **Thermo-physical properties** The thermo-physical properties such as thermal conductivity, the specific heat of the liquid, viscosity, surface tension, and

**Table 2.4.** Comparison of turbine efficiency for the different molecular weight of the fluid.

| Capacity of power | Turbine efficiency for low molecular weight | Turbine efficiency for high molecular weight |
|---|---|---|
| >10 MW | 70–80 | 75–80 |
| 1–5 MW | 50–70 | 75–80 |
| 200–500 kW | 30–50 | 75–80 |
| 10–100 kW | 25–50 | 60–75 |

latent heat of vaporization should be considered while selecting a working fluid. It is advisable to have low specific heat, low viscosity, low surface tension, high thermal conductivity, and high latent heat of vaporization. The ratio of enthalpy between the vaporization to sensible heat should be high enough that the heat required for preheating is less.

(f) **Molecular weight** It is advisable to have low molecular weight for high temperature or high power output and a high molecular weight of the fluid is suitable for low temperature or low power output [38]. The above-mentioned statement is proven by performing many experiments and the range is presented in table 2.4. Even the low molecular weight of the working fluid produces almost the turbine efficiency as 80% in which the fluid of high molecular weight is achieved.

(g) **Thermal stability and compatibility of the fluid with materials** The most critical step in the selection of working fluid is thermal stability and the closeness of the working fluid associated with the material. It is better to have high thermal stability and a longer lifetime. The compatibility of the fluid with the turbine blades has also an impact on the lifetime of the plant. In case the fluid is exposed to chemical decomposition, the periodic replacement should be done with cautious effort otherwise it should have non-condensable gases, which have a deeper impact on the materials. Table 2.5 provides the list of fluids with maximum stability temperature.

### 2.4.2 Low-temperature working fluids

The low-temperature working fluids are mostly suitable for bottoming cycle in a binary or tertiary vapour cycle. The operating temperature of the working fluid is in the range of 30 °C–100 °C to utilize the low heat source either in form of waste heat recovery or in another way. The working fluids have a low boiling point and the generated vapour runs in an organic Rankine cycle. The power obtained from these working fluids is less as compared with medium and high temperature working fluids, since the temperature difference between the generated vapour and the condensed liquid. The generated vapour at 90 °C drives the organic Rankine cycle with an efficiency of 7%–8%. The commonly used working fluids for the low-temperature range are organic fluids such as toluene and methyl chloride and refrigerants such as R11, R113, and R114 [41, 42].

### 2.4.3 Medium-temperature working fluids

The system possesses a temperature from 300 °C to 600 °C is known as a medium temperature system and working fluids used in these temperature limits are known as medium-temperature or intermediate working fluids. None of the working fluids operate in the Rankine cycle with a wide range of temperatures. Hence, different fluids are used to utilize the heat source in the different temperature ranges. The medium temperature working fluids are deployed in the intermediate cycle under the tertiary cycle. The commonly used heat transfer fluid for the intermediate cycle is steam. Other than steam, diphenyl possesses the same thermodynamic properties as steam. These medium-temperature working fluids are used in the tertiary cycle to utilize the heat source from the topping cycle and rejects it to the bottoming cycle or condenser [41–43].

### 2.4.4 High-temperature working fluids

The working fluid possesses a temperature greater than 600 °C is known as a high-temperature working fluid. Mercury is the commonly used working fluid for utilizing the high-temperature source in the topping cycle. The mercury used as working fluid in the topping cycle has a design efficiency of 36%. Since mercury is a toxic fluid, the researchers tried potassium as a working fluid and yielded 53% efficiency. Later they tried with liquid metals such as sodium and the combination of lead and bismuth as the high-temperature working fluid in the topping cycle. The result shows that it performs better than the molten nitrate salts in the topping cycle [41, 42, 44].

### 2.4.5 Novel working fluids for enhancement of cycle efficiency

The novel working fluids on various cycles have a considerable impact on their efficiency. The objective of exploring novel working fluids for the existing cycle is to

**Table 2.5.** Stability temperature of the working fluid.

| Working fluid | Max. stability temperature (°C) | Compatible material |
|---|---|---|
| R227 | 425 | Stainless steel (AISI 316) |
| R23 | 400 | Stainless steel (AISI 316) |
| R236fa | 400 | Stainless steel (AISI 316) |
| R143a | 350 | Stainless steel (AISI 316) |
| R245fa | 300 | Stainless steel (AISI 316) |
| R134a | 368 | Stainless steel |
| R141b | 90 | Stainless steel |
| R13I1 | 102 | Stainless steel |
| R7146 | 204 | Stainless steel |
| R125 | 396 | |
| Methanol | 175–230 | |
| Toluene | 400–425 | |
| R113 | 175–230 | |

improve the system performance by improving the thermo-physical properties of working fluid. A solar ORC system with direct vapour generation (DVG) was examined with seventeen dry and isentropic working fluids. The system consists of a solar collector along with heat storage unit to provide stable power production. As compared with a conventional solar ORC, the system eradicates the intermediate heat exchanger and the HTF. The experimental investigation provides the information about the effect of working fluids on collecting solar radiation. The respective increment in electrical efficiency of all working fluids on this system with respect to different levels of solar radiation as compared with solar ORC with a HTF are shown in table 2.6. The solar ORC with a concept of DVG is possible and it operates with lower temperature and pressure as compared with solar thermal power generation with direct steam generation (DSG). The operating pressure could be lowered based on the working fluid. The proposed system has a benefit of excluding the secondary heat transfer fluid and power requirement of the pumps are reduced. The working fluid in the solar ORC with the concept of DVG had a significant effect on the collection efficiency. The collector efficiency decreases gradually with a rise in critical temperature of the fluid.

Among the seventeen working fluids, the fluid selection for the solar ORC with the concept of DVG was crucial. The cost of the system depends on the operating pressure of the working fluid. From the results shown in table 2.6, it is clear that R123 prefers over the R245 with respect to operating pressure and efficiency. The electrical efficiency of the working fluid R123 was 6.1% with the conditions at critical temperature of 120 °C and average solar radiation level of 800 W m$^{-2}$ [45].

**Table 2.6.** Respective increment in electric efficiency of working fluids in the solar ORC with the concept of DVG as compared with a conventional solar ORC.

| Working fluid | Solar radiation (W m$^{-2}$) | | | |
| --- | --- | --- | --- | --- |
| | 400 | 600 | 800 | 1000 |
| Benzene | 32.47 | 16.03 | 10.53 | 8.17 |
| Cyclohexane | 33.38 | 16.92 | 11.21 | 8.45 |
| Acetone | 33.44 | 16.42 | 10.86 | 8.27 |
| R113 | 34.19 | 17.38 | 11.76 | 8.82 |
| Isopentane | 36.63 | 18.46 | 12.27 | 9.52 |
| R123 | 35.11 | 17.86 | 11.97 | 9.11 |
| R245ca | 37.34 | 18.76 | 12.69 | 9.59 |
| Cis-butene | 36.40 | 18.41 | 12.38 | 9.40 |
| Trans-butene | 37.42 | 19.03 | 12.63 | 9.69 |
| R245fa | 38.97 | 19.75 | 13.27 | 10.08 |
| Butane | 39.04 | 19.87 | 13.21 | 10.08 |
| Isobutene | 31.00 | 15.68 | 10.62 | 8.11 |
| R236ea | 41.69 | 21.25 | 14.32 | 10.97 |
| Isobutane | 42.05 | 21.20 | 14.49 | 10.89 |
| R236fa | 49.63 | 25.08 | 15.77 | 12.76 |

The list of working fluids such as R245fa, R134a, carbon dioxide, n-butane, i-butane and propane were found to be more suitable and efficient up to the temperature range of 250 °C and toluene and water are found to be efficient up for the higher temperature range (up to 340 °C for toluene). In the case of a two-stage solar ORC by using this working fluid, the efficiency was found in the range of 7%–11.8% [46]. The efficiency was supercritical and an ultra-supercritical Rankine cycle was found to be 43% and 49%. The temperature and pressure of steam for both the supercritical and ultra-super critical Rankine cycles are 538 °C and 124 bar and 600 °C and 300 bar respectively. Compared with a conventional cycle, the supercritical cycle possesses a higher number of regenerative units [47].

## 2.5 Power generation from concentrating solar power

On account of industrial growth and rapid urbanization, the energy demand has been drastically increased. Concentrating solar power (CSP) plants can be used for power generation by concentrating the Sunlight on the receiver surface by using the reflectors and transfering the heat to water and convert it to steam drives the steam turbine.

### 2.5.1 Indirect steam generation

Indirect steam generation is defined as the usage of heat transfer fluid to absorb the heat from the receiver and transfer it to water and convert it to steam with the aid of an intermediate heat exchanger. This method of steam generation can be even more fruitful when it gets integrated with thermal energy storage. During the time of excess heat generation, it can be stored in TES and retrieved at a later time for the production of constant power output. Though there are various heat transfer fluids are available at different temperature ranges and applications, synthetic oil and molten salts are the commonly used heat transfer fluids in CSP systems. The central receiver is integrated with TES through the indirect steam generation and is shown in figure 2.22.

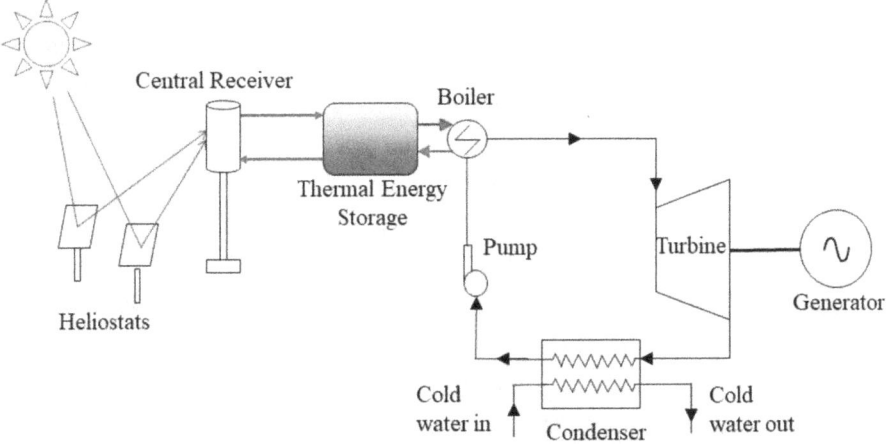

**Figure 2.22.** Central receiver with TES for electricity generation.

## 2.5.2 Direct steam generation

In most of the conventional PTC power plants, synthetic oil is used as the HTF and follows indirect steam generation. The drawback of using synthetic oil is poor heat transfer characteristics, high cost, operating temperature limited up to 390 °C, and degradation of oil owing to continuous heating and cooling. To overcome this drawback, water/steam is used as the HTF, and steam is obtained directly from the solar field and is called a direct steam generation (DSG). In this method, no intermediate heat exchanger is required and it had an impact on the LCoE. The DSG method could generate steam with a temperature of ~550 °C. The advantage of direct steam generation is the steam temperature can reach more than 500 °C. It is better to integrate the CSP plants with TES to provide uniform power output. The concentrated solar radiation on the receiver surface should be sufficient to generate the steam otherwise it should be integrated with TES providing the persistent output during cloudy days. The DSG also eliminates the usage of secondary fluid which had an impact on the investment cost. Table 2.7 shows the list of existing solar power plants operating in the DSG mode. Experimental investigations were performed and compared between DSG and synthetic oil as the HTF in parabolic trough collectors. Levelized cost of electricity shows an 11% reduction in electricity generation cost by DSG as compared with a parabolic trough collector (PTC) with synthetic oil as the HTF. The advantages of DSG are:

- Simple in construction
- Low thermal inertia

**Table 2.7.** Operation of solar power plants using the DSG method.

| Name of the country | Technology | Plant capacity (MW) | Backup/storage method |
|---|---|---|---|
| Pureto Errado 1 (Spain) | Linear Fresnal collector | 1.4 | Thermocline heat storage |
| Thai solar one (Thailand) | Parabolic trough collector | 5 | No storage/backup |
| Kimberlina solar thermal energy power plant (United States) | Linear Fresnel collector | 5 | No storage/backup |
| Sierra Sun Tower (United States) | Power tower | 5 | No storage/backup |
| PS 10 Solar Power Tower (Spain) | Power tower | 11 | Storage type (N/A) |
| PS 20 Solar Power Tower (Spain) | Power tower | 20 | Storage type (N/A) |
| Pureto Errado 2 (Spain) | Linear Fresnel collector | 30 | Thermocline heat storage |
| Ivanpah (United States) | Power, tower | 400 | Natural gas |

- Higher evaporation temperature results in higher conversion efficiency
- In absence of intermediate heat exchangers results in a reduction in capital cost and the improvement in overall efficiency
- Reduction in capital cost
- Reduction in electricity generation cost

Nonetheless the absence of TES or intermediate heat transfer fluid makes the DSG system more complicated during cloudy days and off-sunshine hours. It is difficult to harness the required heat during these times and that ended up in two phases when it comes out from the solar field. It will cause the turbine blade to erosion and corrosion. The attention should be focused on the usage of working fluids with low freezing temperatures otherwise the freezing of working fluids occurs at low ambient temperatures in the receiver tubes and results in the bursting of the absorber tube.

### 2.5.3 Solar-aided power generation

Solar-aided power generation (SAPG) is the process of adding heat from the solar field to the conventional fossil fuel-fired power plant. In such power plants, the extraction of steam (high-grade energy) at various stages of the turbine is replaced with heat produced from solar energy (low-grade energy) to preheat the feed water. It aids to curtail the utilization of fossil fuels, emissions into the atmosphere and it leads to more steam getting expanded in the turbine. Almost two-thirds of the electricity generation around the world is by fossil fuels. Nonetheless, standalone solar thermal power generation has limitations, such as high investment costs and low conversion efficiency. The integration of solar thermal energy into conventional fossil fuel fired power plants also leads to a substantial increment in plant output. In standalone solar thermal power generation, the efficiency of the plant is limited due to the limitation of producing high temperatures from the solar field, and the temperature for heat rejection is fixed. The schematic solar field integrated over the regenerative Rankine cycle is shown in figure 2.21. The major advantages of using SAGP are:

- It can be operated either in power boost (incremental in power output by consuming the same quantity of fuel with the existing system) or fuel reserve mode (produces same power by consuming less quantity of fuel).
- The exergy efficiency is high as compared to standalone solar thermal power generation.
- Low investment cost with less or no difficulty in implementation and more economic benefits (figure 2.23).

The efficiency of the power plant and instant solar share are the two important parameters used to compare the SAPG performance with standalone solar thermal power generation systems. The efficiency of SAPG is calculated by using the formula

$$\eta_{\text{solar}} = \frac{\Delta W_e}{Q_{\text{solar}} + \Delta Q_{\text{boiler}}} \tag{2.7}$$

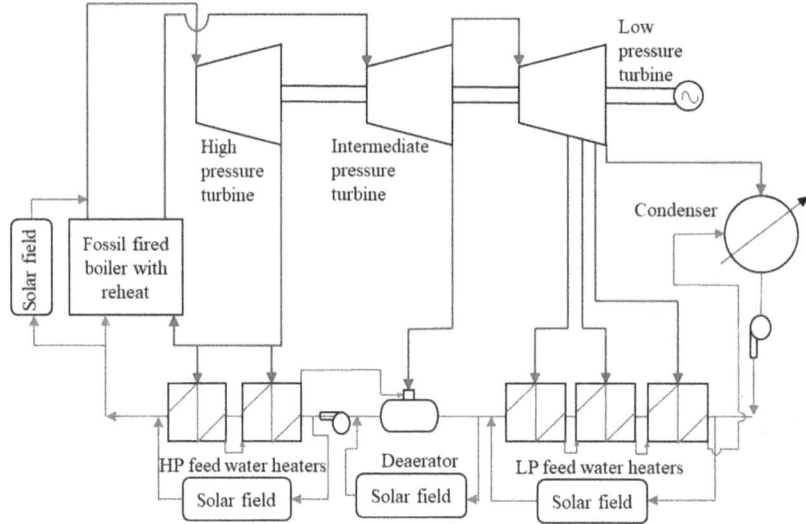

**Figure 2.23.** Solar field integrated with conventional fossil fuel-based power plant.

where $\Delta W_e$ is the power output including the SAPG system.

$Q_{\text{solar}}$ is the input from solar energy.

$\Delta Q_{\text{boiler}}$ is the boiler load calculated after installing the SAPG system.

The solar share is defined as the amount of power generated from incident solar radiation to the total thermal load on the plant. The results reported that SAPG operated in power boost mode, the cycle efficiency was improved by 2.94%. The exergy analysis was evaluated for the SAPG system with a capacity of 600 MW. The result shows that the SAPG system integrated with the regenerative Rankine cycle decreases the exergy loss and improves the exergy efficiency by 10%. Levelised cost of electricity (LCoE) is a major parameter to evaluate its economic investigation of the SAPG plant. LCoE of the SAPG plant operated in power boost mode yields less LCoE as compared with the SAPG plant operated in fuel reserve mode. However, both power boost and fuel reserve mode yields less LCoE as compared with standalone solar thermal power generation. In earlier days, the thermal energy storage system has not been tested with the SAPG system since the maximum incident solar radiation and peak energy demand occurs at the same time. The thermal storage system has been integrated with the SAPG system in later years and the performance has also been evaluated. The result shows that the thermal energy storage system considerably reduces the usage of coal in the regenerative Rankine cycle [48, 49].

## 2.6 4-E analysis of CSP generation

In this part, 4-E (energy-exergy-environmental-economic) analysis of these four concentrators namely the PTC, LFR, central receiver, and PDC are discussed. The objective of doing exergy and energy analysis is to analyze the availability and utilization of solar energy for power generation. Separately, the term exergy gives

**Table 2.8.** Energy and exergy efficiency of concentrating collectors [50–54].

| Name of concentrating collector | Energy efficiency (%) | Exergy efficiency (%) |
|---|---|---|
| Parabolic Trough Collector (PTC) | 23.16 | 32.76 |
| Linear Fresnel Collector (LFR) | 12.17 | 17.21 |
| Central Receiver | 31.24 | 24.5 |
| Parabolic Dish Collector (PDC) | 33.56 | 29.4 |

out the maximum amount of energy that can be obtained from the resources, and energy analysis helps to determine the occurrence of energy loss at various components. This analysis helps to figure out the quantitative part where energy losses have occurred. It also aids to design the solar field more efficiently to produce more output than the conventional design. In environmental aspects, it deals with the reduction in emission of greenhouse gases into the atmosphere and economic aspects deal with the amount spent for unit power electricity generated by each of the four CSP methods. Table 2.8 explains the exergy and energy efficiency of the concentrating collectors. Due to concentrating solar power plant of 1 MW electrical energy output, it could conserve 1 813 000 kg of carbon-dioxide ($CO_2$), 12 250 kg of sulphur-dioxide ($SO_2$), 6230 kg of nitrous oxide ($NO_x$), and 980 kg of particulate matter into the atmosphere [50].

### 2.6.1 Energy analysis

The ratio of net electricity produced to the amount of energy received by the concentrator is known as energy efficiency. The rate of the energy received by the concentrator is given by [50]:

$$\dot{Q}_i = I_{bn}A_{ap}. \tag{2.8}$$

The amount of energy obtained by the receiver is given by:

$$\dot{Q}_r = \dot{Q}_i \eta_{opt}. \tag{2.9}$$

The efficiency of the concentrator is given by:

$$\eta_{I,c} = \frac{\dot{Q}_r}{\dot{Q}_i}. \tag{2.10}$$

The amount of energy transferred to the HTF in the receiver is given by:

$$\dot{Q}_u = \dot{m}_f(h_{f,\,out} - h_{f,\,in}). \tag{2.11}$$

The energy efficiency of the receiver is given by:

$$\eta_{I,r} = \frac{\dot{Q}_u}{\dot{Q}_r}. \tag{2.12}$$

In solar field the efficiency is given by:

$$\eta_{I,\,\text{sf}} = \frac{\dot{Q}_u}{\dot{Q}_i}. \tag{2.13}$$

The power block efficiency is given by

$$\eta_{I,\,\text{PB}} = \frac{PG_{\text{net}}}{\dot{Q}_u}. \tag{2.14}$$

Overall efficiency of the CSP plant is given as:

$$\eta_{I,\,o} = \eta_{I,\,\text{sf}} \times \eta_{I,\,\text{PB}} = \frac{PG_{\text{net}}}{\dot{Q}_i}. \tag{2.15}$$

### 2.6.2 Exergy analysis

The ratio of total power generation to the exergy obtained by the reflector is known as the exergy efficiency of the CSP plant. The rate of exergy obtained by the reflector is given as [50]:

$$\dot{E}x_i = \dot{Q}_i\left[1 - \frac{4T_{\text{amb}}}{3T_{\text{sun}}}(1 - 0.28\ln f)\right] \tag{2.16}$$

where $f$ is the dilution factor $(1.3 \times 10^{-5})$ and it is defined as the mixing ratio of solar radiation from the Sun and radiation emitted by surroundings.

In the receiver, the exergy absorbed is given by:

$$\dot{E}x_r = \dot{Q}_r\left[1 - \frac{T_{\text{amb}}}{T_r}\right]. \tag{2.17}$$

In concentrator, the exergy efficiency is given as:

$$\eta_{\text{II},\,c} = \frac{\dot{E}x_r}{\dot{E}x_i}. \tag{2.18}$$

The rate of exergy usually delivered by the receiver is given as:

$$\dot{E}x_u = \dot{m}_f(\dot{E}x_o - \dot{E}x_i) = \dot{m}_f\left[\left(h_{f,\,\text{out}} - h_{f,\,\text{in}}\right) - T_{\text{amb}}(s_{f,\,\text{out}} - s_{f,\,\text{in}})\right]. \tag{2.19}$$

In receiver, the exergy efficiency is given as:

$$\eta_{\text{II},\,r} = \frac{\dot{E}x_u}{\dot{E}x_r}. \tag{2.20}$$

Exergy efficiency is given by:

$$\eta_{\text{II},\,\text{sf}} = \frac{\dot{E}x_u}{\dot{E}x_i}. \tag{2.21}$$

In power block, the exergy efficiency is given by:

$$\eta_{\text{II, PB}} = \frac{PG_{\text{net}}}{\dot{E}x_u}. \tag{2.22}$$

Overall efficiency of the plant is given as:

$$\eta_{\text{II, }o} = \eta_{\text{I, sf}} \times \eta_{\text{I, PB}} = \frac{PG_{\text{net}}}{\dot{E}x_i}. \tag{2.23}$$

These two analyses, i.e. energy and exergy analyses, are used to find out the irreversibility in the solar power plant. The cycle-tempo is software used to analyse and simulate the performance of the CSP plants. The cycle-tempo software has shown good agreement with simulated values and observed values. It has been noticed that each and every component of the system are accounted for energy and exergy losses. However, the highest losses takes place at two places such as power block and solar field in the PTC power plant. In the case of LFR plants, energy and exergy losses occurred maximum in the solar field. The energy and exergy flow diagrams of 1 MW$_e$ and 50 MW$_e$ LFR and PTC plants are shown in figures 2.24 and 2.25 respectively. The energy losses in the LFR solar field and power block of 1 MW$_e$ were reported as 54.16%, 34.33% whereas the exergy loss in respective fields was found to be 72.73% and 10.98%. The highest energy loss of 1 MW$_e$ LFR plant was observed in the condenser followed by the receiver, concentrator, feed water, and other losses and turbine losses and in terms of percentage were reported as 34.32, 30.72, 23.44, 1.32 and 0.27 respectively. The highest exergy loss took place at the concentrator followed by the receiver, turbine, condenser, and feed water heaters

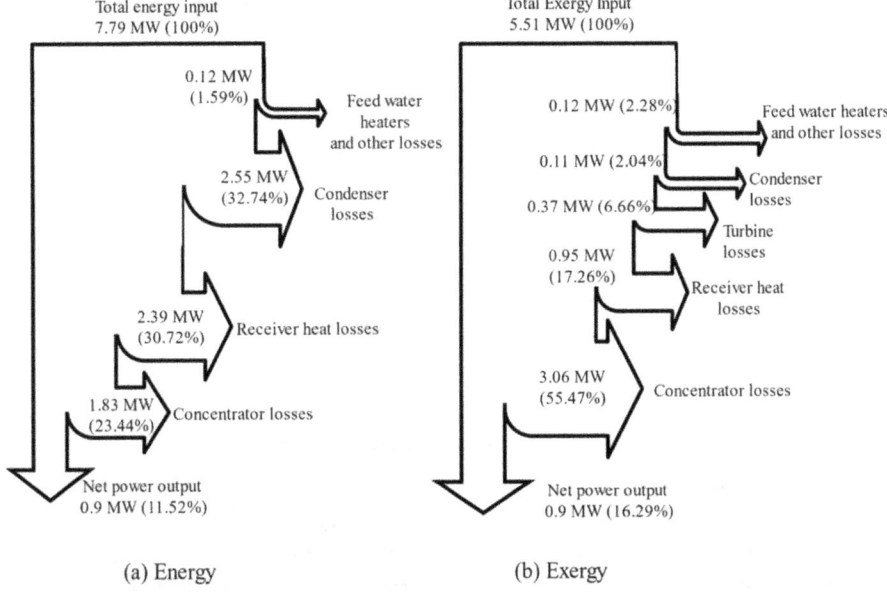

(a) Energy          (b) Exergy

**Figure 2.24.** (a) Energy and (b) exergy flow diagrams of the 1 MW$_e$ LFR power plant.

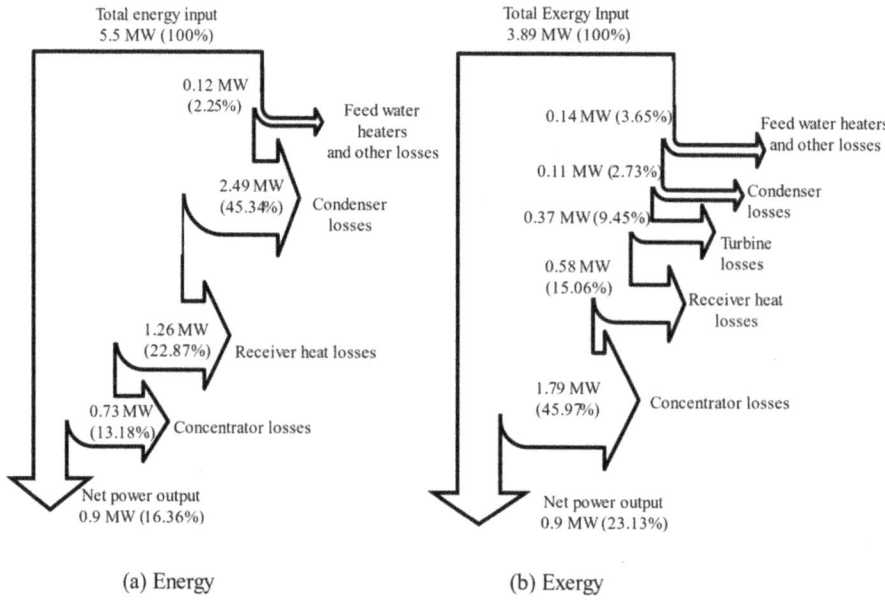

Total energy input
5.5 MW (100%)

0.12 MW
(2.25%)
Feed water
heaters
and other losses

2.49 MW
(45.34%)
Condenser
losses

1.26 MW
(22.87%)
Receiver heat losses

0.73 MW
(13.18%)
Concentrator losses

Net power output
0.9 MW (16.36%)

(a) Energy

Total Exergy Input
3.89 MW (100%)

0.14 MW (3.65%)
Feed water heaters
and other losses

0.11 MW (2.73%)
Condenser
losses

0.37 MW (9.45%)
Turbine
losses

0.58 MW
(15.06%)
Receiver heat
losses

1.79 MW
(45.97%)
Concentrator losses

Net power output
0.9 MW (23.13%)

(b) Exergy

**Figure 2.25.** (a) Energy and (b) exergy flow diagrams of the 1 MW$_e$ PTC power plant.

and other losses are reported in terms of percentage as 55.47, 17.26, 6.66, 2.04, and 2.28 respectively for the 1 MW$_e$ LFR plant.

The comparison of energy and exergy losses for various components in the power plant of 1 MW$_e$ by the LFR and PTC are shown in tables 2.9 and 2.10. In the 1 MW$_e$ PTC plant, the energy loss is higher in the power block as compared with solar field and exergy loss is found to be more in the solar field as compared with other components. The energy loss in the power block and solar fields is reported as 47.58% and 36.06%. The exergy loss takes place at different components of the power plants, such as the concentrator, receiver, turbine, condenser, feed water and other losses, are reported in terms of percentage as 45.97, 15.06, 9.45, 2.73 and 3.65 respectively. It is observed that exergy loss for all components of LFR power plants is higher than in PTC power plants. In particular, the highest exergy loss was reported in the collector field in contrast with other components. The exergy loss occurred due to high-grade solar energy. The exergy efficiency of the power plant components is enhanced by upgrading the reflectivity of the concentrator, optical efficiency, concentration ratio, conversion efficiency, and outlet condition of steam.

### 2.6.3 Environmental impacts of concentrating solar power plants

The concentrating solar power plant does not release particulate matter and greenhouse gases into the atmosphere during its working as compared with conventional power plants. Simply, focussing on CSP plants does not have any impact on the environment. The only concern about the CSP is having disadvantages and an effect on the environment while the making of the materials and

**Table 2.9** Energy and exergy analysis of the 1 MW$_e$ LFR plant

| Plant Components | | Energy (kW) Received | Loss | Energy Loss | Energy Efficiency | Exergy (kW) Received | Loss | Exergy Loss | Exergy Efficiency |
|---|---|---|---|---|---|---|---|---|---|
| Power Block | Turbine | 4485 | 0 | 0.00 | 0.25 | 1661 | 367 | 0.06 | 0.59 |
| | Condenser | 2622 | 2551 | 0.32 | | 114 | 113 | 0.02 | |
| | Feed water heaters and others | 900 | 124 | 0.01 | | 188 | 125 | 0.02 | |
| | Total | 3572 | 2675 | 0.34 | | 1502 | 605 | 0.10 | |
| Solar Field | Concentrator | 7793 | 1827 | 0.23 | 0.45 | 5509 | 3056 | 0.55 | 0.27 |
| | Receiver | 5967 | 2394 | 0.30 | | 2453 | 951 | 0.17 | |
| | Total | 7793 | 4221 | 0.54 | | 5509 | 4007 | 0.72 | |
| Total Power Plant | | 7793 | 6896 | 0.88 | 0.12 | 5509 | 4611 | 0.84 | 0.16 |

2-37

Table 2.10 Energy and exergy analysis of the 1 MW$_e$ PTC plant

| Plant Components | | Energy (kW) | | Energy Loss | Energy Efficiency | Exergy (kW) | | Exergy Loss | Exergy Efficiency |
|---|---|---|---|---|---|---|---|---|---|
| | | Received | Loss | | | Received | Loss | | |
| Power Block | Turbine | 4258 | 0 | 0.00 | 0.25 | 1641 | 367 | 0.09 | 0.59 |
| | Condenser | 2556 | 2493 | 0.45 | | 107 | 106 | 0.02 | |
| | Feed water heaters and others | 731 | 124 | 0.02 | | 174 | 142 | 0.03 | |
| | Total | 3515 | 2616 | 0.47 | | 1515 | 615 | 0.15 | |
| Solar Field | Concentrator | 5498 | 725 | 0.13 | 0.63 | 3886 | 1787 | 0.45 | 0.38 |
| | Receiver | 4773 | 1257 | 0.22 | | 2100 | 585 | 0.15 | |
| | Total | 5498 | 1983 | 0.36 | | 3886 | 2372 | 0.61 | |
| Total Power Plant | | 5498 | 4599 | 0.84 | 0.16 | 3886 | 2987 | 0.77 | 0.23 |

disposal of waste materials. During the construction of the parabolic trough collector plant, 24 g of equivalent $CO_2$ are emitted per kWh. In the coal-fired power plant, the emission of greenhouse gases is reported as 840 g of $CO_2$, 5.8 g of $SO_2$, 2.9 g of $NO_x$, and 0.46 g of particulate matter per unit (kWh) of power production.

### 2.6.4 Economic analysis of concentrating solar power plants

The economic analysis of CSP plants is performed in terms of the levelised cost of electricity, net present value, benefit to cost ratio, and payback period. The levelised cost of electricity (LCoE) depends on the operation and maintenance (O&M) cost and fixed capital cost. The fixed capital cost includes the power block, solar field, infrastructure, and other incidental costs. The O&M cost comprises the equipment cost, solar field replacement components, administration, water treatment, labour, and service cost. The various parameters of economic analysis of CSP power plants are performed as follows [50]:

The levelised cost of electricity is given as:

$$\text{LCoE} = C_{\text{RCap}} + C_{\text{LOM}}. \tag{2.24}$$

The fixed capital cost ($C_{\text{FCap}}$) is given as:

$$C_{\text{FCap}} = \frac{C_{\text{Cap}}}{\text{PP}_{\text{eff}}}. \tag{2.25}$$

Effective power production ($\text{PP}_{\text{eff}}$) is given as:

$$\text{PP}_{\text{eff}} = \text{PP}_{\text{gross}} - \text{PC}_{\text{aux}}. \tag{2.26}$$

The levelised operational and maintenance (O&M) cost ($C_{\text{LOM}}$) is given as:

$$C_{\text{LOM}} = LF(C_{\text{FOM}} + C_{\text{VOM}}). \tag{2.27}$$

The levelising factor (LF) is given by:

$$\text{LF} = \left[ \frac{(1 + i_e)^{\text{lt}} - 1}{i_e(1 + i_e)^{\text{lt}}} \right]\left[ \frac{i(1 + i)^{\text{lt}}}{(1 + i)^{\text{lt}} - 1} \right] \tag{2.28}$$

where dr is the discount rate and $\text{dr}_h$ is the discount rate with hike.

$$\text{dr}_h = \frac{(\text{dr} - e)}{(1 + e)}. \tag{2.29}$$

The fixed O&M ($C_{\text{FOM}}$) is given as:

$$C_{\text{FOM}} = \frac{C_{\text{OM}}}{\text{PP}_{\text{gross}}}. \tag{2.30}$$

The changeable O&M ($C_{\text{COM}}$) includes the supplementary O&M cost other than the fixed O&M and it is given as:

$$C_{\text{COM}} = 0.1 C_{\text{FOM}}. \tag{2.31}$$

**Table 2.11.** LCoE of concentrating solar power systems.

| Parameters | Parabolic trough collector | Linear Fresnel collector | Central receiver | Parabolic dish collector |
|---|---|---|---|---|
| Levelised cost of electricity (LCoE) ($ kWh$^{-1}$) | 0.099 | 0.150 | 0.109 | 0.133 |

The discounted payback period is given as:

$$\text{DPP} = \frac{\ln(B_j - C_j) - \ln\{(B_j - C_j) - \text{dr}C_{\text{cap}}\}}{\ln(1 + \text{dr})}. \tag{2.32}$$

The benefit to cost ratio is given by:

$$\frac{B}{C} = \frac{1}{C_{\text{cap}}}\left[\sum_{j}^{lt} \frac{B_j - C_j}{(1 + \text{dr})^j}\right]. \tag{2.33}$$

The net present value of the system is given as:

$$\text{NPV} = \left[\sum_{j}^{lt} \frac{B_j - C_j}{(1 + \text{dr})^j}\right] - C_{\text{cap}}. \tag{2.34}$$

The levelised cost of electricity (LCoE) for the PTC, LFR, central receiver and PDC are represented in table 2.11.

## 2.7 Recent advancements and future aspects in CSP generation

In recent years the electricity generation from solar photovoltaics is dominant over the CSP plants since the levelised cost of electricity for solar photovoltaics is low compared with CSP. It has a problem of scaling down into communal requirements and the investment cost is high as compared with other power generation systems. Compared to earlier, the cost of CSP components was reduced, however the feed in tariff was slightly higher than solar photovoltaic. Nonetheless, the power generation by CSP is 1.5–3 times greater than solar photovoltaics, CSP is not able to provide electricity at an affordable cost [55]. To overcome the abovementioned drawback, several improvements were persistently made in the solar concentrator, design of the receiver, power block components, heat transfer fluids, and thermal energy storage systems.

### 2.7.1 Advancements in the solar concentrator

The solar concentrator plays a pivot role in enhancing the efficiency of the solar power plants. The slight improvement in reflection has made a significant improvement in the annual performance of the solar power plant. Currently, research work is under progress regarding the development of a thin film reflector that has high reflectivity compared with the silver-polymer mirror. One of the simplest ways to

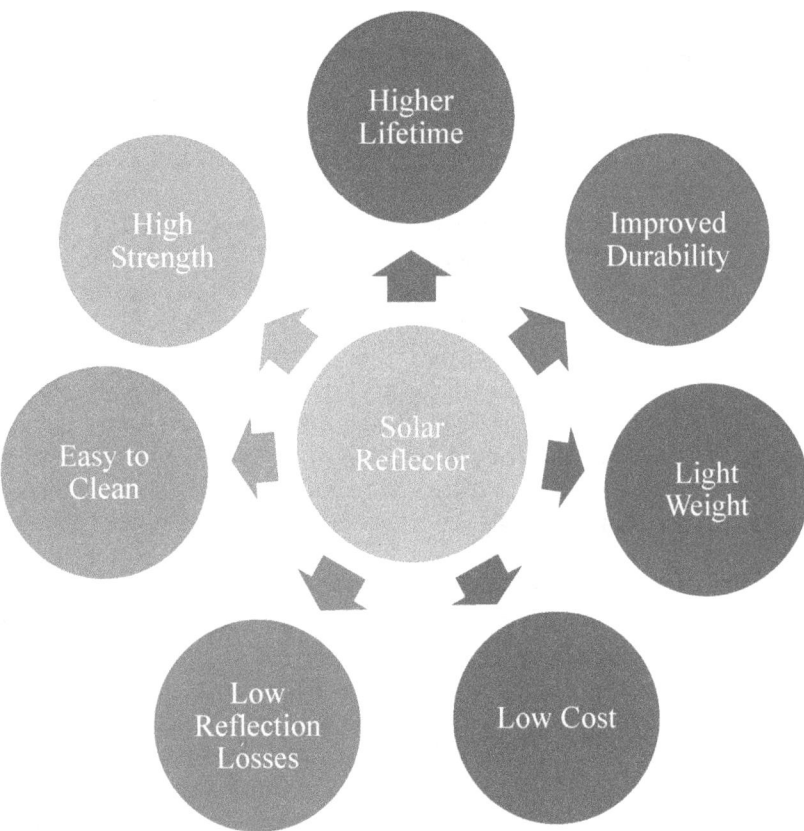

**Figure 2.26.** Characteristics of reflector concentrating solar power systems.

improve reflectance is by reducing the thickness of the mirror. Anti-soil coating on the mirror helps curtail the sedimentation of dust particles on the surface of the concentrator and enhances the efficiency and life of the reflector. The characteristics of the reflector for the CSP system are depicted in figure 2.26.

In a conventional LFR system, the optical and heat losses from the absorber/ receiver are higher since the trapezoidal cavity was used. In CLFR two parallel receivers are utilized in each row of reflectors in which successive mirrors concentrate the incident solar radiation on both sides of the receiver. Further, the heat losses from the receiver can be minimized by the utilization of secondary reflectors and evacuated receiver. The purpose of using a secondary concentrator is to refocus the reflected radiation on the receiver surface. By using this orchestration, the homogeneous heating on the absorber surface could lead up to the high operating temperature of 500 °C.

### 2.7.2 Advancements in heat transfer fluids

Most of the existing CSP plants in the world are using thermic fluids and molten salt as their HTF. Several advancements were made in HTF to improve the performance of the CSP plants. Several HTFs, such as water/steam, liquid metals, compressed gases ($CO_2$, $N_2$, He), with the addition of nanoparticles in the HTF, were examined to improve the performance of the CSP system. The desirable characteristics of the HTF are thermal stability, long life, low vapour pressure, chemical stability and affordable price. The existing central receiver uses molten salt/steam as the HTF operating in the Rankine cycle. The inlet temperature of the turbine at this system is limited to 560 °C. Even further increase the temperature to 1000 °C by exploring various HTF such as falling particles, liquid metal, etc. The Sandia National Laboratory visualizes the concept of the falling particle receiver and the results show that the high operating temperature (1000 °C). The particles falling from the top of the tower to the bottom and get heated up during the traveling time. The heat gets transferred to the working fluid from the solid particles and it also significantly reduces the cost of thermal energy storage. The direct particle receivers, such as free-falling obstructed flow, centrifugal and fluidized designs, possess higher solar to thermal conversion efficiencies since the particles are irradiated directly. The challenges in direct particle receivers are controlling the flow rates and particle loss. However, in an indirect particle receiver, the particle loss may be eliminated but the heat transfer resistance increases between the tubes and particles. The LCoE of the particle receiver is estimated to be 10$ per MW, which is still low as compared with pumped hydro and compressed air energy storage [56]. Advanced research is in progress about the porous structures and the ways of increasing the residence time of the falling particles in the central receiver. The beam down concept has all major equipment on the surface and it implies that installation, operation, and maintenance becomes simple.

In recent years, researchers worked in the area of using supercritical-$CO_2$ as HTF in CSP technologies. A 100 kW CSP plant was developed by the Masdar Institute Solar Platform (MISP) in which the array of 45 reflectors are laid down at the apex of the tower to concentrate the solar radiation to the absorber placed at the ground. The air-Brayton cycle, supercritical-$CO_2$ Brayton cycle and ultra-supercritical cycle are often termed as high-efficiency cycles. In future, implementing the high efficient power cycles may become an alternative to other power generation systems. $CO_2$ has been tested as the HTF/working fluid in the central receiver. The advantages of using $CO_2$ as HTF are non-toxic, non-flammable, high density at the critical point, low-cost fluid, and low corrosive. Furthermore, the work required to compress the $CO_2$ is less when it enters a critical point.

### 2.7.3 Advancements in thermal energy storage systems (TES)

The optics of the CSP and TES are the major areas to look out for in future research. The cost reduction in these areas would make CSP more competitive with solar photovoltaics. The important point is that in the areas of having high DNI both CSP and solar photovoltaics will provide electricity but the advantage of CSP is to store the thermal energy for a longer period of time and thereby generating electricity at low cost as compared with withdrawing energy from batteries. CSP is the most evolving topic for researchers in recent years and it provides an opportunity to integrate the TES with other CSP technologies to minimize the generation cost.

Solar energy is diurnal in and intermittent in nature. To overcome this issue and also reduce the LCoE in CSP plants leads to the integration of TES in CSP plants. Integration of thermal energy storage (TES) into the CSP system helps to generate the constant power output and increase the dispatchability of the plant. Different types of salt mixture used as phase change material (PCM) play a pivotal role because of its desired qualities such as crtitical temperature and high heat storage capacity. There are hundreds of salt mixtures available in the melting temperature range of 120 °C–1000 °C. The commonly used salts as PCM are carbonates, molybdates, chlorides, vanadates, metal alloys, fluorides, hydroxides, nitrates and other salts. In general, hydroxides and nitrate salts have the phase change temperatures at the range of 120 °C–300 °C. Along with the desired melting temperature and large heat storing capacity, the material should possess high thermal conductivity and thermal stability. Nitrate salts alone possess good thermal conductivity and stability along with aforementioned parameters. Molten salt is used as thermal energy storage media for most of the CSP plants.

Solar salts ($NaNO_3$ (60%) + $KNO_3$ (40%)) are relatively inexpensive and the freezing occurs at 220 °C and requires anti freezing or trace heating techniques. The homogenous mixture of two or more salts to reduce the melting temperature and each of the salt or substances melts congruently is known as eutectic salt. Eutectic salt of $NaNO_3$-$KNO_3$ (50%–50%) have melting temperature of 223 °C and heat capacity of 106 kJ kg$^{-1}$. In addition to these properties, it also owns good thermal conductivity, no phase segregation, chemical stability, low corrosion and low cost. The experimental investigation of the performance of $Ca(NO_3)_2$-$NaNO_3$ (30%–70%) salt mixture was performed. The latent heat of this salt mixture was found as 130 kJ kg$^{-1}$ which is the same as solar salt. Nevertheless the cost of $Ca(NO_3)_2$-$NaNO_3$ was half as compared with solar salt. In addition to this mixture, $LiNO_3$ was added to increase the heat capacity. Also, the latent heat of the eutectic mixtures of $KNO_3$-$LiNO_3$ and $KNO_3$-$NaNO_3$-$LiNO_3$ were found to be 178 kJ kg$^{-1}$ and 155 kJ kg$^{-1}$. The primary objectives of using eutectic salt mixtures are to reduce the size of thermal energy storage system. An eutectic salt mixture of $LiNO_3$-$NaCl$ (87%–13%) was studied and the result shows that melting temperature, latent heat and decomposition temperature of the eutectic mixture was 220 °C, 300 kJ kg$^{-1}$ and 400 °C–450 °C, respectively. The compatibility test also conducted and stainless steel 316 was found to be compatible for this eutectic salt mixture [57].

Although, Hitec salts freeze at the temperature of 120 °C and able to work at an operating temperature of 500 °C. To encounter the needs of thermal energy storage over 500 °C, a new ternary eutectic salt mixture was found since nitrate salts are not found to be suitable. The ternary salt mixture consists of $NaCl$, $CaCl_2$, and $MgCl_2$ and its eutectic temperature from phase diagram was 424.05 °C. The actual eutectic temperature of this mixture was 424 °C. Even though the temperature difference was small but the composition has been adjusted to make the eutectic temperature as 424 °C. The new ternary salt mixture was termed as SYSU-C4 and it was prepared by static melting method. The heat capacity of SYSU-C4 was found as 830 J kg-K$^{-1}$ in the solid state and 1190 J kg-K$^{-1}$ in the liquid state. Using the Archimedes principle, the density is calculated for this eutectic salt and it gradually decreases from 2.5 g m$^{-3}$

to 1.9 g m$^{-3}$ with a temperature of 500 °C to 700 °C. Viscosity of the eutectic salt was found to be decreasing from 4 cp to 3 cp as the temperature gets increased. Overall this eutectic salt shows excellent thermophysical properties and thermal stability up to the operating temperature of 700 °C [57]. The most commonly used storage technique in CSP plants is sensible heat-based two tank molten salt storage. In a sensible heat storage system, the amount of heat stored is directly proportional to the quantity of storage media used in the storage system. Latent heat thermal energy storage system has the advantage of storing or releasing energy by changing their phase. However, the latent heat storage system has not been commercially deployed in utility CSP plants. The thermochemical storage system owns a high-energy density with relatively low energy loss. It offers long-term storage with a smaller volume. The recent developments in thermochemical storage are to identify the suitable chemical process for the temperature range of CSP plants [59, 60].

## 2.8 Summary

CSP is one of the prominent technologies for generating electricity to mitigate the usage of conventional fossil fuels. Among the four concentrating solar collectors namely the parabolic trough collector, a linear Fresnel reflector, central receiver and parabolic dish collector, the PTC and central receiver are the most dominant technologies for power generation. Apart from the power generation, the CSP systems can also be used for industrial process heating applications as well. The four CSP plants were elucidated and the novel working fluids concerning temperature and their efficiency on the cycle are presented in detail. This novel working fluid helps to reduce the system design cost and enhance the performance of the system. Hence, the critical step is the selection of a working fluid and it should follow the sequence of steps including the thermo-physical properties of the fluid. The exergy-energy-environmental-economic analysis were also presented to distinguish between the CSP technologies. Finally, the recent advancements and future aspects of the CSP were presented to have a notion to work in this area.

## Nomenclature

| | |
|---|---|
| $A_{ap}$ | Area of the aperture (m$^2$) |
| $C_{CAP}$ | Capital cost (INR) |
| $C_{FCAP}$ | Fixed capital cost (INR) |
| $C_{LOM}$ | Levelised operational and maintenance cost (INR) |
| $C_{FOM}$ | Fixed operational and maintenance cost (INR) |
| $C_{OM}$ | Operational and maintenance cost (INR) |
| $C_{VOM}$ | Variable operational and maintenance cost (INR) |
| $dr_h$ | Discount rate with hike |
| dr | Discount rate |
| $e$ | Escalation |
| $E_{xi}$ | Exergy obtained by reflector (kJ) |
| $E_{xr}$ | Exergy in the receiver (kJ) |
| $E_{xu}$ | Exergy delivered by receiver (kJ) |
| $f$ | Dilution factor |

| | |
|---|---|
| $h_f$ | Enthalpy of fluid (kJ kg$^{-1}$) |
| $I_{bn}$ | Normal beam radiation (W m$^{-2}$) |
| LF | Levelising factor |
| lt | Life time (years) |
| $m_f$ | Mass flow rate of the fluid (kg s$^{-1}$) |
| $PG_{net}$ | Net power generation (kW) |
| $PP_{eff}$ | Effective power production (kW) |
| $PP_{gross}$ | Gross power production (kW) |
| $PC_{aux}$ | Auxiliary power consumption (kW) |
| $Q_i$ | Energy incident on the concentrator (kW) |
| $Q_r$ | Energy received by receiver (kW) |
| $Q_u$ | Energy transferred to HTF in the receiver (kW) |
| $q_{in}$ | Heat addition (kW) |
| $q_{out}$ | Heat rejection (kW) |
| $T_{amb}$ | Ambient temperature (K) |
| $T_{sun}$ | Temperature of sun (K) |
| $W_{turbine,\ out}$ | Work done from turbine (kJ) |
| $W_{pump,\ in}$ | Work done by pump (kJ) |
| $\eta_{th}$ | Thermal efficiency |
| $\eta_{opt}$ | Optical efficiency |
| $\eta_{I,r}$ | Energy efficiency of the receiver |
| $\eta_{I,sf}$ | Energy efficiency of the solar field |
| $\eta_{I,PB}$ | Energy efficiency of the power block |
| $\eta_{I,O}$ | Overall energy efficiency |
| $\eta_{II,C}$ | Exergy efficiency of concentrator |
| $\eta_{II,r}$ | Exergy efficiency of receiver |
| $\eta_{II,sf}$ | Exergy efficiency of solar field |
| $\eta_{II,PB}$ | Exergy efficiency of power block |
| $\eta_{II,O}$ | Overall exergy efficiency |

## Abbreviations

| | |
|---|---|
| CO | Condenser |
| CLFR | Compact Linear Fresnel Collector |
| CSP | Concentrating Solar Power |
| DPP | Discounted Payback Period |
| DSG | Direct Steam Generation |
| DVG | Direct Vapour Generation |
| ECOM | Economiser |
| EVA | Evaporator |
| GE | Generator |
| HRSG | Heat Recovery Steam Generator |
| HTF | Heat Transfer Fluid |
| HS | Heat Source |
| IEA | International Energy Agency |
| LCoE | Levelised Cost of Electricity |
| LFR | Linear Fresnel Collector |
| MDM | Octamethyltrisiloxane |

| MIL101 | Material Institut Lavoisier composed of trimeric chromium (III) octahedral clusters interconnected by 1,4 benzenedocarboxylates |
|---|---|
| MISP | Masdar Institute Solar Platform |
| MWCNT | Multi-Walled Carbon Nano-Tube |
| ORC | Organic Rankine Cycle |
| O&M | Operation and Maintenance |
| PDC | Parabolic Dish Collector |
| PTC | Parabolic Trough Collector |
| PU | Pump |
| RE | Recuperator |
| SAPG | Solar-Aided Power Generation |
| SHR | Superheater |
| TB | Turbine |
| TES | Thermal Energy Storage |
| TTD | Terminal Temperature Difference |

# References

[1] Islam T, Huda N, Abdullah A B and Saidur R 2018 A comprehensive review of state-of-the-art concentrating solar power (CSP) technologies: current status and research trends *Renew. Sustain. Energy Rev.* **91** 987–1018

[2] Pitz-paal R 2020 Concentrating solar power *Future Energy* (Amsterdam: Elsevier) pp 413–30

[3] International Energy Agency 2020 *Concentrating Solar Power* https://iea.org/reports/concentrating-solar-power-csp

[4] Cac G 2013 Concentrated solar power plants: review and design methodology *Renew. Sustain. Energy Rev.* **22** 466–81

[5] Alva G, Lin Y and Fang G 2018 An overview of thermal energy storage systems *Energy* **144** 341–78

[6] Moroz O G L, Frolov B and Burlaka M 2014 Turbomachinery flowpath design and performance analysis for supercritical CO2 *Turbine Technical Conference and Exposition GT2014*, pp 1–8, https://doi.org/10.1115/GT2014-25385

[7] Alghoul M A, Sulaiman M Y, Azmi B Z and abd Wahab M 2005 Review of materials for solar thermal collectors *Anti-Corrosion Methods Mater.* **52** 199–206

[8] Nathan G J *et al* 2018 Solar thermal hybrids for combustion power plant: a growing opportunity *Prog. Energy Combust. Sci.* **64** 4–28

[9] Van Sark W and Corona B 2020 Concentrating solar power *Technological Learning in the Transition to a Low-Carbon Energy System* (New York: Academic) pp 221–31

[10] Khandelwal D K, Ravi Kumar K and Kaushik S C 2019 Heat transfer analysis of receiver for large aperture parabolic trough solar collector *Int. J. Energy Res.* **43** 4295–311

[11] Malan A and Kumar K R 2021 Coupled optical and thermal analysis of large aperture parabolic trough solar collector *Int. J. Energy Res.* **45** 4630–51

[12] Malan A and Kumar K R 2021 A comprehensive review on optical analysis of parabolic trough solar collector *Sustain. Energy Technol. Assess.* **46** 101305

[13] Peiró G, Gasia J, Miró L, Prieto C and Cabeza L F 2017 Influence of the heat transfer fluid in a CSP plant molten salts charging process *Renew. Energy* **113** 148–58

[14] González-Roubaud E, Pérez-Osorio D and Prieto C 2017 Review of commercial thermal energy storage in concentrated solar power plants: steam vs. molten salts *Renew. Sustain. Energy Rev.* **80** 133–48

[15] Giaconia A, Iaquaniello G, Metwally A A, Caputo G and Balog I 2020 Experimental demonstration and analysis of a CSP plant with molten salt heat transfer fluid in parabolic troughs *Sol. Energy* **211** 622–32

[16] Qazi S 2017 Solar thermal electricity and solar insolation *Standalone Photovoltaic (PV) Systems for Disaster Relief and Remote Areas* (Amsterdam: Elsevier) pp 203–37

[17] Vant-Hull L L 2012 Introduction to concentrating solar power technology *Concentrating Solar Power Technology. Principles, Developments and Applications* ed K Lovegrove and W Stein (Woodhead Publishing Limited) pp 3–7

[18] Pang L, Wang T, Li R and Yang Y 2017 Two-stage solar power tower cavity-receiver design and thermal performance analysis *AIP Conf. Proc.* 1850 (June)

[19] Garcia R F, Carril J C, Catoira A D and Gomez J R 2012 An efficient parabolic dish engine based on rankine cycle *Renew. Energy Power Qual. J.* **1** 312–16

[20] Abbas M, Boumeddane B, Said N and Chikouche A 2011 Dish stirling technology: a 100 MW solar power plant using hydrogen for Algeria *Int. J. Hydrogen Energy* **36** 4305–14

[21] Pal R K and Kumar K R 2021 Thermo-hydrodynamic modeling of flow boiling through the horizontal tube using Eulerian two-fluid modeling approach *Int. J. Heat Mass Transf.* **168** 120794

[22] Pal R K and Kumar K R 2021 Two-fluid modeling of direct steam generation in the receiver of parabolic trough solar collector with non-uniform heat flux *Energy* **226** 120308

[23] Pal R K and Kumar K R 2021 Investigations of thermo-hydrodynamics, structural stability, and thermal energy storage for direct steam generation in parabolic trough solar collector: a comprehensive review *J. Clean. Prod.* 311 127550

[24] Santos J J C S, Palacio J C E, Reyes A M M, Carvalho M, Freire A J R and Barone M A 2018 Concentrating solar power *Advances in Renewable Energies and Power Technologies* (Amsterdam: Elsevier) pp 373–402

[25] Aboelwafa O, Fateen S K, Soliman A and Ismail I M 2017 A review on solar Rankine cycles: working fluids, applications, and cycle modifications *Renew. Sustain. Energy Rev.* **82** 868–85

[26] Thaddaeus J *et al* 2020 Overview of recent developments and the future of organic Rankine cycle applications for exhaust energy recovery in highway truck engines applications for exhaust energy recovery in highway truck engines ABSTRACT *Int. J. Green Energy* **17** 1005–21

[27] Mondejar M E *et al* 2017 Prospects of the use of nanofluids as working fluids for organic Rankine cycle power systems *Energy Procedia* **129** 160–7

[28] Cavazzini G, Bari S, McGrail P, Benedetti V, Pavesi G and Ardizzon G 2019 Contribution of metal-organic-heat carrier nanoparticles in a R245fa low-grade heat recovery organic rankine cycle *Energy Convers. Manag.* 199 111960

[29] Bellos E and Tzivanidis C 2017 Parametric analysis and optimization of an organic Rankine cycle with nanofluid based solar parabolic trough collectors *Renew. Energy* **114** 1376–93

[30] Habibi H, Zoghi M, Chitsaz A, Javaherdeh K and Ayazpour M 2018 Thermo-economic performance comparison of two configurations of combined steam and organic Rankine cycle with steam Rankine cycle driven by Al2O3-therminol VP-1 based PTSC *Sol. Energy* **180** 116–32 2019

[31] Loni R, Askari Asli-Ardeh E, Ghobadian B, Najafi G and Bellos E 2018 Effects of size and volume fraction of alumina nanoparticles on the performance of a solar organic Rankine cycle *Energy Convers. Manag.* **182** 398–411 2019

[32] Refiei A, Loni R, Najafi G, Sahin A Z and Bellos E 2020 Effect of use of MWCNT/oil nanofluid on the performance of solar organic Rankine cycle *Energy Reports* **6** 782–94

[33] Reyes-Belmonte M A, Sebastián A, González-Aguilar J and Romero M 2017 Performance comparison of different thermodynamic cycles for an innovative central receiver solar power plant *AIP Conf. Proc.* 1850

[34] Wang Z, Hu Y, Xia X, Zuo Q, Zhao B and Li Z 2020 Thermo-economic selection criteria of working fl uid used in dual-loop ORC for engine waste heat recovery by multi-objective optimization *Energy* **197** 117053

[35] Bao J and Zhao L 2013 A review of working fluid and expander selections for organic Rankine cycle *Renew. Sustain. Energy Rev.* **24** 325–42

[36] Wang Z, Tian H, Shi L, Shu G, Kong X and Li L 2020 Fluid selection of transcritical rankine cycle for engine waste heat recovery based on temperature match method *Energies* **13** 1830

[37] Yu H, Feng X and Wang Y 2016 Working fluid selection for organic Rankine cycle (ORC) considering the characteristics of waste heat sources *Ind. Eng. Chem. Res.* **55** 1309–21

[38] Li J, Alvi J Z, Pei G, Ji J, Li P and Fu H 2016 Effect of working fluids on the performance of a novel direct vapor generation solar organic Rankine cycle system *Appl. Therm. Eng.* **98** 786–97

[39] Tchanche B, Papadakis G, Lambrinos G and Frangoudakis A 2008 Criteria for working fluids selection in low-temperature solar organic Rankine cycles *EUROSUN 2008* pp 1–7

[40] Linke P, Papadopoulos A I and Seferlis P 2015 Systematic methods for working fluid selection and the design, integration and control of organic Rankine cycles—a review *Energies* **8** 4755–801

[41] Sukathme S P 1978 Solar thermal power generation, 1978. *Proc. Indian Acad. Sci.* **109** pp 521–31

[42] Vignarooban K, Xu X, Arvay A, Hsu K and Kannan A M 2015 Heat transfer fluids for concentrating solar power systems—a review *Appl. Energy* **146** 383–96

[43] McWhirter J D 1996 Multiple Rankine topping cycles https://osti.gov/servlets/purl/206551

[44] Pacio J, Fritsch A, Singer C and Uhlig R 2014 Liquid metals as efficient coolants for high-intensity point-focus receivers: implications to the design and performance of next-generation CSP systems *Energy Procedia* **49** 647–55

[45] Alvi J Z, Feng Y, Wang Q, Imran M and Pei G 2021 Effect of working fluids on the performance of phase change material storage based direct vapor generation solar organic Rankine cycle system *Energy Reports* **7** 348–61

[46] Aboelwafa O, Fateen S E K, Soliman A and Ismail I M 2017 A review on solar Rankine cycles: working fluids, applications, and cycle modifications *Renew. Sustain. Energy Rev.* **82** 868–85 2018

[47] Salazar-Pereyra M, Lugo-Leyte R, Bonilla-Blancas A E and Lugo-Méndez H D 2016 Thermodynamic analysis of supercritical and subcritical rankine cycles *Proc. ASME Turbo Expo* **8** 1–8

[48] Qin J, Hu E and Li X 2020 Solar aided power generation: a review *Energy Built Environ.* **1** 11–26

[49] Ravi Kumar K, Krishna Chaitanya N V V and Sendhil Kumar N 2021 Solar thermal energy technologies and its applications for process heating and power generation—a review *J. Clean. Prod.* **282**

[50] Kumar K R and Reddy K S 2012 4-E analyses of line-focusing stand-alone concentrating solar power plants *Int. J. Low-Carbon Technol.* **7** 82–96

[51] Xu C, Wang Z, Li X and Sun F 2011 Energy and exergy analysis of solar power tower plants *Appl. Therm. Eng.* **31** 17–18 3904–13

[52] Javanshir A and Sarunac N 2017 Thermodynamic analysis of a simple organic Rankine cycle organic Rankine cycle *Energy* **118** 85–96

[53] Tyagi S K, Wang S, Singhal M K, Kaushik S C and Park S R 2007 Exergy analysis and parametric study of concentrating type solar collectors *Int. J. Therm. Sci.* **46** 1304–10

[54] Application P 2013 Exergetic analysis and optimisation of a parabolic dish collector for low power application *Cent. Renew. Sustain. Energy Stud.* **53** 1689–99

[55] Jelley N and Smith T 2015 Concentrated solar power: recent developments and future challenges *Proc. Inst. Mech. Eng. A* **229** 693–713

[56] Liu M *et al* 2016 Review on concentrating solar power plants and new developments in high temperature thermal energy storage technologies *Renew. Sustain. Energy Rev.* **53** 1411–32

[57] Zhou D and Eames P 2017 A study of a eutectic salt of lithium nitrate and sodium chloride (87–13%) for latent heat storage *Sol. Energy Mater. Sol. Cells* **167** 157–61

[58] Xiaolan W, Ming S, Qiang P, Jing D and Jianping Y 2014 A new ternary chloride eutectic mixture and its thermo-physical properties for solar thermal energy storage *The 6th Int. Conf. on Applied Energy—ICAE2014* 61 1314–17

[59] Ho C K 2016 A review of high-temperature particle receivers for concentrating solar power *Appl. Therm. Eng.* **109** 958–69

[60] Ray A K, Rakshit D, Ravi Kumar K and Gurgenci H 2020 Silicon as high-temperature phase change medium for latent heat storage: a thermo-hydraulic study *Sustain. Energy Technol. Assess.* **46** 101249 2021

# Chapter 3

## Storage of electricity generated from the renewable sources using electrochemical energy conversion devices

**Roushan Nigam Ramnath Shaw, Ravi Sankannavar, G M Madhu, A Sarkar, K R V Subramanian and Raji George**

In recent years, various electrochemical approaches have been studied for renewable energy storage and conversion process, while several bottleneck problems are yet to be addressed to utilize these approaches effectively. Predominantly, these energy storage systems are utilized to store electricity generated from renewable energies in chemical energy and discharge it in the form of electricity when required. Among the electrochemical energy storage and conversion systems studied, water electrolyzers, redox flow batteries and flow capacitors provide very promising ways to store renewable energies. Thus, in this chapter, we discuss these systems with a focus on both electrochemical and thermodynamics analysis.

## 3.1 Introduction

Due to the ever-increasing consumption and depletion of fossil fuels, there is a huge concern for environmental safety and monitoring. The increased dependency of human life on fossil fuels has been compensated with huge consumptions of fossil fuels in a way that it is nearing exhaustion. The source of energy was necessary from time immemorial; there were great discoveries of oil, natural gas fields and coals. The widespread distribution of electricity from coal-powered power plants became the greatest energy source, so why shift from traditional sources of energy to some non-conventional? The answer to this question can be summarized in two words, i.e. abundance and degradation. We know that fossil fuels are very limited, and if we do not find any other alternative energy sources, we might drive humanity into a big crisis. The combustion of fossil fuels produces $NO_x$, $SO_x$ gases and fine particulates, which contributes to global warming and many other environmental hazards. Both of these factors motivate us to find and develop alternate energy sources that are

doi:10.1088/978-0-7503-3711-3ch3

abundant and do not degrade the environment. Renewable energy sources such as solar and wind are found to be the most suitable to replace depleting fossil fuels. Not so long ago, the invention of energy storage and conversion devices for small-scale and grid-level applications have modernized our day-to-day life. The grid-scale commercialization of energy devices like supercapacitors, batteries and photovoltaic cells have become very advantageous due to the fact that they can store electrical energy and the same energy can be released during peak power consumption (Zheng *et al* 2013, Argyrou *et al* 2018, Fang *et al* 2018, Ren *et al* 2018, Luta and Raji 2019). We can get energy from various technologies like photovoltaic, voltaic, hydro-electric, biomass, geothermal, tidal, wind, and hydrogen energy from renewable energy resources. But one may ask how do we store all the energy generated from renewable energy resources and use it later? Because this is an important question, especially if one cannot get continuous supply during the night or when the wind does not blow. The answer to this question may be to store the electricity generated from renewable sources in chemical form. Thus eco-environmental and sustainable energy storage and conversion techniques became crucial in the present days for modern society. In this regard, electrochemical energy conversion and storage technique has contributed significantly to the utilization of renewable energies. This electrochemical technique has the advantage that conversion from electrical energy to chemical energy and vice-versa share the same charge carrier, i.e., electron. The benefit of having a common charge carrier for conversion from one to another limits the energy losses in the energy conversion process. Electrochemical energy conversion and storage devices include water electrolyzers, fuel cells, batteries, electroreduction of $CO_2$ and supercapacitors. Among the electrochemical energy conversion and storage devices, water electrolyzers and redox flow batteries (RFBs) are found to be the most suitable to store electricity generated from renewable sources on a large scale. These electrochemical energy conversion systems are the key to a more sustainable future and convert renewable electricity in the form of chemical energy such as hydrogen gas or as chemically charged species. Thus electrochemical engineering plays a significant role in addressing the issues related to the utilization of renewable energies.

In the past, considerable research has been done to utilize water electrolyzers and redox flow battery technologies techniques to store renewable energies. In water electrolyzers, electrical energy generated from renewable sources has been used to split water into hydrogen and oxygen gases, and these gases are stored separately. Thus the renewable energies can be stored in the chemical form, i.e., hydrogen. During the water electrolysis, a hydrogen evolution reaction occurs at the anode of the electrolyzer, and the produced hydrogen gas will be stored as hydrogen energy. Later, the stored hydrogen gas can be used in the hydrogen-oxygen fuel cells to generate back electricity in the odd hours. The most important uniqueness of the combination of the water electrolyzer and fuel cell technologies is that both technologies do not emit any greenhouse gases. Hence, these technologies are considered sustainable, and hence hydrogen gas is considered a clean energy carrier in the present case. Thus, water electrolyzer technology is a promising option to store renewable energies. The overview of the water electrolysis process for the utilization of renewable energies to produce clean hydrogen gas is presented in figure 3.1, along with the various applications of hydrogen.

However, most of the water technologies studied so far suffer from energy losses that are major due to the sluggishness of the oxygen evolution reaction that takes place at the anode of the water electrolyzer. Hence, several efforts have been made to decrease the energy loss due to the anodic reaction for designing and developing efficient electrocatalysts for the oxygen evolution reaction. In comparison, redox flow batteries are a type of batteries and represent another class of electrochemical energy conversion and storage systems (figure 3.2). In these RFBs, redox couples are employed to store the energy from the renewable sources in the liquid electrolyte solution forms that circulate through flow cell during charging and discharging processes. The chemistry involved in the RFBs has been well established and show proven ability in the power sector due to their very low degradation of electrolytes, long-duration energy output and large-scale deployments. RFBs are predominantly used for grid-level energy storage because of their high-energy efficiencies even at reasonably high current density (Wei *et al* 2016, Zhang *et al* 2018). This type of battery architecture is based on redox-active pairs in separate solutions that are contained within external storage tanks, termed the anolyte (negolyte) and the catholyte (posolyte). Power is generated upon the flow of the anolyte and catholyte within a central electrochemical cell, comprising an ion-exchange membrane.

The chemical energy stored in the redox components is converted to electrical energy transmitted through the external circuit (Wang *et al* 2020). These systems have high cycling efficiencies compared to other conventional batteries, making their shelf life arguably high (Hu *et al* 2017, Luo *et al* 2019). Currently, the all-vanadium redox flow batteries (AVRFBs) are being used commercially to some extent, and they have been extensively investigated in the literature. The large-scale

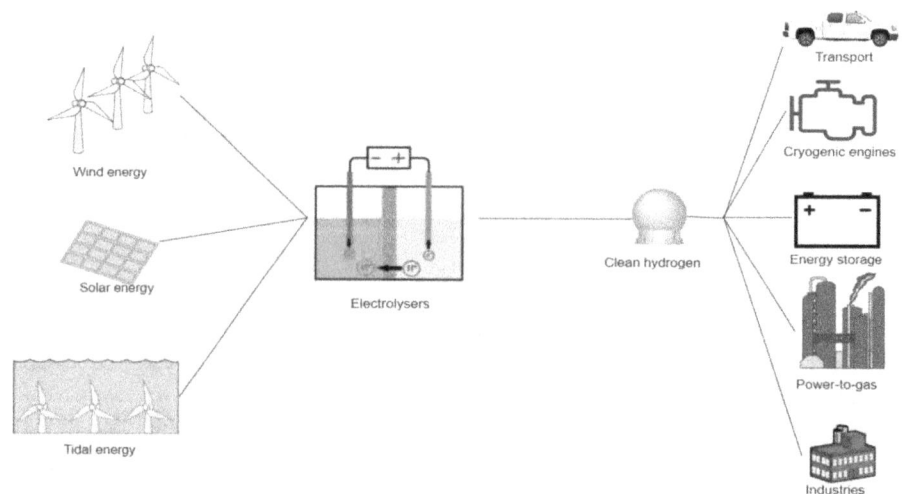

**Figure 3.1.** Overview of the utilization of renewable energies to produce clean hydrogen gas using water electrolyzers and its various applications.

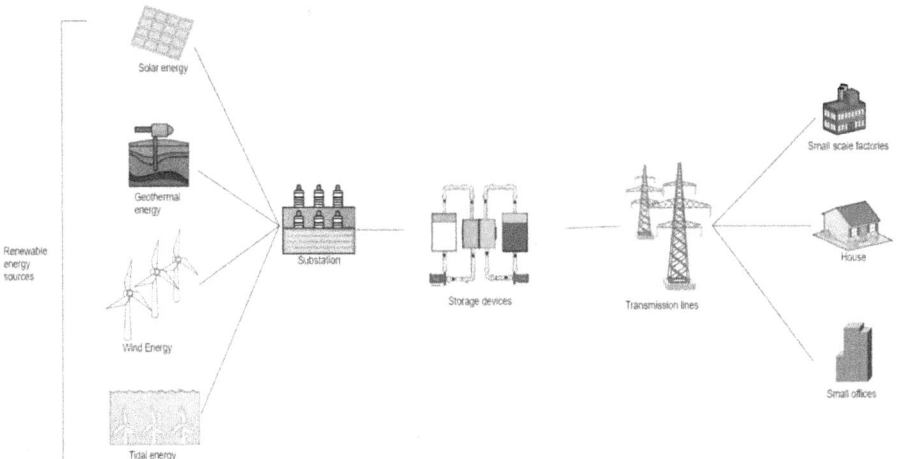

**Figure 3.2.** Overview of the utilization of renewable energies to charge the electrolytes of redox flow battery (RFB) and applications of the electricity from the RFB.

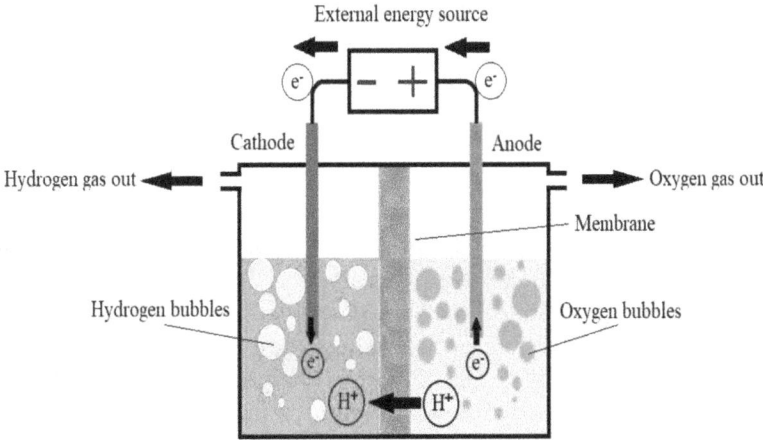

**Figure 3.3.** Schematic representation of the water electrolysis process for hydrogen and oxygen gas production.

commercialization of VRFBs is still under consideration due to reasons like the high cost of the electrolyte component, the toxicity of vanadium when exposed to the environment and the expertise required for its maintenance. Although VRFBs pose environmental concerns, other redox couples are feasible in an RFB.

Here we also emphasize the importance of electrochemical flow capacitor (EFC) (Presser *et al* 2012; Presser *et al* 2013a; Presser *et al* 2013b which can solve the disadvantages of batteries and supercapacitors and combine their advantages of high energy, high power, long life and ease of charge–discharge (rate performance). The working of an EFC has been explained in recent studies (Presser *et al* 2012; Presser *et al* 2013a; Presser *et al* 2013b). A carbon-based slurry (carbon + electrolyte) is filled in containers outside the flow cell. On charging, it is pumped into the flow cell and charge is stored in the carbon particles electrostatically. Previous studies have reported the use of various carbon materials like phenolic resin-derived activated carbon beads, non-spherical carbide-derived carbon powder, and carbon black-added activated carbon, which were all in the micron-size range. It has been proven that there is a drastic improvement in the electrical properties as well as the storage capacity of a carbon material when its size is reduced to a nanosize range (Zhang *et al* 2013, Pumera 2011, Dai *et al* 2012). The specific surface area increases as the size decreases and a high specific surface area is a desirable characteristic of an electrode material for any energy storage device. The well-known nanocarbon forms, such as carbon nanotubes, graphene, and fullerenes, have demonstrated their efficiency as electrode materials in devices (Guo and Li 2011; Simon and Gogotsi 2008; Zhai *et al* 2011; Kamat *et al* 2013). In a recent study (Tomai *et al* 2017), a promising approach to capacitance enhancement in EFCs, based on the impregnation of redox-active quinone derivatives in nanoporous carbon for flowable slurry electrolytes was demonstrated. The resulting slurry electrolytes exhibited good flowability, high power capability and an energy density of approximately 20 W h kg$^{-1}$, 2.5 times higher than that of the unmodified carbon slurry. In another recent study (Akuzum *et al* 2018), an approach was adopted to flow cell design that can enable a significantly improved power output (~10×) for electrochemical flow capacitors (EFCs), even at large flow channel gaps. Reticulated vitreous carbon (RVC) electrodes of various average pore sizes (0.43–2 mm) were integrated into EFC flow cell fixtures with channel gaps of 5 mm. Electrochemical testing under flow conditions showed a 10-fold improvement in the power density with the RVC integration (290 W m$^{-2}$, 580 W kg$^{-1}$) for the same slurry composition. RVCs having an average pore size of 0.55 mm showed the best performance out of all studied cases with improved coulombic efficiency and good specific capacity (85 F g$^{-1}$) under flowing conditions. The successful incorporation of the carboxymethyl cellulose sodium salt (CMC-Na)-assisted carbonaceous suspension electrode in aqueous media for the electrochemical flow capacitor concept (Torop *et al* 2019) and demonstration of the electrochemical charge storage in flowable electrodes using a cation conductive membrane as separator in a double-pipe flow-electrode module was shown. During continuous operation of the system, the capacitance of the flow electrode reached to 0.3 F L$^{-1}$ providing the energy and current densities of 7 mWh kg$^{-1}$ and 56 mW L$^{-1}$, respectively.

Li-ion batteries have transformed portable electronics and will play a key role in the electrification of transport. The highest energy storage possible with Li-ion batteries is insufficient for the long term needs of society. To go beyond the horizons of Li-ion batteries, the rechargeable Li-air battery (based on aqueous or non-aqueous electrolyte) is a promising research area. In the Li-air battery, oxygen is the fuel. A Li-air battery with a non-aqueous electrolyte has been in the research limelight (Bruce *et al* 2011, Lee *et al* 2011). Major challenges have to be solved if Li-air batteries are to succeed.

In this chapter, the working principles and thermodynamics of water electrolyzers and RFBs/flow capacitors will be discussed in detail with respect to the electro-chemical energy conversion and storage of renewable energies.

## 3.2 Water electrolyzers

### 3.2.1 Renewable energy utilization for hydrogen production

In the search for processes for preventing catastrophic global warming and utilizing renewable energy sources for the production of hydrogen on a large-scale, the water-splitting process is considered economical and practical. Water splitting is the process of breaking down a water molecule into hydrogen ($H_2$) and oxygen ($O_2$) gases (equation (3.1) and figure 3.3). It is predominantly utilized to establish the hydrogen economy. However, hydrogen production by water splitting requires a considerable amount of energy, making it incompatible with currently available hydrocarbon-based $H_2$ production technologies. Therefore, any research and development efforts to reduce the energy to be applied for water splitting and on improving its efficiency of process are encouraging.

$$2H_2O + energy/heat \rightarrow 2H_2 + O_2. \tag{3.1}$$

Numerous studies have shown that renewable energy sources (solar and wind) are far away from meeting the requirements of the world's energy demand due to their seasonal variation in the power supply. In this sense, a number of technologies are suggested for the production of $H_2$-gas by splitting water (equation (3.1)), such as water electrolysis, photocatalytic water splitting, photoelectrochemical water splitting, water biophotolysis, thermal decomposition of water (Funk and Reinstrom 1966). These $H_2$ production techniques are generally distinguished based on the sources of energy supplied for splitting water. In general, water electrolysis is the most suitable to make $H_2$ gas for both residential and industrial applications due to its high degree of reliability. Although water electrolysis is being used far a long time, research and development efforts are still required for its effective use for $H_2$ production as efficient and economical approaches as water electrolysis would be a breakthrough that could underpin a hydrogen economy. In this regard, further details of the water electrolysis process are discussed below.

In 1789, van Troostwijk and Deinman observed the decomposition of water into gases by an electric discharge and, in the 1980s, William Nicholson and Anthony Carlisle found that $H_2$ and $O_2$ gases are generated when battery's terminals were immersed in water (Trasatti 1999). After this invention of water splitting, months later, Johann Wilhelm Ritter fabricated a set-up to collect these gases separately. From these humble beginnings, water electrolysis as technology has progressed substantially in the past two centuries. As mentioned earlier, the present-day research predominantly focuses on the utilization of water electrolysis technology for the storage of renewable energy in the form of $H_2$ fuel. For this reason, the electricity derived from renewable sources may be used for water electrolysis. In this respect, it is important to study factors that influence the water electrolysis process.

### 3.2.2 Thermodynamics of water electrolyzer

As mentioned earlier, water splitting (equation (3.1)) requires energy to split the water molecule. The minimum standard Gibbs free energy of $\Delta G° = 237.13$ kJ is applied as the electrical energy input to produce molecular hydrogen and oxygen gases from one mole of water at standard conditions. Thus, the minimum cell potential ($E_{cell}^{\circ}$) necessary for one mole of water electrolysis can be calculated using

$$\Delta G^{\circ} = -nFE_{cell}^{\circ} \tag{3.2}$$

here $\Delta G^{\circ}$, $n$, $F$, and $E_{cell}^{\circ}$ are the standard Gibbs free energy (237.13 kJ), the number of electrons participating in a splitting of one mole of water (2), Faraday's constant (96 485 C mol$^{-1}$), and standard cell potential for water electrolysis (V), respectively. Thus, the minimum cell potential required for water electrolysis at standard operating conditions comes to be 1.229 V. When electrical energy is applied across the water electrolyzer, the oxygen evolution reaction (OER), equation (3.3) and hydrogen evolution reaction (HER), equation (3.4) take place at the anode (positive) and cathode (negative) electrodes, respectively, and these reactions are given as follows:

$$2H_2O \rightarrow O_2 + 4H^+ + 4e^-; E_{OER}^{\circ} = 1.229 \text{ V } vs. \text{ SHE} \tag{3.3}$$

$$4H^+ + 4e^- \rightarrow 2H_2; E_{HER}^{\circ} = 0.000 \text{ V } vs. \text{ SHE} \tag{3.4}$$

$$\text{Overall cell reaction: } 2H_2O \rightarrow 2H_2 + O_2; E_{cell}^{\circ} = -1.229 \text{ V } vs. \text{ SHE.} \tag{3.5}$$

Here $E_{OER}^{\circ}$ and $E_{HER}^{\circ}$ represent the standard reduction electrode potentials of OER and HER at standard conditions of 25 °C and 1 atm with 1 M concentration of all the reactants and products. Henceforth, the abbreviation SHE is used for the standard hydrogen electrode (SHE). The net electrochemical cell potential for water electrolysis (equation (3.5)) is also given by $E_{cell}^{\circ} = E_{HER}^{\circ} - E_{OER}^{\circ} = 0.000 - 1.229 = -1.229$ V $vs.$ SHE. However, when water electrolysis is carried out at non-standard conditions, the Nernst equation is used for calculating the cell potential, which is given by the following equation:

$$E_{cell} = E_{cell}^{0} + \frac{2.303RT}{4F} \log \left( \frac{p_{H_2}^2 p_{O_2}}{p_{H_2O}} \right) \tag{3.6}$$

where $R$, $T$ and $F$ represent the gas law constant (8.314 J mol$^{-1}$ K$^{-1}$), temperature (K) and Faraday's constant (96 485 C mol$^{-1}$), respectively. Whereas $p_{H_2}$, $p_{O_2}$, and $p_{H_2O}$ are the partial pressures of $H_2$ and $O_2$ gases and water vapour, respectively. However, in practice, water electrolysis usually requires a higher potential than the theoretical cell potential of 1.229 V $vs.$ SHE due to the energy losses from resistances such as the anode and cathode reactions, electrolyte, separator, and connection circuit. As a result, the actual potential required for water electrolysis is higher than $E_{cell}^{\circ} = 1.229$ V $vs.$ SHE. The energy losses can be minimized by improving the OER and HER kinetics by using suitable electrocatalysts, decreasing the gap between electrodes, using selective permeable membrane, etc. It is also to be noted here that the cell potential for water electrolysis can also be decreased by increasing the operating temperature. However, additional complications arise at high temperatures, such as increased corrosion and heat exchange problems. Hence, minimizing energy losses by ramping operating conditions like temperature may not be feasible unless using waste heat from industries. The energy losses in electrochemical processes represent the overpotential, and it is defined as 'the additional potential (beyond the thermodynamic requirement) needed to drive a reaction at a certain rate' (Bard and Faulkner 2001). Thus, considerable research has been done in the past to minimize the overpotential of water electrolyzers (Carmo *et al* 2013). It is reported that the overpotential for water electrolysis is around 50 to 75% of the $E_{cell}^{\circ}$ (Matsumoto and Sato 1986).

Main thermodynamic functions such as enthalpy ($\Delta H$), entropy ($T\Delta S$), and Gibbs free energy ($\Delta G$) are related by equation (3.7)

$$\Delta G = \Delta H - T\Delta S. \tag{3.7}$$

The thermodynamic functions are plotted against temperature in the range of 0–1000 °C as it can be seen from figure 3.4 that there is a discontinuity at 100 °C as marked the figure. This discontinuity is because water changes its phase from liquid to vapour phase at 100 °C, and the enthalpy change accompanying vaporization is 40.7 kJ mol$^{-1}$. The $\Delta G$ is 0 at 100 °C because the liquid phase of water is in equilibrium with the vapour phase. The entropy and enthalpy increase as the temperature increases. Consequently, the Gibbs free energy, which is given by equation (3.7) decreases as per thermodynamics if $\Delta G < 0$; then the reaction is promising, which can easily be shown by below equation (3.8)

$$\Delta G = -RT \ln K \tag{3.8}$$

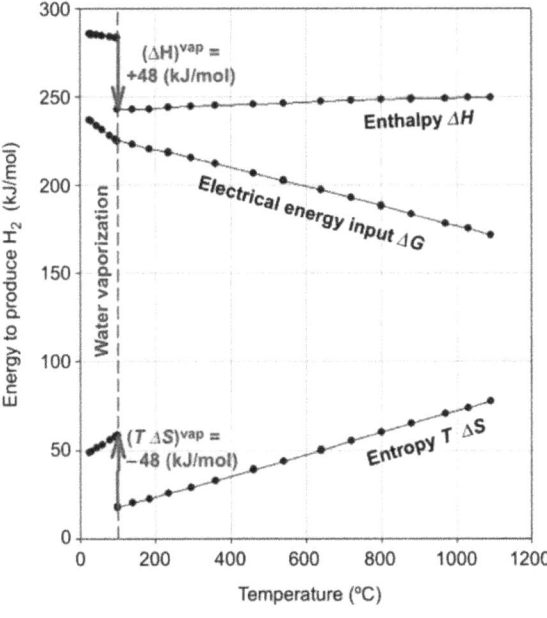

**Figure 3.4.** Thermodynamic data for the water electrolysis process. This figure is adopted from Millet (2015).

where $K$ is an equilibrium constant that has to be greater than 1 (i.e., $K > 1$); thus, it shows that with an increase in temperature, less electricity is required when compared to electrolysis at room temperature (Millet 2015).

## 3.3 Redox flow batteries

A redox flow battery is a type of battery in which electrolytes flow through the cell, converting electrical energy to chemical energy and vice-versa. As we aim towards achieving zero carbon emission by 2050 and go away from fossil fuels, we use renewable energy sources, but the nature of renewable energy sources increases fluctuation in an electrical grid hence the need to store the energy deepens. To stabilize the grid and store a surplus of electricity, we incorporate electrochemical energy storage devices, and one of the leading technologies that help us achieve this goal is redox flow batteries. While traditional batteries are suitable for high power mobile devices like laptops, smartphones, etc., flow batteries can store power for a longer period and we might have noticed that batteries in our smartphone tend to degrade which we notice when the battery life of smartphone degrades which is not the case with flow batteries, it lasts long for more than a decade or two and this makes it a suitable candidate for large-scale application.

### 3.3.1 Construction and principle of redox flow batteries

An RFB consists of two half cells, electrolyte tanks, pumps, piping and an ion-exchange membrane. Each pump circulates an anolyte and catholyte into a container where the reactions occur; oxidation takes place at negative electrode $A^-/A^+$ and reduction takes place at positive electrode $B^+/B^-$ simultaneously. The cell is separated into two parts by an ion exchange membrane that prevents the intermixing of the anolyte and catholyte but is still a porous material that can transfer ions. Within the half cells, the electrolytes are chemically oxidized and reduced to convert electrical energy to chemical energy and vice versa.

### 3.3.2 Working principle of vanadium redox flow batteries

Here we take the example of a vanadium flow battery (figure 3.5). When an electrical power source-like energy from solar energy is applied to the half cells to charge the RFB, the electrolytes are passed through the cell, which converts electrical energy to chemical energy which are then passed back to the electrolyte reservoirs, which charges the battery. At the cathode, half cell catholyte $V^{4+}$ gets oxidized to $V^{5+}$ this gives off an $H^+$ which gets diffused across the ion exchange membrane to the other half cell here electron from external circuit reduces $V^{3+}$ to $V^{2+}$ thus the batteries are charged.

Another component of the RFBs that influences the system's performance is the membrane, which is used to separate two electrolytes and ionic conductions. However, it should not let the reactive species crossover, which should be resistant to chemical attacks. Membranes can be classified into two types as (i) porous membranes and (ii) dense ion-conducting membranes (Ye *et al* 2017). Further, dense ion conduction membranes can be classified into two types: the cation exchange membrane and the anion exchange membranes, which can conduct both cations and anions. Both types of membranes are described below in brief.

The porous membranes—diaphragm cannot prevent mixing if the pore size is more than the ion size; recently available microfiltration membranes and ultra-filtration membranes are used for this process; the latter can be used for batteries with a polymeric redox couple. For vanadium flow batteries, a polyacrylonitrile membrane was prepared by the phase inversion process, membranes which are made by this process consists of large pores and a dense thing selective layer. In order to improve the wetting property, the nitrile group of membranes are hydrolyzed by immersing in 10 wt% NaOH solution (Zhang *et al* 2013).

The cation exchange membrane consists of polymers substituted with acidic groups. These groups are sulfonic or phosphonic acids for strong ion exchange materials and carboxylic acid groups for weak ion exchange materials. Due to the difference in hydrophilicity of the backbone and functional group, it shows phase difference. These have advantages like high conductivity, high chemical stability and structural variety. Disadvantages, like susceptibility to ion crossover, plagues it. Anion exchange membranes are positively charged functional groups and repel cation. This inhibits the crossover of vanadium ions. It is found that they have higher conductivity than pure membranes because it absorbs some sulfuric acid when immersed in electrolytes, as shown by Yun *et al* (2015). Further, they have tested the suitability of anion exchange membranes for vanadium/cerium redox flow batteries, and they found out that trimethylammonium functionalized poly

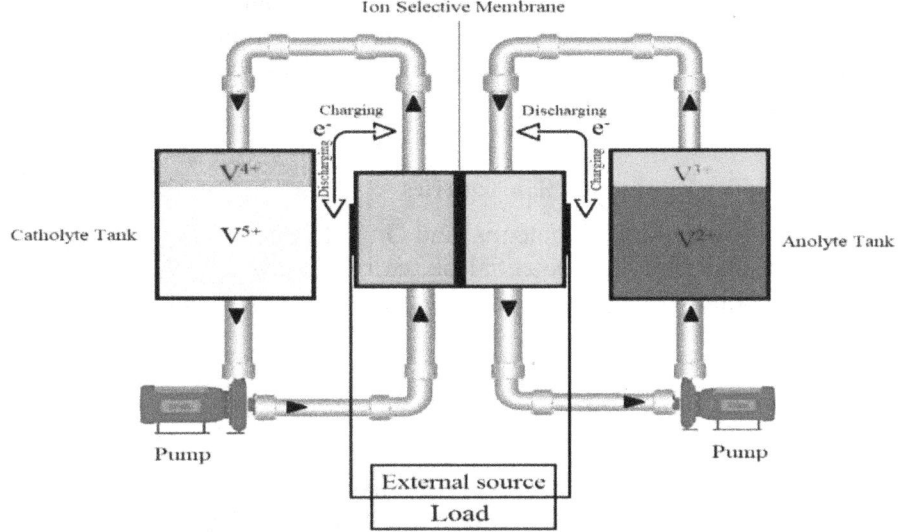

**Figure 3.5.** Schematic representation of the vanadium redox flow battery for renewable energy conversion showing charging and discharging processes.

(ether ketone)-based anion exchange membrane is chemically and mechanically stable, which can prevent the mixing of active cycles.

### 3.3.3 Electrochemistry of redox flow batteries

When an external load is attached to the cell, the chemical energy stored in the electrolytes gets converted to electrical energy due to the spontaneous reactions, which drives the external load. When an external load is connected to the battery, the opposite phenomena of charging occur, i.e., $V^{2+}$ gets oxidized to form $V^{3+}$ which release electrons that drive the load and $V^{5+}$ are reduced to form $V^{4+}$ these electrolytes leave the cell in their discharged state. This process happens until both the electrolytes come back to their discharged state (figure 3.5). Electrochemical reactions occurring at positive and negative electrodes in the vanadium RFB are presented below.

$$\text{Positive electrode reaction: } VO^{2+} + H_2O \rightleftarrows VO_2^+ + 2H^+ + e^- \tag{3.9}$$

$$\text{Negative electrode reaction: } V^{3+} + e^- \rightleftarrows V^{2+}. \tag{3.10}$$

Here $\rightleftarrows$ indicates charging and discharging processes. The electrolytes associated with vanadium flow batteries are toxic and corrosive; for a flow battery, a very large amount of electrolyte is required. This implies the storage of a large amount of corrosive and toxic substance; another main challenge is cost and energy density; the cost associated with the electrolyte and membrane is very high.

Though there are environmental hazards related to redox flow batteries, it is an excellent source of energy. In the process of improving the efficiency of RFBs, there are recent developments in organic electrolytes like AQDS (9,10-Anthraquinone-2,7-disulphonic acid), TEMPO (2,2,6,6-Tetramethylpiperidine 1-oxyl), which might solve both of the above-mentioned issues. The membrane problem can be solved by combining cation-exchange and anion-exchange membranes either by spray coating or electrocoating to form a bipolar membrane.

### 3.3.4 Thermodynamics of redox flow batteries

The relation of equilibrium cell potential and Gibbs free energy is given by equation (3.2), and the total electrode potential of the cell is given by $E_{cell} = E_{positive} - E_{negative}$. The Gibbs free energy can be calculated using equation (3.11) if the energies of each species are known.

$$\Delta_f G_{i,b} = \Delta_f G_{i,b}^{\circ} + RT \ln\left(\frac{b_i \gamma_{i,b}}{b^{\circ}}\right) F. \tag{3.11}$$

For the calculation of Gibb's free energy, one may use the suitable equation to find the activity coefficients like NRTL, extended UNIQUAC, Pitzer correlations, and the Bromley-Zemaitis model. These activity coefficients are necessary and must be calculated for high ionic strength (Hall *et al* 2020).

## 3.4 Flow capacitors

### 3.4.1 Use of nanomaterials in the electrochemical flow capacitor

Figure 3.6(a) shows a schematic diagram for the electrochemical flow capacitor. The electrochemical flow capacitor will be assembled as shown in figure 3.6(b) with a modification to the current collector design for more area of contact between the flowing slurry and current collectors. Inside each sub-compartment, there are two current collector plates inter-connected, as shown in figure 3.6(c). The current collectors are made of stainless steel. All containers and the flow cell are made of an insulating material, and PTFE (poly tetrafluoroethylene) was used as the separator. Peristaltic pumps will be used for pumping and withdrawing the slurry.

In the intermittent flow, only incremental volumes were to be fed inside the main chamber, while in full flow, the complete slurry volume was fed in one shot. The composite-based slurry will be prepared using graphene prepared as above, oleylamine, sodium lauryl sulphate, isopropanol and ionic electrolyte 1-butyl-1-methylpyrrolidiniumbis(trifluoromethylsulfonyl) imide, while the electrolyte is 50% v/v ionic electrolyte mixed with 50% v/v propylene carbonate with added additives, i.e., 5 M camphor sulphonic acid. The volume ratio of the ionic electrolyte in the entire slurry is 40% while the remaining constituents will make up the 60%. 100 ml slurries is prepared and the typical amounts of the various constituents in the slurry were as follows: 20 ml of 1-butyl-1-methylpyrrolidinium bis(trifluoromethylsulfonyl)imide, 20 ml of propylene carbonate, 2 ml of 5 M camphor sulphonic acid, 1.2 g of graphene composites, 3 ml of oleylamine, 12 mg of sodium lauryl sulphate and 36 ml of isopropanol.

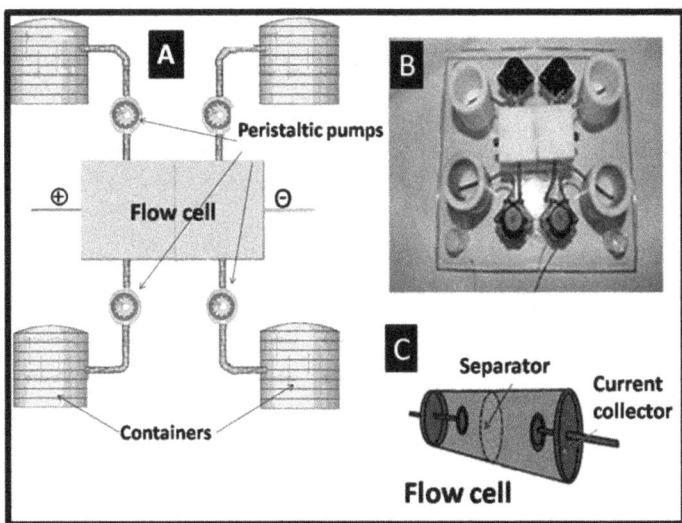

**Figure 3.6.** (A) A schematic diagram of the EFC, (B) a lab-scale EFC device set-up, (C) modification of the current collector. Reprinted from Sasi *et al* 2015, Copyright (2015), with permission from Royal Society of Chemistry.

The EFC device performance would be analyzed using chronoamperometry with composite slurries, while the graphite slurry will be used as the control. The flow device capacitance in Farads (F) and device specific capacitance in F g$^{-1}$ would be calculated from the chronoamperometry charging curve using equations (3.12$a$) and (3.12$b$). The energy density (from the chronoamperometry charging curve) and power density (from the chronoamperometry discharging curve) would be calculated using equations (3.13$a$), (3.13$b$) and (3.14). Cycling study of the EFC would be carried out for a specified no. of cycles in a LiOH electrolyte with an applied voltage of 1.5 V for 500 s of charging and discharging.

$$C = (2/\Delta E) \lceil idt \qquad (3.12a)$$

where $C$ is the device capacitance, $E$ the potential applied, $i$ is the current and $m$ is the weight of active material.

$$C_{sp} = (2/\Delta Em) \lceil idt \qquad (3.12b)$$

where $C_{sp}$ is the specific capacitance.

$$\text{Energy density} = \lceil idtv/3600 \, m. \qquad (3.13a)$$

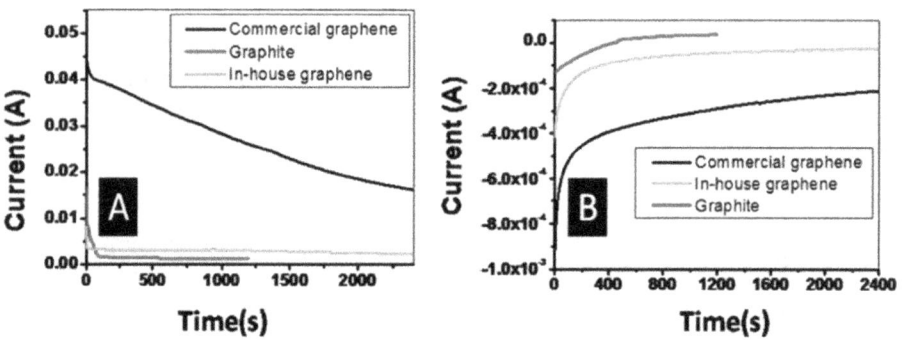

**Figure 3.7.** Intermittent EFC performance: (A) charging behaviour and (B) discharging behavior. Reprinted from Sasi *et al* 2015, Copyright (2015), with permission from Royal Society of Chemistry.

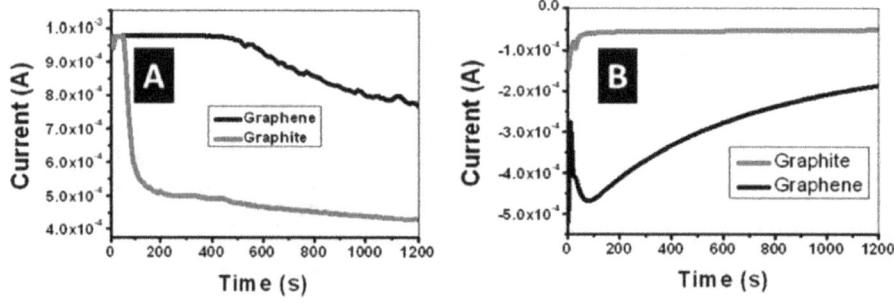

**Figure 3.8.** Full flow EFC device performance: (A) charging behaviour and (B) discharging behavior. Reprinted from Sasi *et al* 2015, Copyright (2015), with permission from Royal Society of Chemistry.

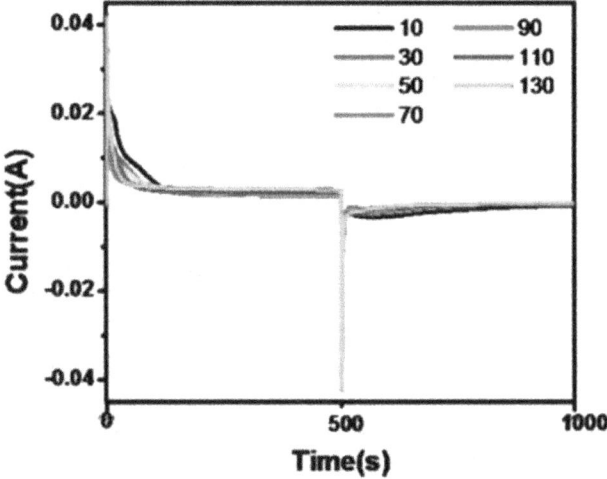

**Figure 3.9.** Cycling study of the EFC. Reprinted from Sasi *et al* 2015, Copyright (2015), with permission from Royal Society of Chemistry.

The other symbols have already been described in previous equations.

$$\text{Volumetric energy density} = \lceil\ idtv/3600 \times \text{vol} \qquad (3.13b)$$

$$\text{Power density} = (i1V + i2V + i3V + i4V + i5V + i6V + i7V + i8V \\ + i9V + i10V)/10\ m \qquad (3.14)$$

where $i1$, $i2$,... are the random current values from the $i$–$t$ discharge graph.

The graphs in figures 3.7, 3.8 and 3.9 show device performance of the EFC when graphene nanoplatelets were used (Sasi *et al* 2015). The device performance is expected to be even better when using graphene composites because of their higher electrical conductivity, surface area and favorable electronic band structure.

## 3.5 Summary

In this chapter, we have briefly discussed the working, construction, thermodynamics and electrochemistry of electrochemical devices, water electrolyzer and redox flow batteries and have seen how these electrochemical devices can make usage of current renewable energy resources more effectively. Unfortunately, the majority of energy losses are associated with slow kinetics of anodic and cathodic reactions and membranes; thus, there is no highly efficient redox flow battery or water electrolyzer for susceptible renewable energy conversion. Although there is a very high amount of academic and industrial research going into the design and development of efficient electrocatalysts, better ion conducting membranes, environment friendly electrolytes for both devices, still an opportunity exists for research to develop efficient renewable energy conversion devices. In addition, of course, cost reduction is also necessary, which will further push the developments of highly integrated systems that can take advantage of these electrochemical devices for renewable energy utilization. Irrespective of the limitations associated with these energy conversion devices, water electrolyzers and redox flow battery/flow capacitors can become an essential means of energy storage to achieve clean and sustainable energy. Thus, flexible, efficient, and environmentally friendly electrochemical energy conversion and storage devices have the scope to meet the energy demand.

# References

Akuzum B, Hudson D D, Eichfeld D A, Dennison C R, Agartan L, Gogotsi Y and Kumbur E C 2018 *J. Electrochem. Soc.* **165** A2519–27

Argyrou M C, Christodoulides P and Kalogirou S A 2018 Energy storage for electricity generation and related processes: technologies appraisal and grid scale applications *Renew. Sustain. Energy Rev.* **94** 804–21

Bard A J and Faulkner L R 2001 *Electrochemical Methods: Fundamentals and Applications* 2nd edn (New York: Wiley)

Bruce P G *et al* 2011 *Mat. Res. Soc. Bull.* **36** 506

Carmo M, Fritz D L, Mergel J and Stolten D 2013 *Int. J. Hydrogen Energy* **38** 4901–34

Dai L, Chang D W, Baek J-B and Lu W 2012 *Small* **8** 1130–66

Fang Y, Chen Z, Xiao L, Ai X, Cao Y and Yang H 2018 *Small* **14** 1703116

Funk J E and Reinstrom R M 1966 *Ind. Eng. Chem. Proc. Des. Dev.* **5** 336–42

Guo C X and Li C M 2011 *Energy Environ. Sci.* **4** 4504–7

Hall D M, Grenier J, Duffy T S and Lvov S N 2020 *J. Electrochem. Soc.* **167** 110536

Hu B, DeBruler C, Rhodes Z and Liu T L 2017 *J. Am. Chem. Soc.* **139** 1207–14

Kamat P V, Guldi D M and D'Souza F 2013 *Fullerenes and Nanotubes: The Building Blocks of Next Generation Nanodevices: Proceedings of the International Symposium on Fullerenes, Nanotubes and Carbon Nanoclusters* (Philadelphia PA: The Electrochemical Society)

Lee J S *et al Adv. Energy. Mat.* **1** 2011 34

Luo J, Wu W, Debruler C, Hu B, Hu M and Liu T L 2019 *J. Mater. Chem.* A **7** 9130–6

Luta D N and Raji A K 2019 *Energy* **166** 530–40

Matsumoto Y and Sato E 1986 *Mater. Chem. Phys.* **14** 397–426

Millet P 2015 Hydrogen production by polymer electrolyte membrane water electrolysis *In Woodhead Publishing Series in Energy, Compendium of Hydrogen Energy* pp 255–86

Presser V, Dennison C R, Campos J, Knehr K W, Kumbur E C and Gogotsi Y 2012 *Adv. Energy Mater.* **2** 895–902

Presser V, Dennison C R, Campos J, Knehr K W, Kumbur E C and Gogotsi Y 2013a *Electrochim. Acta* **111** 888–97

Presser V, Dennison C R, Campos J, Knehr K W, Kumbur E C and Gogotsi Y 2013b *Electrochim. Acta* **98** 123–30

Pumera M 2011 *Energy Environ. Sci.* **4** 668–74

Ren W, Chen X and Zhao C 2018 *Adv. Energy Mater.* **8** 1801413

Sasi S, Murali A, Nair S V, Nair A S and Subramanian K R V 2015 *J. Mater. Chem.* A **3** 2717–25

Simon P and Gogotsi Y 2008 *Nat. Mater.* **7** 845–54

Trasatti S 1999 *J. Electroanal. Chem.* **476** 90–1

Tomai T, Saito H and Honma I 2017 *J. Mater. Chem.* A **5** 2188–94

Torop J, Summer F, Zadin V, Koiranen T, Janes A, Lust E and Aabloo A 2019 *Eur. J. Phys E.* **42** 8

Wang H, Sayed S Y, Luber E J, Olsen B C, Shirurkar S M, Venkatakrishnan S, Tefashe U M, Farquhar A K, Smotkin E S and McCreery R L *et al* 2020 *ACS Nano* **14** 2575–84

Wei X, Duan W, Huang J, Zhang L, Li B, Reed D, Xu W, Sprenkle V and Wang W 2016 *ACS Energy Lett.* **1** 705–11

Yun S, Parrondo J and Ramani V 2015 *ChemPlusChem* **80** 412–21

Zhai Y, Dou Y, Zhao D, Fulvio P F, Mayes R T and Dai S 2011 *Adv. Mater.* **23** 4828–50

Zhang J, Jiang G, Xu P, Kashkooli A G, Mousavi M, Yu A and Chen Z 2018 *Energy Environ. Sci.* **11** 2010–15

Zhang H, Zhang H and Li X 2013 *ECS Trans.* **53** 65–8

Zhang Q, Uchaker E, Candelaria S L and Cao G 2013 *Chem. Soc. Rev.* **42** 3127–71

Zheng H, Li S, Zang C and Zheng W 2013 Coordinated control for grid integration of PV array, battery storage, and supercapacitor *2013 IEEE Power & Energy Society General Meeting* pp 1–5 (Piscataway, NJ: IEEE)

**IOP** Publishing

Thermodynamic Cycles for Renewable Energy Technologies

**K R V Subramanian and Raji George**

# Chapter 4

# Thermodynamic cycles for renewable energy utilization

**Entesar H Betelmal**

Environmental issues and lack of energy resources lead to the utilization of industrial waste heat in thermodynamic applications to improve thermodynamic cycles' performance and keep pace with climate change.

An essential for thermodynamics application is the analysis of power generation cycles. Exergy analysis is a powerful method in the optimization of thermodynamic processes. This chapter examines the modified thermodynamic cycles incorporating renewable energy to compare the different cycle efficiencies. We then evolve the exergy balance equation for it to then be applied to each cycle component. Furthermore, we discuss future technologies for modified Rankine, Stirling, Kalina, and Brayton cycles using a new working fluid, and we discuss the different working fluids and electrical energy storage. We also consider the OTEC Rankine with new working fluids.

The continued pursuit for higher thermal efficiencies has resulted in some new modifications to the basic thermodynamic cycle. Steam is the most used working fluid within the vapor power cycle due to its many preferable merits, these include low cost, availability, and high enthalpy of vaporization. This chapter is devoted to the discussion of steam and gas power plants, exploring steam power plants with renewable energy applications and industrial waste heat utilization, these technologies are current trends in the energy world. Regardless of the type of sources used to supply heat to the steam, the steam goes through the same basic cycle. Therefore, all cycle processes analyze in the same manner.

**OBJECTIVES OF CHAPTER**

- Analyze power cycles that use different working fluids and heat sources.
- Analysis of thermodynamic cycles concerning renewable energies and various working fluids.

doi:10.1088/978-0-7503-3711-3ch4
4-1

- Analysis of decrease in global discharge of $CO_2$ in the world by using renewable energy.
- Analyze the second law of thermodynamics in vapor power cycles.
- Modification of the basic Rankine, Stirling, Brayton, Kalina, and OTEC Rankine cycles to increase thermal efficiency.
- Analyze the exergy rate and exergy destruction of thermodynamic cycles in different applications.

## 4.1 General considerations in energy sources

Over past decades, global warming has increased with the Earth's population's growth due to increased demand for power supply. Consequently, it has increased interest in renewable energy, mostly solar and wind energy, to reduce carbon dioxide emissions from the combustion of fossil fuels. Exhaust gases are the primary cause of global warming. They can cause severe damage to our health and the environment even though conventional energy sources are significant for a country's economic growth.

Renewable energy obtained from natural sources is referred to as clean energy, which is continuously replenished, like sunlight, wind, rain, tides, waves, and geothermal heat. Biomass also falls into this category, as shown in figure 4.1.

Carbon dioxide ($CO_2$) is the most present greenhouse gas, but other air pollutants such as methane are also emitted in mass and contribute to global warming, with clean energy increasing, it has now been evaluated at 19.3% of global energy consumption since 2015, and this continues to increase. Furthermore, carbon dioxide emissions from fossil fuels were nearly unvarying in 2016 for the third year in a row because of declining coal use worldwide.

The clean energy industry has changed positively, and its costs have fallen dramatically, technologies have enhanced and become more effective with solutions to integrating renewables into electric grids progressing. Therefore, the capacity of renewable energy in the power sector will grow with time. Energy, entropy, and exergy analysis are commonplace in thermodynamics applications, and they are heavily applied in this chapter, and we take an in-depth view to provide a better understanding of these concepts. It also covers the basic principles and practical applications of thermodynamic cycles. Some illustrative examples are presented to highlight the importance of the aspects of energy, entropy, and exergy and their roles in thermal engineering.

## 4.2 Exergy concept

The energy enters through the system (100% of the input exergy) as shown in figure 4.2 and exits the system at various points along the way. Figure 4.3 shows an exergy flow diagram of the Rankine cycle at a particular operating condition. This flow diagram shows where the exergy losses occur in the process and how the exergy is destroyed. This analysis is very useful when considering the system design. It also gives information about the possibilities of improving the performance and where to direct efforts for improvement.

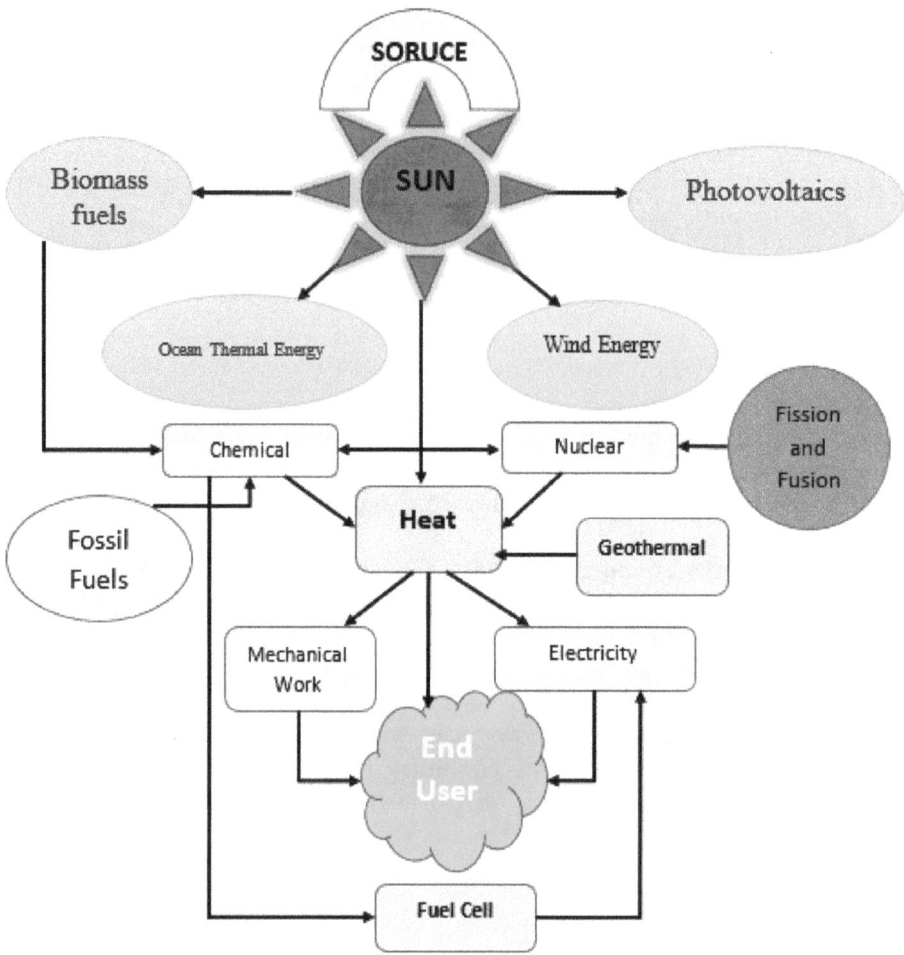

**Figure 4.1.** Energy sources and conversion processes.

## 4.3 Rankine cycle

Figure 4.4 shows the steam power plant, it is the source that provides electricity. The Rankine cycle is a type of thermodynamic cycle and an essential operation of all power plants and used in these plants for converting thermal energy into mechanical energy, then transformed into electric power by a generator. This cycle was developed in 1859 by Scottish engineer William J M Rankine.

The Rankine cycle is utilized to evaluate the performance of a steam turbine. In this process, fuel in the boiler is used to produce heat to convert the working fluid (water) undergoing phase change into steam and then expands through a steam turbine producing useful work. The ideal Rankine cycle does not have any internal losses in each of the four components, which means any losses are neglected.

Figure 4.3 consists of the following four processes:

1–2 Isentropic compression within a pump.

**Figure 4.2.** Exergy = energy − losses.

2–3 Constant pressure heat addition in a large heat exchanger, called a boiler.

3–4 Isentropic expansion within a turbine.

4–1 Constant pressure heat rejection in a heat exchange called a condenser.

The Rankine cycle runs using water as the working fluid which enters the pump at state 1 as saturated water to isentropically compress to the boiler's desired pressure. In state 2, water flows into the boiler as a compressed liquid and leaves as superheated vapor at state 3. The heat delivered to the boiler originates from the heat source or energy source, this is achieved with steam as the medium, as water is heated by an isobaric process converting the water into steam.

The boiler and the superheater (where the steam is superheated) are known as the steam generator. The superheated vapor expands from state 3 to state 4 through turbine stages converting mechanical energy to electrical energy producing useful work by rotating the shaft linked to an electric generator. The steam exits the turbine

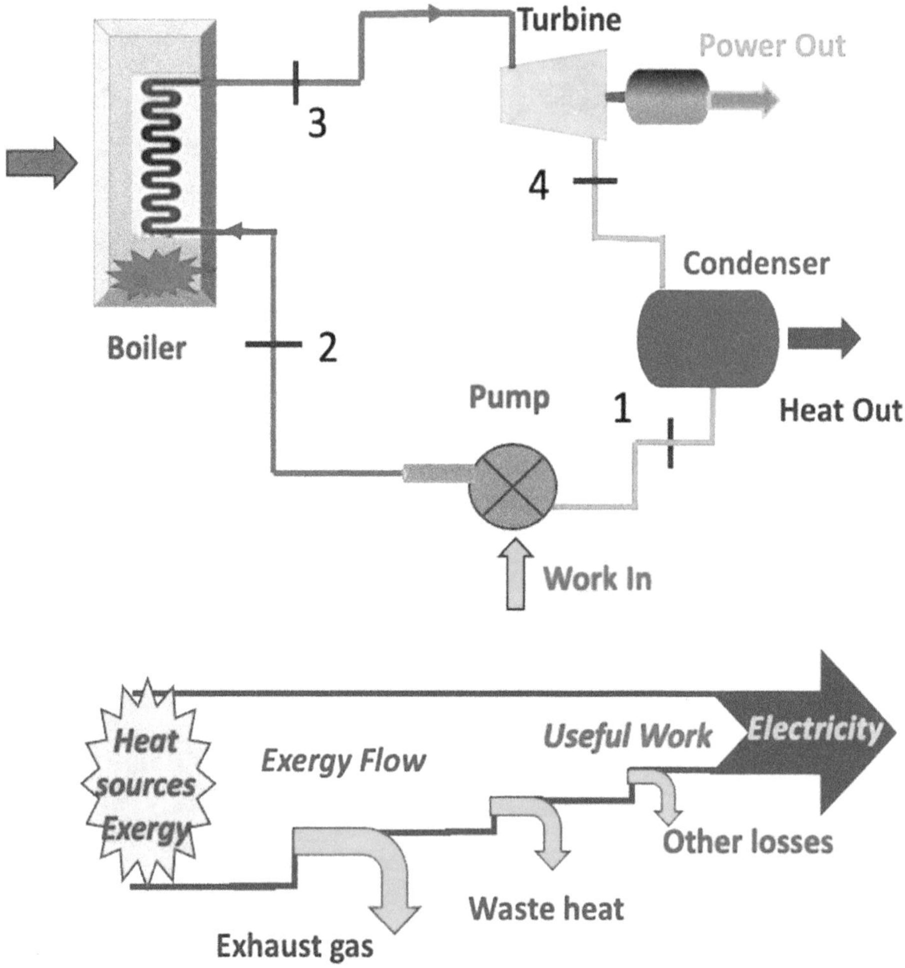

**Turbine**

Power Out

3

4

Condenser

Boiler

2

Pump

1

Heat Out

Work In

Heat sources Exergy

Exergy Flow

Useful Work

Electricity

Other losses

Waste heat

Exhaust gas

**Figure 4.3.** Schematic diagram of the Rankine cycle and exergy flow diagram.

**Figure 4.4.** Steam power plant.

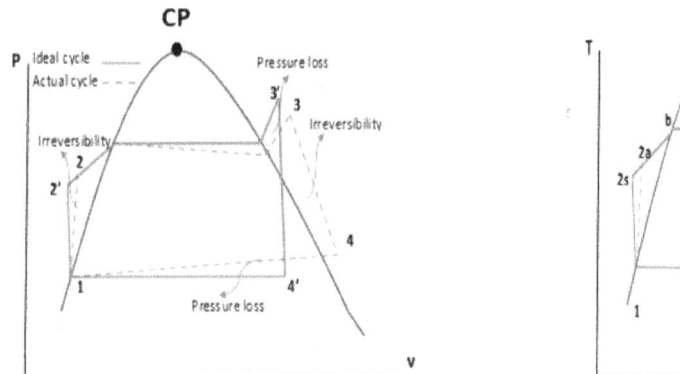

**Figure 4.5.** Rankine cycle $P$–$v$ and $T$–$s$ diagram.

at low pressure and temperature at state 4 as a saturated liquid-vapor mixture with high quality (x) and flows into a large heat exchanger called a condenser. Steam is condensed at constant pressure by rejecting heat from the steam to the atmosphere, a lake, or a river. Saturated steam leaves the condenser and enters the pump again and the cycle is repeated. The power plants are cooled by air instead of water. Such a method of cooling, which is also used in car engines, is called dry cooling.

The equation below presents how to define the *thermal efficiency of power cycles.*

$$\eta = \frac{W_{\text{net}}}{Q_{\text{in}}} = \frac{W_{34} - W_{51}}{Q_{13}} \tag{4.1}$$

To determine the values of heat and work of the Rankine cycle, we apply the first law of thermodynamics to every process of the cycle. The process within a turbine is assumed to be an isentropic expansion in order to calculate quality (x). The Rankine cycle components (pump, boiler, turbine, and condenser) are steady-flow devices, and any changes in the kinetic and potential energy of the steam can be neglected as the changes are minor relative to the work and heat transfer values.

Figure 4.5 illustrates the Rankine cycle on $P$–$v$ and $T$–$s$ diagram,

Where Ideal Rankine Cycle 1–2′–b–3′–4′–1.

Actual Rankine Cycle 1–2–b–3–4–1.

As shown in figure 4.5, the critical point (CP) is at the curve's highest point. The curved lines to the left of the CP are saturated-liquid lines, and the region to the left of these lines are called sub-cooled liquid regions. Similarly, curved lines to the right of the CP are saturated vapor lines, and the region to the right of these lines are called superheat vapor regions. The $T$–$s$ diagram shows the internally reversible heat transfer processes. The area under process 2–3 shows the heat transferred to the water in the boiler, and the area under the process curve 4–1 shows the heat rejected in the condenser. The net work done during this cycle is equal to the difference between these two areas. During the isentropic compression process, the water temperature rises slightly because of a trivial decrease in water's specific volume. The isentropic process is represented in the line between states 1 and 2 on the $T$–$s$ diagram.

The Rankine cycle is ideal; there is no loss associated with the compression and expansion processes. In a real steam power cycle, the compression and expansion processes are non-isentropic, In other words, this cycle is irreversible, and the energy is lost. Additionally, entropy is increased during the compression and expansion processes. The real cycle requires more power for the pump, and the turbine generates less power. Because of water-droplet formation, the actual steam turbine efficiency will be limited. Erosion of turbine blades by droplet impingement reduces the turbine blades' life and efficiency. To avoid this problem superheating the steam is requiring. As shown in figure 4.5 ($T$–$s$ diagram), state 3′ is in the saturated vapor region, so when the steam is expanded it will be very wet. By superheating the steam, the state will be in state 3 as shown in figure 4.5 and drier steam is produced after the expansion process.

### 4.3.1 Energy analysis of real cycle

In a real Rankine cycle, all processes are irreversible causing a drop in the overall efficiency. Turbine and pump irreversibilities are included in calculating the overall cycle efficiency, where subscript ($a$) indicates actual values and subscript ($s$) shows isentropic values. The continuity equation, the first law of thermodynamic of the Rankine cycle, can be developed using equations [1].

$$\sum m_{in} = \sum m_{out} \tag{4.2}$$

$$Q + W = \sum m_{out} h_{out} - \sum m_{in} h_{in}. \tag{4.3}$$

The subscripts in and out represent the inlet and exit states, $Q$ and $W$ are the net heat and work inputs.

The energy equation in each of the components is illustrated below.

**Pump:** Pump pressurize the liquid water from prior going to the boiler. Assuming no heat transfer with the surroundings, the work done by the pump to compress the water can be calculated by the following equation.

The energy balance in the pump is

$$W_{pump} = \dot{m}(h_2 + h_1) = V(P_2 - P_1) \tag{4.4}$$

$$\eta_p = \frac{W_s}{W_a} = \frac{V(P_{out} - P_{in})}{W_a}. \tag{4.5}$$

**Turbine:** The steam from the boiler, which has an elevated temperature and pressure, expands through the turbine to produce useful work and then it is discharged to the condenser with relatively low pressure. The heat transfer between the turbine and the surroundings can be neglected, the energy balance in the turbine is

$$W_t = \dot{m}(h_3 + h_4) \tag{4.6}$$

$$\eta_t = \frac{W_a}{W_s} = \frac{(h_{in} - h_{out})_a}{(h_{in} - h_{out})_s} \tag{4.7}$$

where $\eta_{th}$ is the thermal efficiency of the system and $W_t$ is the turbine's work done.

If $\eta_t$ and $\eta_p$ are known, we can determine the actual enthalpy after the compression and expansion processes from the isentropic processes' values. In the real cycle, each stage of the Rankine cycle has irreversibility, reducing the cycle efficiency.

**Boiler:** Liquid water enters the boiler and is heated to a superheated state. The heat $q_{in}$ that is supplied to the boiler comes from the energy source. The energy balance in the boiler is

$$Q_{in} = \dot{m}(h_3 + h_2). \tag{4.8}$$

**Condenser:** Steam leaves the turbine is condensed to liquid water in the condenser. The energy balance in the condenser is

$$Q_{out} = \dot{m}(h_4 + h_1). \tag{4.9}$$

In this state, the heat is withdrawn from the cycle. This is an important Rankine cycle process from a technological viewpoint since the pump requires a liquid medium to work efficiently.

We can obtain the energy balance for the whole cycle by applying the first law of thermodynamics and summarizing the four energy equations above as the following equation.

$$\left(Q_{in} - Q_{out}\right) - \left(W_{t,\,out} - W_{p,\,in}\right) = 0. \tag{4.10}$$

Most of the fuel energy is wasted as most advanced boilers transform only 40% of this energy into usable steam energy. There are two main reasons for this loss:

- The combustion gas temperatures are higher than the highest vapor temperatures, where the range of the combustion gas temperature is between 1000 °C and 2000 °C. The heat transfer between the large temperature difference causes the increased entropy.
- Combustion (oxidation) at technically sensible temperatures is highly irreversible.

The condensation process occurs at a temperature higher than the cooling medium's temperature because the heat transfer surface in the condenser has a finite value that causes the generation of entropy.

The efficiency of the condensers is reduced due to the deposition of dirt during operation with cooling water.

**EXAMPLE 1** In a thermal power plant working on a simple non-ideal Rankine cycle as shown below, the boiler pressure is 5.5 MPa and the condenser pressure is

75 kPa. Heat input to the steam in a furnace is kept at 950 K and the heat is rejected to the surroundings at 295 K. The maximum temperature in the cycle is 400 °C. The turbine isentropic efficiency is 85% and the pump isentropic efficiency is 80%. Determine the thermal efficiency of the cycle.

**Properties** The properties during the processes in the cycle is determined using data from steam tables.

**Analysis** A non-ideal Rankine cycle and steady conditions are considered. No changes in kinetic and potential energy. The schematic of the Rankine cycle and the $T$–$s$ diagram is shown in figure 4.6.

**Solution**

1) $P_1 = 75$ kPa, $T_1 = 92$ °C, $h_1 = 385.4$ kJ kg, $s_1 = 1.216$ kJ kg$^{-1}$ K$^{-1}$, $v_1 = 0.001\ 028$

2) $s_1 = s_{2s} = 1.216$ kJ kg$^{-1}$ K$^{-1}$    $h_{2s} - h_1 = v_1(P_2 - P_1)$

$h_{2s} = h_1 + v_1(P_2 - P_1) \implies 385.4 + 0.001028 * (5500 - 75)$

$h_{2s} = 391.031$ kJ kg$^{-1}$

$\eta_p = 0.8 \implies 0.8 = \frac{h_{2s} - h_1}{h_{2a} - h_1} = \frac{391.031 - 385.4}{h_{2a} - 385.4} \implies h_{2a} = 392.4$ kJ kg$^{-1}$

$s_{2a} = 1.220$

3) $p_3 = 5.5$ MPa, $T_3 = 400$ °C, $h_3 = 3187.6$ kJ kg$^{-1}$ K$^{-1}$, $s_3 = 6.5937$ kJ kg$^{-1}$ K$^{-1}$

4) $s_{4s} = 6.5937$ kJ kg$^{-1}$ K$^{-1}$, $s_{4sf} = 1.216$ kJ kg$^{-1}$ K$^{-1}$, $s_{4sfg} = 6.214$ kJ kg$^{-1}$ K$^{-1}$

$$x = \frac{(s_{4s} - s_{4sf})}{s_{4sfg}} \qquad x = \frac{6.5937 - 1.216}{6.214} = 0.865$$

$h_{4s} = h_f + x\, h_{fg}$  $h_f = 317.62$ kJ kg$^{-1}$ K$^{-1}$  $h_{fg} = 2318.4$ kJ kg$^{-1}$ K$^{-1}$

$h_{4s} = 385.4 + (0.865 * 2276.4)$  $h_{4s} = 2302.9$ kJ kg$^{-1}$ K$^{-1}$

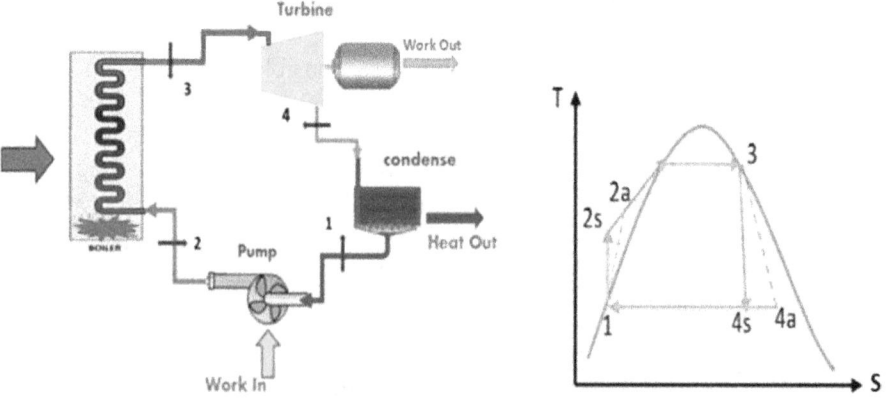

**Figure 4.6.** The simple Rankine cycle and $T$–$s$ diagram.

$$\eta_t = \frac{(h_3 - h_{4a})}{(h_3 - h_{4s})}$$

$$h_{4a} = 2435.6 \text{ kJ kg}^{-1}$$

$$s_{4a} = s_{4sf} + x s_{4sfg} \Rightarrow s_{4a} = 1.216 + (0.865 * 6.238) = 6.61 \text{ kJ kg}^{-1} \text{ K}^{-1}.$$

The cycle thermal efficiency is,

$$\eta_{th} = \frac{w_t - w_p}{q_H} = \frac{(h_3 - h_{4a}) - (h_{2a} - h_1)}{(h_3 - h_{2a})}$$

$$\eta_{th} = \frac{(3187.6 - 2435.6) - (392.4 - 385.4)}{(3487.08 - 329.11125)} = 0.266.$$

$$\eta_{th} = 26.6\%$$

**EXAMPLE 2** An ideal reheat-regenerative Rankine cycle with one open feedwater heater, one closed feedwater heater, and one reheater is shown in figure 4.7. Calculate the part of the flow drawn out of the turbines and the cycle efficiency.

*Analysis*
- All processes in the cycle are steady state.
- No changes in kinetic and potential energy.

The enthalpies at different points in the cycle and specific work of the pump can be determined by using the water tables.

*Solution*
1) saturated water state $P_1 = 10$ kPa $h_1 = 191.83$ kJ kg$^{-1}$ $v_1 = 0.001\,01$ m$^3$ kg$^{-1}$
2) compressed water $P_2 = 1$ MPa

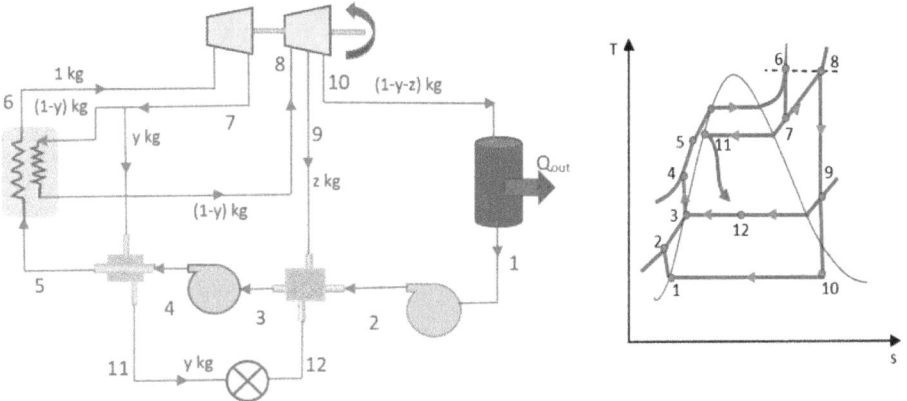

**Figure 4.7.** An ideal reheat-regenerative Rankine cycle and *T–s* diagram.

$$w_p = v_1 (P_2 - P_1)$$
$$= 0.001\,01(1000 - 10) = 1.0 \text{ kJ kg}^{-1}$$
$$h_2 = h_1 + w_p = 192.83 \text{ kJ kg}^{-1}$$

3) saturated water state   $P_3 = 1$ MPa   $h_3 = 762.81$ kJ k$^{-1}$   $v_3 = 0.001\,13$ m$^3$ kg$^{-1}$

4) compressed water   $P_4 = 16$ MPa

$$w_p = v_3 (P_4 - P_3)$$
$$= 0.001\,13(16000 - 1000) = 16.9 \text{ kJ kg}^{-1}$$
$$h_4 = h_3 + w_p = 762.81 + 16.9 = 779.71 \text{ kJ kg}^{-1}$$

5) saturated water state   $P_5 = 16$ MPa   $h_5 = 1650.1$ kJ kg$^{-1}$

6) superheated vapor   $P_6 = 16$ MPa   $T_6 = 600$ °C
   $h_6 = 3569.8$ kJ kg$^{-1}$ K$^{-1}$   $s_6 = 6.6988$ kJ (kg K)$^{-1}$

7) superheated vapor state   $P_7 = 5$ MPa   $s_7 = s_6 = 6.6988$ kJ (kg K)$^{-1}$
   $h_7 = 3222.4$ kJ kg$^{-1}$ K$^{-1}$

8) superheated vapor state   $P_8 = 5$ MPa   $T_8 = 600$ °C (given)
   $h_8 = 3665.6$ kJ kg$^{-1}$ K$^{-1}$   $s_8 = 7.2731$ kJ (kg K)$^{-1}$

9) superheated vapor state   $P_9 = 1$ MPa   $s_9 = s_8 = 7.2731$ kJ (kg K)$^{-1}$
   $h_9 = 3138.1$ kJ kg$^{-1}$ K$^{-1}$

10) saturated mixture state   $P_{10} = 10$ kPa   $s_{10} = s_8 = 7.2731$ kJ (kg K)$^{-1}$
    $x_8 = (s_8 - s_f$ at 10 kPa$)/s_{fg}$ at 10 kPa $= 88.3\%$
    $h_{10} = h_f$ at 10 kPa$+ x_8 h_{fg}$   at 10 kPa $= 2304.76$ kJ kg$^{-1}$

11) saturated water   $P_{11} = 5$ MPa $h_{11} = 1154.2$ kJ kg$^{-1}$

12) at throttling device (The Trap)   $h_{12} = h_{11} = 1154.2$ kJ kg$^{-1}$

Assume $y$ kg of water steam is extracted from the high-pressure turbine per kg of mass entering the boiler and assume $z$ kg of steam that is extracted from the low-pressure turbine for each kg of mass entering the boiler, using the energy balance of the closed and open feedwater heaters to calculate $y$ and $z$.

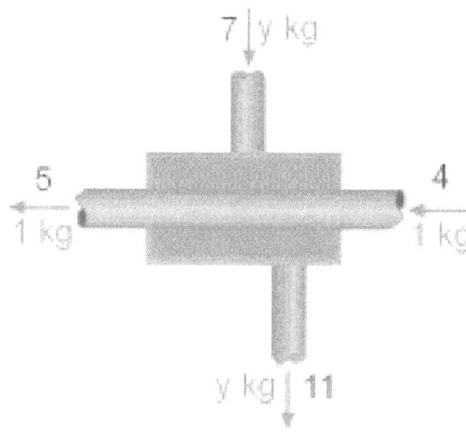

Closed feedwater heater:

$$y(h_7 - h_{11}) = 1(h_5 - h_4)$$
$$y = (1650.1 - 779.71)/(3222.4 - 1154.2)$$
$$y = 0.42$$

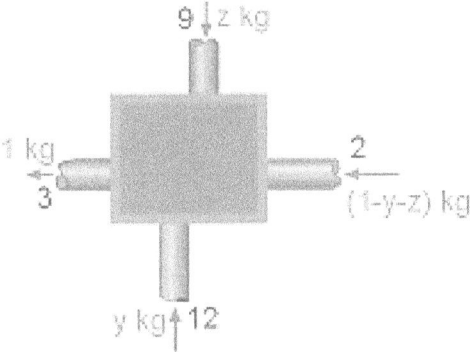

Open feedwater heater:

$$h_3 = zh_9 + (1 - Y - z)h_2 + Yh_{12}$$
$$z = 0.056$$

The high-pressure and low-pressure turbines power output for 1 kg of steam flowing through the boiler is:

$$w_t = (h_6 - h_7) + (1 - Y)(h_8 - h_9) + (1 - Y - z)(h_9 - h_{10})$$
$$w_t = (3569.8 - 3222.4) + (1 - 0.42) * (3665.6 - 3138.1) + (1 - 0.42 - 0.056)$$
$$* (3138.1 - 2304.7)$$
$$w_t = 1090.0 \text{ kJ kg}^{-1}$$

The pump input works in (1) is

$$w_p = (1 - y - z)1.0 + 16.9$$
$$w_p = 17.4 \text{ kJ kg}^{-1}$$

The total network output for 1 kg mass of water flowing through the boiler is

$$w_{\text{net}} = w_t - w_p$$
$$w_{\text{net}} = 1090.0 - 17.4 = 1072.6 \text{ kJ kg}^{-1}$$

The total heat input to the cycle is the heat input in the boiler plus the heat input from the reheater.

$$q_{\text{in}} = (h_6 - h_5) + (1 - Y)(h_8 - h_7)$$
$$q_{\text{in}} = (3569.8 - 1650.1) + (1 - 0.42) * (3665.6 - 3222.44)$$
$$q_{\text{in}} = 2176.7 \text{ kJ kg}^{-1}$$

The cycle thermal efficiency is

$$\eta_{th} = w_{net}/q_{in} = 1072.6/2176.7 = 49.3\%.$$

### 4.3.2 Exergy analysis

The real cycle differs from the ideal Rankine cycle because of the irreversibility in various components. Two significant factors of irreversibility are fluid friction and heat loss. An exergy analysis is a powerful method that measures the availability of energy to do work. The exergy is a large indication of the useful work that resources can do in a particular environment. The exergy concept shows the quality of the energy as an additional use of its consumption during the conversion or transfer of energy.

One of the uses of exergy is its balance in thermal analysis systems. This loss identification and qualification allows for the evaluation and improvement of thermal system design.

Exergy destruction analysis is a measure of resource degradation and the location of energy losses. These methods can calculate most cases of thermodynamic deficiency. Actual work and reversible work equations are often formulated in the exergy function for open and closed systems. The dead state of the system occurs when it is in equilibrium with the environment. This equilibrium occurs in standard atmospheric conditions (298.15 K and 1.013 25 bar (1 atm)). The following equation can derive the second law of thermodynamics of any component.

$$\phi_{heat} + W = \sum \phi_{out} - \sum E\phi_{in} + I \tag{4.11}$$

$$\phi = (U - U_0) + P_0(V - V_0) - T_0(S - S_0) \tag{4.12}$$

$$E = U + KE + PE. \tag{4.13}$$

Here $\phi$ is the exergy, $E$ is the energy, $V$ is the volume, and $S$ is the entropy. The exergy efficiency of component ($i$) is defined as the ratio of used exergy of component ($i$) to the available exergy of the same component.

$$\eta_{ex,\ component} = \frac{\phi_{i,\ used}}{\phi_{i,\ available}}. \tag{4.14}$$

The total exergy efficiency of system is the ratio of its total used exergy to total available exergy. It can be determined as:

$$\eta_{ex,\ cycle} = \frac{\phi_{total,\ used}}{\phi_{total,\ available}}. \tag{4.15}$$

The following equation can calculate the exergy per mass flow rate in each component:

$$\varphi_i = h_i + h_0 - T_0(s_i - s_0). \tag{4.16}$$

Based on the exergy balance presented in equation (4.16), the exergy destruction was calculated for each component of the system using the following equation.

$$\dot{I}_i = \sum \dot{m}_{\text{in}}\varphi_{\text{in}} - \sum \dot{m}_{\text{out}}\varphi_{\text{out}} + \dot{\phi}_{Qi} - \dot{\phi}_{Wi} \tag{4.17}$$

where,

$\dot{I}_i$ is the exergy destruction rate in component $i$,

$\varphi_{\text{in}}$ and $\varphi_{\text{out}}$ are the exergy rate associated with the inlet and outlet flow.

$\dot{\phi}_{Qi}$, $\dot{\phi}_{Wi}$ are the exergy rate by work and heat transfer through the boundary,

The exergy rate by heat transfer is calculated based on equation (4.17).

$$\dot{\phi}_{Qi} = Q_i\left(1 - \frac{T_0}{T_s}\right) \tag{4.18}$$

where $T_0$ is the environment temperature, and $T_s$ is the source or sink temperature. Another expression to calculate the exergy destruction rate of a component ($I_i$) based on the entropy generation rate is presented in equation (4.19) as follows.

$$I_i = T_0 S_{\text{gen}i} \tag{4.19}$$

where

$S_{\text{gen}i}$ is the entropy generation rate in the component ($i$).

The exergy analysis method aims to evaluate the exergy losses (Irreversibilities) associated with a process. Hence, it is required to calculate the irreversibility in all the power cycle components to estimate exergy efficiency. The exergy destruction in each of the components is calculated for the specified dead state ($P_0$, $T_0$).

**Boiler**

$$\varphi_{b,\,\text{ph}} = \dot{m}_{\text{in}}[h_{\text{out}} + h_{\text{in}} - T_0(s_{\text{out}} - s_{\text{in}})] \tag{4.20}$$

$$\varphi_{b,\,\text{fuel}} = Q_b\left(1 - \frac{T_0}{T_s}\right) \tag{4.21}$$

$$\varphi_b = \varphi_{b,\,\text{ph}} + \varphi_{b,\,\text{fuel}} \tag{4.22}$$

$$I_b = Q_b\left(1 - \frac{T_0}{T_s}\right) - m_s[(h_{\text{out}} - h_{\text{in}}) - T_0(s_{\text{out}} - s_{\text{in}})]. \tag{4.23}$$

**Steam Turbine**

The irreversibility in the steam turbine is given by

$$\varphi_t = \dot{m}_{\text{in}}[h_{\text{out}} + h_{\text{in}} - T_0(s_{\text{out}} - s_{\text{in}})] \tag{4.24}$$

$$I_t = T_0 \dot{m}_{\text{in}}(s_2 - s_1). \tag{4.25}$$

**Condenser:**

Mass flow rate of cooling water circulated to condense the steam to liquid water, that is obtained from the energy equation balance.

$$m_{\text{cw}}c_{\text{pw}}(T_{w,\,\text{in}} - T_{w,\,\text{out}}) = m_s(h_2 - h_3) \tag{4.26}$$

$$\varphi_{c,\,\mathrm{ph}} = \dot{m}_{\mathrm{in}}[h_{\mathrm{out}} + h_{\mathrm{in}} - T_0(s_{\mathrm{out}} - s_{\mathrm{in}})] \tag{4.27}$$

$$\varphi_{c,\,w} = Q_{\mathrm{cold}}\left(1 - \frac{T_0}{T_s}\right) \tag{4.28}$$

$$\varphi_c = \varphi_{b,\,\mathrm{ph}} + \varphi_{b,\,\mathrm{fuel}}. \tag{4.29}$$

Irreversibility in the condenser

$$I_c = T_0\left[m_s(s_2 - s_3) - \left(m_{\mathrm{cw}}c_{\mathrm{pw}}\ln\frac{T_{wo}}{T_{wi}}\right)\right]. \tag{4.30}$$

**Pump**
   Irreversibility rate in the boiler feed pump

$$\varphi_p = \dot{m}_{\mathrm{in}}[h_{\mathrm{out}} + h_{\mathrm{in}} - T_0(s_{\mathrm{out}} - s_{\mathrm{in}})] \tag{4.31}$$

$$I_p = m_s T_0(s_2 - s_1). \tag{4.32}$$

Total Irreversibility:

$$I_{\mathrm{total}} = \left[I_{\mathrm{bolier}} + I_{\mathrm{turbine}} + I_{\mathrm{condenser}} + I_{\mathrm{pump}}\right]. \tag{4.33}$$

**EXAMPLE 3** Refrigerant-134a enters an adiabatic compressor as a saturated vapor at $-26\,°C$ and $0.039\,59\,\mathrm{kg\ s^{-1}}$ and leaves the compressor at 800 kPa and 50 °C. We need to find the power input, isentropic efficiency, rate of exergy destruction, and second-law efficiency.
   *Assumptions*   (1) The process is a steady flow since there are no changes with time. (2) Kinetic and potential energy changes are negligible. (3) There is no heat transfer since the compressor is adiabatic.
   *Analysis*   (a) Using R-134a tables to get the properties of the refrigerant at the inlet and exit states of the compressor:
   *Properties*   Using steam tables: $h_1 = 2197.2\ \mathrm{kJ\ kg^{-1}}$; $s_1 = 7.2136\ \mathrm{kJ\ kg^{-1}\ K^{-1}}$.
   State 2 is saturated liquid $\Rightarrow x_2 = 0$ and $p_2 = p_1$.
**Solution:** $x_1 = 1$, $T_1 = -26\,°C$, $m_1 = 0.03959\ \mathrm{kg\ s^{-1}}$

$h_1 = 234.68\ \mathrm{kJ\ kg^{-1}}$; $s_1 = 0.9514\ \mathrm{kJ\ kg^{-1}\ K^{-1}}$; $v_1 = 0.18946\ \mathrm{m^3\ kg^{-1}}$
$P_2 = 800\ \mathrm{kPa}$   and   $T_2 = 50\,°C$
$h_2 = 286.69\ \mathrm{kJ\ kg^{-1}}$; $s_2 = 0.9802\ \mathrm{kJ\ kg^{-1}\ K^{-1}}$; $s_1 = s_2 \rightarrow h_{2s} = 277.53\ \mathrm{kJ\ kg^{-1}}$

The actual work input is

$$W_{\mathrm{act}} = \dot{m}(h_2 - h_1) = 0.03959 * (286.69 - 234.68)$$
$$W_{\mathrm{act}} = 2.06\ \mathrm{kW}.$$

The isentropic power input is

$$\dot{W}_{isen} = \dot{m}(h_{2s} - h_1) = 0.03959 * (277.53 - 234.68)$$
$$\dot{W}_{isen} = 1.696 \text{ kW}.$$

The isentropic efficiency is

$$\eta_c = \frac{W_{isen}}{W_{act}} = \frac{1.696}{2.059} = 0.8238 = 82.4\%.$$

The exergy destruction is

$$I_c = \dot{m}T_0(s_2 - s_1) = 0.03959 * 300 * (0.9802 - 0.9514)$$
$$I_c = 0.3417 \text{ kW}.$$

The reversible power is calculated as:

$$W_{rev} = W_{act} - I_c = 1.717 \text{ k}.$$

The second-law efficiency is calculated as:

$$\eta_{c,\,II} = \frac{W_{rev}}{W_{act}} = \frac{1.717}{2.059} = 0.8341 = 83.4\%.$$

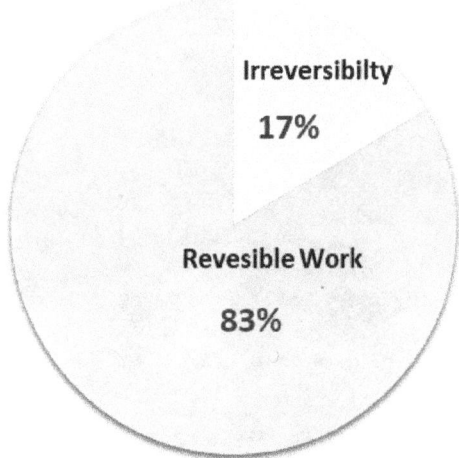

**EXAMPLE 4** A steam turbine that has an isentropic efficiency of 0.88 is supplied with steam at 12 MPa and 700 °C and leaves at 0.6 MPa. Determine:
   a) Reversible work and irreversibility
   b) Second-law efficiency.

**Properties** At 12 MPa and 700 °C from superheated table, we get
$s_1 = s_2 = 7.0757$ kJ kg$^{-1}$ K$^{-1}$ $h_1 = 3858.4$ kJ kg.

The two properties that we know in state 2 are entropy and pressure, $s_2 = 7.0757$ kJ kg$^{-1}$ K$^{-1}$. K and $P_2 = 600$ kPa → so we still have superheated vapor Therefore, $T_2 = 225.2$ °C and $h_2 = 2904.1$ kJ kg$^{-1}$.

**Analysis** Kinetic and potential energy changes can be ignored.

Process 2–3 is constant pressure and heat is added. In process 4–1, heat is rejected, so they are internally reversible and externally irreversible.

**Solution:**

a) From the energy equation, find work per unit mass. State 1–2
$w_a = h_1 - h_2 = 954.3$ kJ kg$^{-1}$

And $w_{rev} = (h_1 - T_0 s_1) - (h_2 - T_0 s_2) = (h_1 - h_2) = 954.3$ kJ kg$^{-1}$

$$\eta_s = 0.88 = \frac{w_a}{w_s} = \frac{w_a}{954.3} \implies w_a = 839.8 \text{ kJ kg}^{-1}.$$

From 1st law: $w_a = (h_1 - h_2) \Rightarrow h_2 = 3018.6$ kJ kg$^{-1}$

$h_2$  *and*  $p_2 \Rightarrow s_2 = 7.2946$ kJ kg$^{-1}$ K$^{-1}$  and  $T_2 = 279.4$ °C.

Then: $w_{rev} = (h_1 - T_0 s_1) - (h_2 - T_0 s_2) = 905$ kJ kg$^{-1}$.
The irreversibility is given by

$$I = w_{rev} - w_{irrev} = (905 - 839.8) = 65.2 \text{ kJ kg}^{-1}.$$

b) The second law efficiency is

$$\Rightarrow \eta_{II} = \frac{w_a}{w_{rev}} = \frac{839.8}{905} = 0.928 = 92.8\%.$$

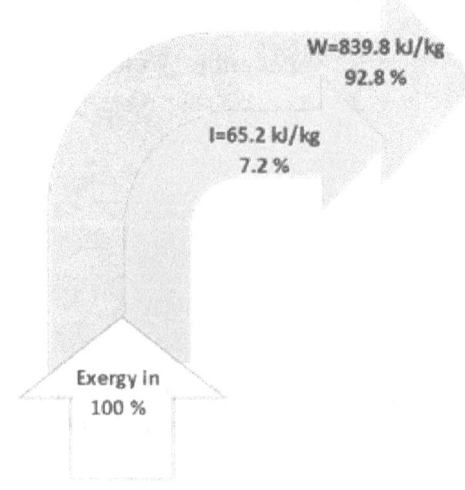

W=839.8 kJ/kg
92.8 %

I=65.2 kJ/kg
7.2 %

Exergy in
100 %

**EXAMPLE 5** Calculate the value of exergy destruction associated with each process of a simple Rankine cycle that produces power of 210 MW. The inlet condition to the turbine is 10 MPa and 500 °C and the condenser pressure is 10 kPa, assuming a source temperature of 1500 K and a sink temperature of 290 K.

> ***Properties*** Refer to the saturation water table, the values of $h_1$
> and    $v_1$ at $P_1 = 10$ kPa are

$$h_1 = h_f = 191.81 \text{ kJ kg}^{-1} \quad v_1 = v_f = 0.00101 \text{ m}^3 \text{ kg}^{-1}.$$

The specific work input to the cycle

$$w_p = v_1(P_2 - P_1).$$

Substitute $v_1$, $P_1$ and $P_2$

$$w_p = 0.00101(10{,}000 - 10) = 10.09 \text{ kJ kg}^{-1}.$$

Express the enthalpy of steam of state 2.

$$h_2 = h_1 + w_p = 191.81 + 10.09 = 201.9 \text{ kJ kg}^{-1}.$$

Refer to superheated water table, the values of $h_3$    and $s_3$ at $P_3 = 10$ MPa and $T_3 = 500$ °C are

$$s_3 = s_4 = 6.5995 \text{ kJ kg}^{-1} \text{ K}^{-1} \quad h_3 = 3375.1 \text{ kJ kg}^{-1}.$$

Similarly, get the values of $s_f$ and $s_{fg}$ from saturated water pressure table at $P_4 = 10$ kPa

$$s_f = 0.6492 \text{ kJ kg}^{-1} \text{ K}^{-1} \quad s_{fg} = 7.4996 \text{ kJ kg}^{-1} \text{ K}^{-1}$$

the quality of steam ($x_4$) at the end of the heat rejection process is

$$x_4 = \frac{s_4 - s_f}{s_{fg}}.$$

Substitute by specific entropy of saturated liquid is $s_f$, specific entropy of vaporization is $s_{fg}$ and specific entropy at state 4 is $s_4$

$$x_4 = \frac{6.5995 - 0.6492}{7.4996} = 0.7934.$$

Express the specific enthalpy of steam of state 4.
From the saturated table at $P_4 = 10$ kPa

$$h_4 = h_f + x_4 h_{fg} = 191.81 + (0.7934 * 2392.1) = 2089.7.$$

Specific heat input to the cycle is $q_{in}$, specific enthalpy of steam at state 3 is $h_3$

$$q_{in} = 3375 - 201.9 = 3173.2 \text{ kJ kg}^{-1}.$$

Express the specific heat output of the Rankine cycle.

$$q_{out} = h_4 - h_1$$
$$q_{out} = 2089.7 - 191.81 = 1897.9 \text{ kJ kg}^{-1}.$$

Express the exergy destruction for the process $I$

Process 1–2, and 3–4, are isentropic processes, therefore, the value of exergy destruction during these processes is zero (0).

Process 2–3

$$I_{2-3} = T_0\left(s_3 - s_2 + \frac{q}{T_s}\right)$$
$$I_{2-3} = 290\left(6.5995 - 0.6492 + \frac{-3173.2}{1500}\right).$$

The exergy destruction during process 2–3 is

$$I_{2-3} = 1112.1 \text{ kJ kg}^{-1}.$$

Process 4–1

$$I_{4-1} = T_0\left(s_1 - s_4 + \frac{q}{T_s}\right)$$
$$I_{4=1} = 290\left(0.6492 - 6.5995 + \frac{1897.9}{290}\right).$$

The exergy destruction during process 4–1 is

$$I_{4-1} = 172.3 \text{ kJ k}^{-1}.$$

## 4.4 Organic Rankine cycle

The organic Rankine cycle is a system based on the normal Rankine cycle, which is used when the high temperatures needed to convert water to steam are not available. Instead of using water, we use an organic working fluid that has a lower boiling point than water and needs low-temperature heat sources to produce power. This cycle can generate electric and thermal power using different sources, like renewable sources, traditional fuels, or waste heat from power plants (figure 4.8).

The organic Rankine cycle has a function like a steam Rankine cycle. However, it uses a different working fluid, a high molecular mass organic fluid is used as a working medium. It operates on low temperatures and low pressures comparative to conventional Rankine cycles [2]. The selected working fluids efficiently exploit low-temperature heat sources (70 to 400 °C) to produce electricity. A heat source in the evaporator vaporizes the organic fluid, the organic fluid vapor expands and is then condensed in a heat exchanger finally, the condensate flows back to the evaporator, subsequently, completing the thermodynamic cycle.

Organic fluids are used in an organic Rankine cycle, fluids such as n-pentane or toluene instead of water. The advantage of these working fluids is allowing the use of low temperature heat sources, like solar ponds that normally operate at around 70–90 °C. This lower temperature range causes a decrease in the cycle efficiency,

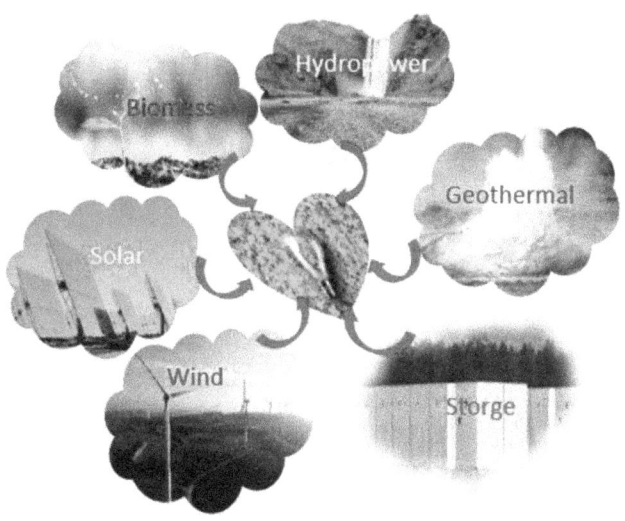

**Figure 4.8.** Renewable energy.

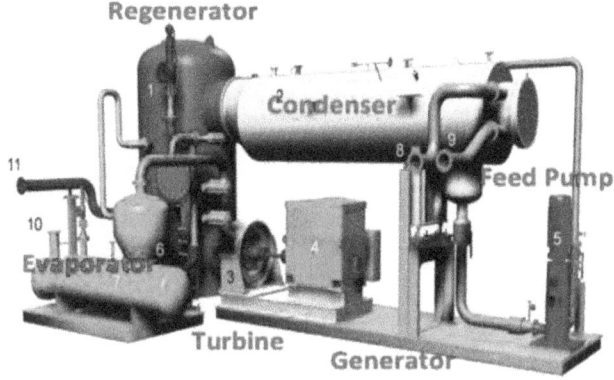

**Figure 4.9.** Organic Rankine cycle. Source: BIOS Bioenergiesysteme GmbH, Graz, Austria; www.bios-bioenergy.at

however the lower) temperature is valuable for the lower heat cost (figure 4.9). Figure 4.10 shows the simple organic Rankine cycle.

The thermal efficiency of the organic Rankine cycle [3] may be expressed as:

$$\eta_{th} = W_{net,\,out}/Q_{in} = W_{net,\,out}/(m_{in}(h_2 - h_1)) \tag{4.34}$$

$$W_{net,\,out} = W_{turbine} - W_{pump}. \tag{4.35}$$

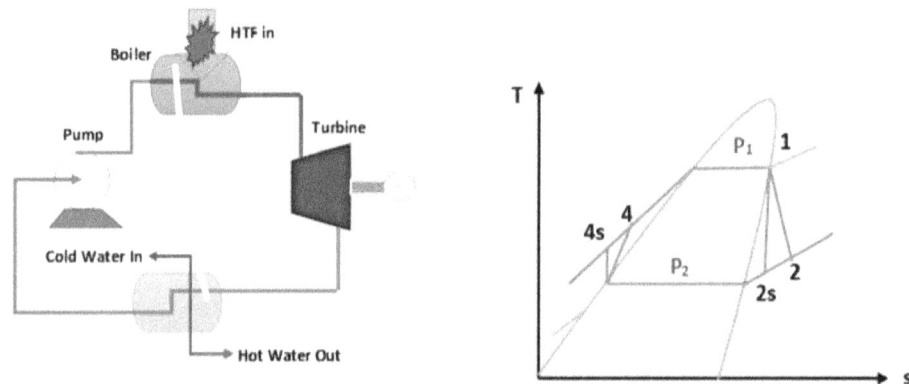

**Figure 4.10.** (a) Simple organic Rankine cycle. (b) $T$–$s$ diagram.

**Table 4.1** Basic properties of working fluids. Source: ethermo.us/Refrigerant.

| Working fluid | Critical temperature (°C) | Critical pressure (kPa) | Critical density (kg m$^{-3}$) |
|---|---|---|---|
| R134a | 101 | 4059 | 511.90 |
| R227ea | 102.8 | 2925 | 594.25 |
| R245fa | 154 | 3651 | 516.08 |
| R123 | 183.68 | 3661.8 | 550 |
| R600 | 152 | 3797 | 228 |
| Toluene | 319 | 4126 | 291 |
| Iso-butane | 134.7 | 3650 | |
| Iso-pentane | 187.2 | 3629 | |

### 4.4.1 Working fluid

Working fluid selection of the organic Rankine cycle is a crucial consideration, as several criteria need to be considered, such as environmental sustainability, ozone depletion potential (ODP), global warming potential (GWP), and safety (non-flammable, non-toxic, and non-corrosive). Besides, the vapor pressure in the boiler, critical temperature, and thermal stability should be considered.

Table 4.1 shows the basic properties of some working fluids [4].

There have been several investigations on the types of working fluids and the research has concluded with the results on some types of working fluid properties given in figure 4.11. Three types of working fluids are used in the organic Rankine cycle: wet, dry, and isentropic types, which have different slopes of the vapor saturation curves in the $T$–$s$ diagram as shown in figure 4.11. The wet fluids have a negative slope concerning the vapor saturation curve, while the dry fluids have a positive slope. The isentropic fluids have a vertical.

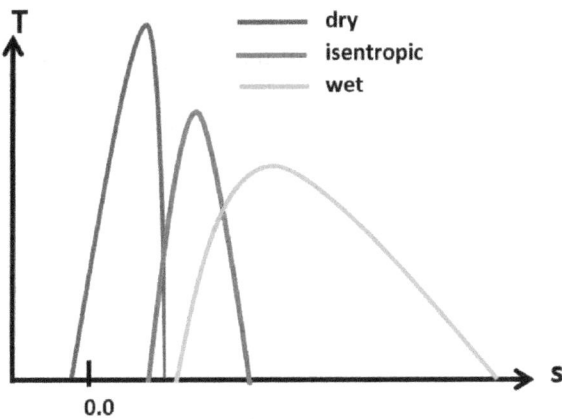

**Figure 4.11.** Three types of organic Rankine cycle working fluids: dry, isentropic, and wet.

### 4.4.2 Effects of different working fluids and its thermophysical properties on exergy analysis

The choice of working fluid is crucial since it has a major influence on power output, thermal efficiency, heat transfer area, and exergy analysis. The second-law efficiency (or exergy efficiency) is a useful tool that is used to investigate and compare working fluids with each other. The second-law efficiency illustrates the ratio of the total exergy output to exergy input. The importance of second-law efficiency lies in its ability to give clearer information on system losses and performance. The second law efficiency is a substantial parameter that evaluates which working fluid is better for a given heat source and heat sink temperature. Working fluids with different critical temperatures have different thermal efficiencies for the same operating condition. The working fluids influence the performance of organic Rankine cycles like internal and external exergy efficiencies, the thermophysical properties of the working fluid have a small influence on internal exergy efficiency which increases with the increase of evaporation temperature for all working fluids but have significant effects on external exergy efficiency. Furthermore, the working fluid with lower critical temperature has a higher evaporation temperature, which leads to larger overall exergy efficiency [5].

The overall exergy efficiency of the organic Rankine cycle can be expressed as

$$\eta_{ex} = \frac{W_{net}}{\phi_{s,\,in}} \tag{4.36}$$

or

$$\eta_{ex} = \eta_{ex,\,e} * \eta_{ex,\,i} \tag{4.37}$$

where $\phi_{s,\,in}$ is the inlet exergy of the heat source and defined as:

$$\phi_{s,\,in} = \dot{m}_s[(h_{s,\,in} - h_{s,\,0}) - T_0(s_{s,\,in} - s_{s,\,0})]. \tag{4.38}$$

The characteristics of the heat transfer processes for the same heat source are different due to the physical properties of different working fluids; hence the outlet condition of the source is different. Therefore, it is appropriate to choose the inlet exergy of the heat source as the total exergy input of the system when choosing the working fluid.

$$\eta_{ex,\,e} = \frac{\phi_{working\ fluid}}{\phi_{s,\,in}} \tag{4.39}$$

$$\eta_{ex,\,i} = \frac{W_{net}}{\phi_{working\ fluid}} \tag{4.40}$$

where $\phi_{working\ fluid}$ is the exergy augmentation of the working fluid in the evaporator, which can be calculated by

$$\phi_{working\ fluid} = q_{evp} - T_0(s_{out} - s_{in}). \tag{4.41}$$

The entropy changes of the working fluid from outlet and inlet state can be calculated approximately as

$$s_{out} - s_{in} = \overline{C_P}\ln\frac{T_1}{T_3} + \frac{\gamma_e}{T_1} \tag{4.42}$$

where $\gamma_e$ is evaporation latent heat [kJ kg$^{-1}$ K$^{-1}$].

**EXAMPLE 6** Consider a power plant operating on an organic Rankine cycle using R-410A as a working fluid, it has a boiler operating at 3 MPa and superheating the working fluid to 180 °C, with a condenser at a pressure of 800 kPa. Determine all four energy transfers and the thermal efficiency of the cycle.
  *Properties* Kinetic and potential energy changes are negligible.
  *Analysis* organic Rankine cycle, working fluid R-410A.

**Solution**
  From the saturated R-410a P = 800 kPa to find $h_{f,800\ kPa}$

State 1: Interpolating at 0 °C to 5 °C  $\qquad \frac{h_f - 57.76}{65.41-57.76} = \frac{800-798.7}{933.9-798.7}$

$$h_1 = h_f \qquad = 57.833\ kJ\ kg^{-1}$$
$$\frac{v_f - 0.000855}{0.000870 - 0.000855} = \frac{800 - 798.7}{933.9 - 798.7}$$
$$v_1 = v_f \qquad = 0.000\ 855\ kJ\ kg^{-1}$$

from the superheated R-410a
  at 3 MPa,180 °C $h_3 = 445.09\ kJ\ kg^{-1}$ $s_3 = 1.3661\ kJ\ kg^{-1}\ K^{-1}$.

From the saturated R-410a $s_f$ at 800 kPa

$$\frac{s_f - 0.2264}{0.2537 - 0.2264} = \frac{800 - 798.7}{933.9 - 798.7} s_f = \text{kJ kg}^{-1} \text{K}^{-1}$$

$$\frac{h_{fg} - 221.37}{215.13 - 221.37} = \frac{800 - 798.7}{933.9 - 798.7} h_{fg} = \text{kJ kg}^{-1} \text{K}^{-1}.$$

From the R-410a superheated table at $P = 800$ kPa and $s_4 = 1.3661$ kJ kg$^{-1}$ K$^{-1}$ by interpolated $h_4 = 385.97$ kJ kg$^{-1}$.

State 1: control volume: **Pump**

$$w_p = (h_2 - h_1) = v_1(P_2 - P_1)$$
$$w_{1-2} = 0.000855(3000 - 800) = 1.88122 \text{ kJ kg}^{-1}$$
$$h_2 = w_p + h_1 = 1.88122 + 57.833$$
$$h_2 = 59.71422 \text{ kJ kg}^{-1}.$$

State 2: control volume: **Boiler**

The enthalpy and entropy at state 3 are determined from the given temperature and pressure with the superheated values from R-410a.

$$q_H = h_3 - h_2 = 445.09 - 59.714$$
$$q_H = 385.38 \text{ kJ kg}^{-1}.$$

State 3: control volume: **Turbine**

$$w_t = (h_3 - h_4) = 445.09 - 385.97 = 59.12 \text{ kJ kg}^{-1}.$$

State 4: control volume: **Condenser**

Heat lost in the condenser is $q_L = h_4 - h_1 = 385.97 - 57.833 = 328.21$ kJ kg$^{-1}$

$$q_L = 221.3 \text{ kJ kg}^{-1}.$$

The thermal efficiency then is:

$$\eta_{th} = \frac{w_{net}}{q_H} = \frac{w_t - w_p}{q_H} = \frac{59.12 - 1.88122}{385.45} = 0.148$$
$$\eta_{th} = 14.8\%.$$

## 4.5 Solar Rankine cycle

It is a novel thermodynamic cycle that may have great potential to improve energy conversion efficiency and significantly reduce the global discharge of $CO_2$ by using solar energy as the system's energy source. Solar radiation is divided into two components, direct radiation, and diffuse radiation. Direct or beam radiation is the radiation that arrives at the ground without being scattered by clouds. Diffuse radiation, however, is known as scattered radiation and has been affected by molecules in the air [6]. The total radiation received, the sum of the beam, and diffuse radiation are called global radiation. The engine uses a heat source, such as

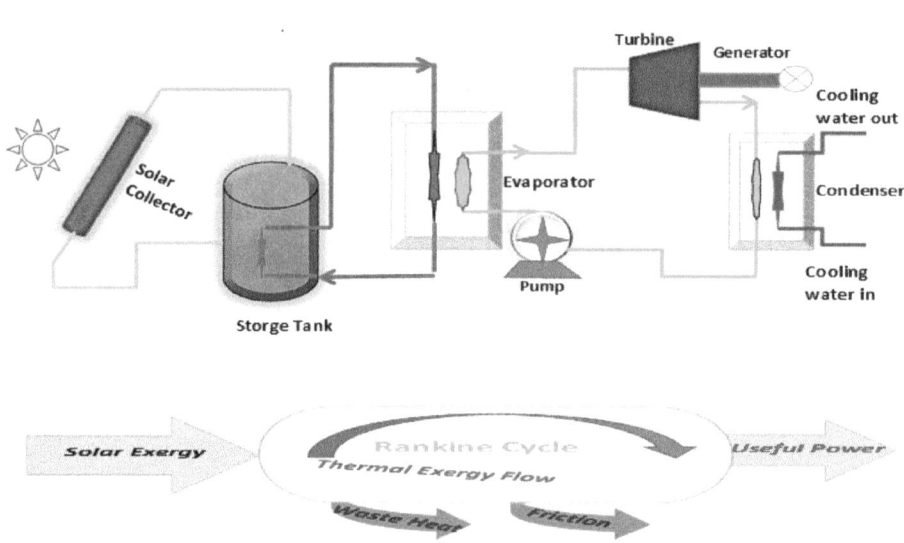

**Figure 4.12.** Thermal energy storage for solar-powered organic Rankine cycle and exergy flow diagram.

concentrated solar radiation as shown in figure 4.12, to provide energy to a fluid in a closed cycle; this, in turn, drives a turbine that produces electricity. This cycle is the so-called solar Rankine cycle engine.

### 4.5.1 Nanofluids as working fluid for solar Rankine cycle

The selection of working fluid for solar Rankine cycles is significant due to the significant impact the working fluid has, especially for low-temperature applications since the working fluid affects the cycle efficiency. Recently, the nanofluids have been proposed as working fluids for the organic Rankine cycle as a consequence of their boosted thermal properties, which is beneficial to the design of the Rankine cycle components and performance [7, 8]. A nanofluid containing pentane and silver nano-particles has been used as a working fluid for a solar Rankine cycle, using the nanofluid as a working fluid permits the use of smaller heat exchangers and increases efficiencies compared to the conventional working fluid. Although using nanofluids has various challenges that need more research including investigating the stability of nanofluids, transmigration of nanoparticles to the gas phase, performance in expanders and settling in heat exchangers, with the information available we can evaluate the possible benefits that nanofluids can bring to the organic Rankine cycle. Several correlations have been developed in recent years to predict the thermophysical properties (density, heat capacity $C_p$, thermal conductivity $k$, and dynamic viscosity) of nanofluids [9]. The properties of the nanoparticles and volume concentration used by linear correlations are used for density and specific heat capacity, giving satisfactory results mostly at low temperatures and low volume concentrations. The nanofluid can be considered as a pure fluid hence we can use the equations of continuity, motion, and energy.

As a consequence, the dimensionless correlations of pure fluid heat transfer can apply to nanofluids by using nanofluid thermophysical properties instead. To calculate the thermophysical and thermodynamic properties of nanofluids as specific heat, thermal conductivity, viscosity, and density, using the law of mixtures, we use the following equation [10].

$$\alpha_{\text{total}} = \alpha_{\text{particles}} + \alpha_{\text{base fluid}} \tag{4.43}$$

where $\alpha$ illustrates a specific thermophysical property of the nanofluid.

The thermal conductivity is related to thermal diffusivity and nanofluid density as follows:

$$\lambda = \alpha \rho C_p \tag{4.44}$$

where $C_p$ is the specific heat, $\alpha$ is the thermal diffusivity, $\lambda$ and $\rho$ represent the thermal conductivity and density, respectively. The nanofluids specific heat is calculated as:

$$C_{p,\,\text{nf}} = \frac{(1 - \phi)(\rho C_P)_{\text{bf}} + \phi(\rho C_P)_p}{(1 - \phi)\rho_{\text{bf}} + \phi\rho_p} \tag{4.45}$$

where 'nf' and 'bf' indicate the nanofluid and the basic fluid, respectively.

$\phi$ is the nanofluid particle concentration,

$\rho$ is density.

The density of nanofluids is calculated as:

$$\rho_{\text{nf}} = \phi_p\rho_p + (1 - \phi)\rho_{\text{bf}}. \tag{4.46}$$

### 4.5.2 Energy analysis

Two principal assumptions are considered: (1) the system reaches a steady state, and (2) pipe pressure drop and heat loss to the environment is neglected. Equations used to perform the energy analysis are

$$\textbf{Evaporator}: \dot{Q}_{\text{Ev}} = \dot{m}_{\text{ORC}}(h_{\text{out, ORC}} - h_{\text{in, ORC}}) \tag{4.47}$$

$$\textbf{Turbine:} \dot{W}_t = \dot{m}_{\text{ORC}}(h_{\text{out, ORC}} - h_{\text{in, ORC}}) \tag{4.48}$$

$$\textbf{Condenser:} \dot{Q}_{\text{cd}} = \dot{m}_{\text{ORC}}(h_{\text{out, ORC}} - h_{\text{in, ORC}}) \tag{4.49}$$

$$\textbf{Pump:} \dot{W}_p = \dot{m}_{\text{ORC}}(h_{\text{out, ORC}} - h_{\text{in, ORC}}) \tag{4.50}$$

$$\textbf{Thermal efficiency}: \eta_{\text{ORC}} = \frac{\dot{W}_t - \dot{W}_p}{\dot{Q}_H}. \tag{4.51}$$

### 4.5.3 Exergy analysis

The energy balance equation is fundamentally related to the quantity, not the quality of energy. In thermodynamics, the quality of a specified quantity of energy is described by its exergy since analyzing the exergy is considered as extended energy analysis. Furthermore, exergy analysis is a useful tool because it provides an accurate and robust analysis of thermodynamic systems; it accounts for the energy stream that can do work. Exergy analysis can apply to a system, subsystem, or component. All cases give different information that is useful for system improvement. The real inefficiencies in a thermodynamic system are the exergy destruction occurring in the system boundaries and the value of exergy transferred to the surrounding system. The reason for exergy loss is chemical reactions, fluid friction, throttling of flow, mixing dissimilar flow, and heat transfer through finite temperature differences. The total system exergy is formed mainly of four components: chemical, physical, potential, and kinetic exergy. The analysis reference temperature $T_0$ is 298.15 K, and pressure $p_0$ is 1.013 bar.

The following equations represent the exergy analysis that is employed to evaluate the performance of the system based on the irreversibility that is occurring in every component of the ORC system (as non-isentropic expansion and compression) [11].

$$\textbf{Fuel: } \phi_f = Q_H\left(1 - \frac{T_0}{T_H}\right) \tag{4.52}$$

$$\textbf{Evaporator: } \phi_{Ev} = \dot{m}_{ORC}(h_{out,\,ORC} - h_{in,\,ORC} - T_0(s_{out,\,ORC} - s_{in,\,ORC})) \tag{4.53}$$

$$I_{Ev} = \phi_f - \phi_{Ev} = \dot{Q}_H\left(1 - \frac{T_0}{T_H}\right) - \dot{m}_{ORC}(\varphi_{out,\,ORC} - \varphi_{in,\,ORC}) \tag{4.54}$$

$$\textbf{Turbine: } \phi_t = \dot{m}_{ORC}(h_{out,\,ORC} - h_{in,\,ORC} - T_0(s_{out,\,ORC} - s_{in,\,ORC})) \tag{4.55}$$

$$I_t = \dot{m}_{ORC} * T_0(s_{out,\,ORC} - s_{in,\,ORC}) \tag{4.56}$$

$$\textbf{Condenser: } \varphi_C = \dot{m}_{ORC}[h_{out,\,ORC} + h_{in,\,ORC} - T_0(s_{out,\,ORC} - s_{in,\,ORC})] \tag{4.57}$$

$$\textbf{Pump: } \phi_p = \dot{m}_{ORC}(h_{out,\,ORC} - h_{in,\,ORC} - T_0(s_{out,\,ORC} - s_{in,\,ORC})) \tag{4.58}$$

$$I_p = \dot{m}_{ORC} * T_0(s_{out,\,ORC} - s_{in,\,ORC}). \tag{4.59}$$

The exergy efficiency of the ORC system calculates the effectiveness of a cycle relative to its performance in reversible processes and is calculated below:

$$\eta_{II,\,ORC} = \frac{\dot{W}_t - \dot{W}_p}{\phi}. \tag{4.60}$$

The irreversibility ratio of each component of the solar driven ORC system is defined as:

$$IR = \frac{I_{component}}{I_{total}}. \tag{4.61}$$

**EXAMPLE 8** Consider an engine operating on a solar Rankine cycle and Refrigerant 134a is its working fluid. The saturated vapor enters the turbine at 60 °C and exits as a saturated liquid-vapor mixture and enters the condenser. The condenser pressure is 6 bar and the working fluid exits the condenser as a saturated liquid. The R-134a fluid is then pressurized in the pump before entering the constant solar heater. The energy input to the collectors from solar radiation is 0.4 kW m$^{-2}$. Find the minimum possible solar collector surface area, in m$^2$, per kW of power produced by the plant.

*Properties* Obtain the following properties of R-134a at 60 °C from saturated R-134a tables. $T_1 = 60$ °C $h_1 = h_g = 278.5$ kJ kg$^{-1}$ $s_1 = s_g = 0.907$ kJ kg$^{-1}$ K$^{-1}$.

Obtain the following properties of R-134a at 6 bar from saturated R-134a tables. $v_3 = v_f = 0.0008$ kJ kg$^{-1}$ $h_3 = h_f = 81.5$ kJ kg$^{-1}$.

*Analysis* There are no pressure drops in the condenser; and R-134a leaves the condenser as the saturated liquid. The model used is a Rankine cycle and all processes are steady state with no changes in potential and kinetic energy. The cycle diagram is shown in figure 4.13.

*Solution* The turbine process is a reversible isentropic process at 60 °C.

$$s_1 = s_2.$$

From saturated table at 6 bar

$$s_1 = s_2 = s_f + x_2 s_{fg}$$
$$0.907 = 0.308 + x_2 * 0.614 x_2 = 0.975$$
$$h_2 = h_f + x_2 h_{fg}$$
$$h_2 = 81.5 + 0.975 * 181 = 257.97 \text{ kJ kg}^{-1}.$$

Therefore, the **pump** work

$$w_p = -v_f(p_4 - p_3) = -0.0008 * (1682.8 - 600)$$
$$w_p = -0.86624 \text{ kJ kg}^{-1}.$$

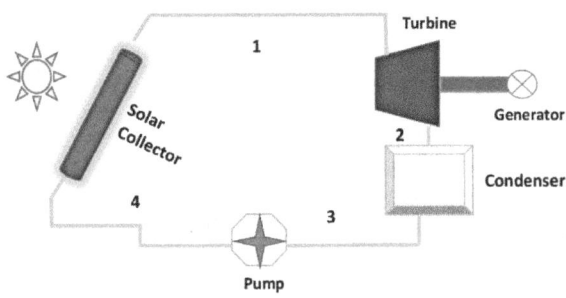

**Figure 4.13.** The schematic diagram of the solar power plant.

Enthalpy at state 4

$$h_4 = h_3 - w_p = 81.5 + 0.86624$$
$$h_4 = 82.36 \text{ kJ kg}^{-1}.$$

At the **turbine** entrance, we have $T_3 = 60 \text{ °C}$, $x_3 = 1$.
Thus, we find $w_t = (h_1 - h_2) = (278.5-257.97)$

$$w_t = 20.53 \text{ kJ kg}^{-1}.$$

The energy input in a Rankine cycle is isobaric.
The heat input is given by

$$q_{in} = (h_1 - h_4) = (278.5-82.36)$$
$$q_{in} = 196.14 \text{ kJ kg}^{-1}.$$

Heat rejected in the **condenser**

$$q_{out} = (h_2 - h_3) = (257.97-81.5)$$
$$q_{out} = 176.47 \text{ kJ kg}^{-1}.$$

Let us assume that we develop 1 kW of network, the required heat power input is.
The network is $W_{cycle} = \dot{m}(w_t - w_p)$

$$1 = \dot{m}(20.53-0.86624)$$

$$\dot{m} = 0.0508 \text{ kg s}^{-1}.$$

The amount of heat addition $Q_{in} = \dot{m} * q_{in} = 0.0508 * 196.14$

$$Q_{in} = 9.963 \text{ kW}.$$

So the required surface area to develop this 1 kW of net power is

$$\frac{Q_{in}}{A} = 0.4 = \frac{9.963}{A} = 0.4$$
$$A = 24.907 \text{ m}^2.$$

## 4.6 Geothermal energy application

Geothermal energy is renewable energy obtained using natural heat from the interior of the Earth. Geothermal energy sources can be found with a wide range of temperatures and can be used for power generation.

The temperature of geothermic heat sources vary from 50 °C to 350 °C, the organic Rankine cycle can therefore optimally use this application as a heat source [12, 13]. However, it is crucial to keep in mind that for low-temperature geothermal sources (typically less than 100 °C), the efficiency of the cycle will be low and depends on heat sink temperature (defined by the ambient temperature).

### 4.6.1 Type of geothermal heat sources

Hot spring is hot water or steam, that can flow outside naturally, by umping or by impulses, usually using two wells for the operation. Geothermal wells are a low- and medium-temperature geothermal energy, having temperatures that can reach 150 °C. Volcanic heat is depending on the ambient temperatures and the electricity can be generated with heat sources above 90 °C (figure 4.14).

**EXAMPLE 9** Consider an engine operating on an ideal Rankine cycle that uses geothermal hot water as the energy source and R-134a as a working fluid. Saturated vapor R-134a exits from the boiler at a temperature of 85 °C, and the condenser temperature is 40 °C. Determine the cycle thermal efficiency.

*Properties* Obtain the following properties of R-134a fluid at 85 °C from saturated R-134a tables. $P_2 = 2928.2$ kPa $h_2 = h_g = 279.50$ kJ kg$^{-1}$ $s_2 = s_g = 0.8812$ kJ kg$^{-1}$ K$^{-1}$.

Obtain the following properties of R-134a at 40 °C from saturated R-134a tables. $P_4 = 1017.1$ kPa $v_4 = v_f = 108.28$ kJ kg$^{-1}$ $h_4 = h_f = 108.28$ kJ kg$^{-1}$ $h_g = 271.31$ kJ kg$^{-1}$ $s_f = 0.394\,93$ kJ kg$^{-1}$ K$^{-1}$ $s_g = 0.915\,52$ kJ kg$^{-1}$ K$^{-1}$.

*Analysis* All processes of the ideal Rankine cycle are steady state with no changes in potential and kinetic energy. The $T$–$s$ diagram is shown in figure 4.15.

**Solution**

Pump (use R-134a table)

Determine the pump work

$$-w_p = h_2 - h_1 = \int_1^2 vP \approx v_1(P_2 - P_1)$$
$$-w_p = 0.000873(2926.2 - 1017.0) = 1.67 \text{ kJ kg}^{-1}$$
$$h_2 = h_1 - w_p = 256.54 + 1.67 = 258.21 \text{ kJ kg}^{-1}$$

Boiler

$$q_H = h_3 - h_2 = 428.10 - 258.21 = 169.89 \text{ kJ kg}^{-1}$$

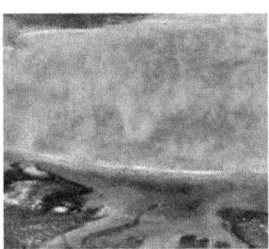

**Figure 4.14.** Geothermal sources. Source: Rank®, www.rank-orc.

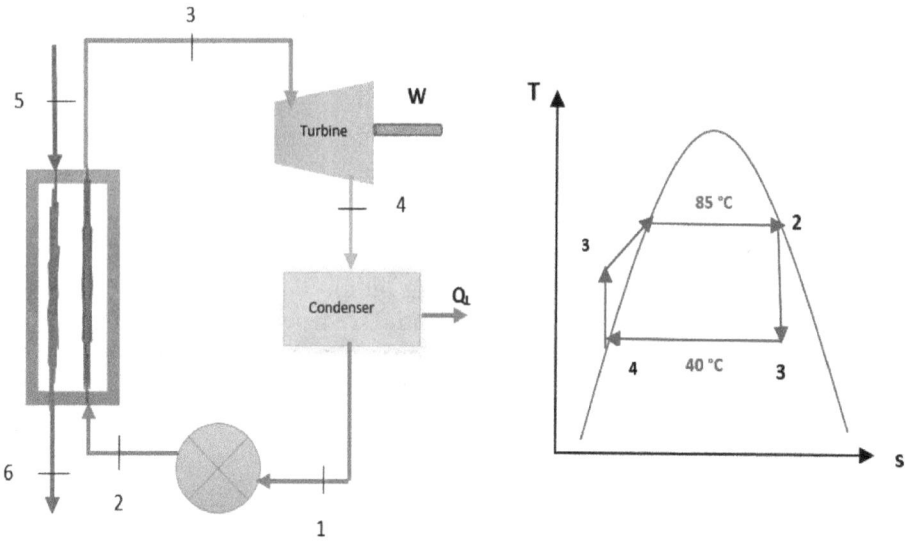

**Figure 4.15.** Rankine cycle and *T–s* diagram.

Turbine

$$s_4 = s_3 = 1.6782 = 1.1909 + x_4 \times 0.5214 \gg x_4 = 0.9346$$
$$h_4 = 256.54 + 0.9346 \times 163.28 = 409.14$$

*Energy equation:*

$$w_t = h_3 - h_4 = 428.1 - 409.14 = 18.96 \text{ kJ kg}^{-1}$$
$$w_{\text{net}} = w_t - w_p = 18.96 - 1.67 = 17.29 \text{ kJ kg}^{-1}$$
$$\eta_{\text{th}} = \frac{w_{\text{net}}}{q_H} = \frac{17.29}{169.89} = 0.102$$
$$\eta_{\text{th}} = 10\%.$$

**EXAMPLE 10** A single-flash geothermal power plant is shown in figure 4.16. A saturated liquid at 230 °C is supplied from the geothermal resource. 230 kg s$^{-1}$ of geothermal liquid is extracted from the production well and is flashed by an isenthalpic flashing process; the pressure will be 500 kPa, at which the steam is separated from the liquid in the separator and sent to the turbine. The steam leaves the turbine at 10 kPa with a moisture content of 5% and is then condensed in the condenser, it is routed to the reinjection well along with the liquid coming off the separator.

Determine
a) The turbine power output and the cycle thermal efficiency.

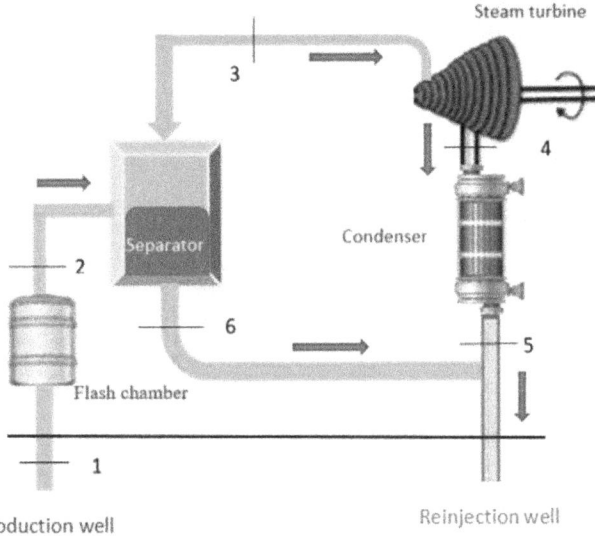

**Figure 4.16.** Schematic of a single-flash geothermal power plant.

b) The exergy of the geothermal liquid at the exit of the flash chamber, the exergy destruction, and the second-law efficiency for the flash chamber, the turbine, and the entire plant.

***Properties*** $T_2 = 230$ °C $P_2 = 500$ kPa $m = 230$ kg s$^{-1}$.
***Analysis*** The model used is a single-flash geothermal power plant and all processes are steady state with no changes in potential and kinetic energy. The diagram is shown in figure 4.16.
***Solution***
State 1 Obtain the following values from the saturated water-temperature tables at temperature $T_1 = 230$ °C.

$$h_1 = h_f = 990.14 \text{ kJ kg}^{-1} \quad s_1 = s_f = 2.6100 \text{ kJ kg}^{-1} \text{ K}^{-1}.$$

From state 1 enthalpy is constant $h_1 = h_2$.
Calculate the dryness fraction.

$$h_1 = h_f + x h_{fg}.$$

State 2 Obtain the following values from the saturated water-pressure tables at pressure

$$P_2 = 500 \text{ kPa}$$

$$h_f = 640.09 \text{ kJ kg}^{-1} \quad h_{fg} = 2108 \text{ kJ kg}^{-1}$$

$$s_f = 1.8603 \text{ kJ kg}^{-1} \text{ K}^{-1} \quad s_{fg} = 4.9603 \text{ kJ kg}^{-1} \text{ K}^{-1}$$

$$h_2 = h_f + x h_{fg}$$

$$990.14 = 640.09 + x * 2108$$

$$x_2 = 0.1661.$$

Calculate the entropy at state 2

$$s_2 = s_f + x_2 s h_{fg} = 1.8604 + 0.1661 * 4.9603$$

$$s_2 = 2.6841 \text{ kJ kg}^{-1} \text{K}^{-1}.$$

Calculate the mass flow rate at state 2 $\dot{m}_1 = 230 \text{ kg s}^{-1}$

$$m_2 = x_2 \dot{m}_1 = 0.1661 * 230 = 38.203 \text{ kg s}^{-1}.$$

State 3 Obtain the following values from the saturated water-pressure tables at pressure $P_3 = 500$ kPa

$$h_3 = h_g = 2748.1 \text{ kJ kg}^{-1} \quad s_3 = s_g = 6.8207 \text{ kJ kg}^{-1} \text{K}^{-1}.$$

State 4 Obtain the following values from the saturated water-pressure tables at pressure

$$P_4 = 10 \text{ kPa}$$

$$h_f = 191.81 \text{ kJ kg}^{-1} \quad h_{fg} = 2392.1 \text{ kJ kg}^{-1}$$

$$s_f = 0.6492 \text{ kJ kg}^{-1} \text{ K}^{-1} \quad s_{fg} = 7.4996 \text{ kJ kg}^{-1} \text{K}^{-1}.$$

Calculate the enthalpy:
$$h_4 = h_f + x h_{fg} \quad x \text{ is the dryness fraction} = 0.95$$

$$h_4 = 191.81 + 0.95 * 2392.1 = 2464.29 \text{ kJ kg}^{-1} \text{K}^{-1}.$$

Calculate the entropy:

$$s_2 = s_f + x s h_{fg} = 0.6492 + 0.95 * 7.4996$$
$$s_4 = 7.7739 \text{ kJ kg}^{-1} \text{K}^{-1}.$$

State 5 Obtain the following values from the saturated water-pressure tables at pressure
$P_3 = 500$ kPa.

$$h_5 = h_f = 640.09 \text{ kJ kg}^{-1} \quad s_5 = s_f = 1.8604 \text{ kJ kg}^{-1} \text{K}^{-1}.$$

Calculate the mass flow rate at state 6:

$$\dot{m}_6 = \dot{m}_1 - \dot{m}_3 = 230 - 38.203 = 191.79 \text{ s}^{-1}.$$

Calculate the work down by the turbine:

$$W_t = \dot{m}_3(h_3 - h_4) = 38.203 * (2748.1 - 2464.3)$$
$$W_t = 10842 \text{ kW}.$$

Obtain the following values from the saturated water tables at $T_0 = 25$ °C.

$$h_0 = h_f = 104.83 \text{ kJ kg}^{-1} \quad s_0 = s_f = 0.3672 \text{ kJ kg}^{-1} \text{K}^{-1}.$$

Calculate the input energy from the turbine:

$$Q_{\text{in}} = \dot{m}_1(h_1 - h_0) = 230 * (990.14 - 104.83)$$
$$Q_{\text{in}} = 203621.3 \text{ kW}.$$

Calculate thermal efficiency of the plant:

$$\eta_{\text{th}} = \frac{W_t}{Q_{\text{in}}} = \frac{10842}{203621.3} = 0.053$$
$$\eta_{\text{th}} = 5.3\%.$$

Calculate the specific exergy at state 1:

$$\varphi_1 = h_1 - h_0 - T_0(s_1 - s_0)$$
$$\varphi_1 = 990.14 - 104.83 - [(25+273) * (2.6100 - 0.3672)]$$
$$\varphi_1 = 216.53 \text{ kJ kg}^{-1}.$$

Calculate the specific exergy at state 2:

$$\varphi_2 = h_2 - h_0 - T_0(s_2 - s_0)$$
$$\varphi_2 = 990.14 - 104.83 - [(25+273) * (2.6841 - 0.3672)]$$
$$\varphi_2 = 194.44 \text{ kJ kg}^{-1}.$$

Calculate the specific exergy at state 3:

$$\varphi_3 = h_3 - h_0 - T_0(s_3 - s_0)$$
$$\varphi_3 = 2748.1 - 104.83 - [(25+273) * (6.8207 - 0.3672)]$$
$$\varphi_3 = 719.10 \text{ kJ kg}^{-1}.$$

Calculate the specific exergy at state 4:

$$\varphi_4 = h_4 - h_0 - T_0(s_4 - s_0)$$
$$\varphi_4 = 2464.3 - 104.83 - [(25+273) * (7.7739 - 0.3672)]$$
$$\varphi_4 = 151.05 \text{ kJ kg}^{-1}.$$

Calculate the specific exergy at state 6:

$$\varphi_6 = h_6 - h_0 - T_0(s_6 - s_0)$$
$$\varphi_6 = 640.09 - 104.83 - [(25+273) * (1.8604 - 0.3672)]$$
$$\varphi_6 = 89.97 \text{ kJ kg}^{-1}.$$

Calculate the exergy of the geothermal liquid at state 6:

$$\phi_6 = \dot{m}_6 \varphi_6 = 191.79 * 89.97$$
$$\phi_6 = 17255 \text{ kW}.$$

Calculate the exergy destruction of the flash chamber:

$$I_{FC} = \dot{m}_1(\varphi_1 - \varphi_2) = 230 * (216.53 - 194.44)$$
$$I_{FC} = 5080 \text{ kW}.$$

Calculate the efficiency of the flash chamber:

$$\eta_{FC} = \frac{\varphi_2}{\varphi_1} = \frac{194.44}{216.53} = 0.898.$$

Therefore, the efficiency of the flash chamber is

$$\eta_{FC} = 89.9\%.$$

Calculate the exergy destruction of the turbine:

$$I_t = \dot{m}_3(\varphi_3 - \varphi_4) - \dot{W}_t = 38.203 * (719.10 - 151.05) - 10842$$
$$I_t = 10854 \text{ kW}.$$

Calculate the efficiency of the turbine:

$$\eta_t = \frac{W_t}{\dot{m}_3(\varphi_3 - \varphi_4)} \text{turbine} \quad is = \frac{10842}{38.203 * (719.10 - 151.05)} = 0.50.$$

Therefore, the efficiency of the

$$\eta_{FC} = 50.0\%.$$

Calculate the exergy input of the plant.

$$\phi_{in, \text{ plant}} = \dot{m}_1 \varphi_1 = 230 * 216.53.$$

Therefore, the exergy input of the plant is

$$\phi_{in, \text{ plant}} = 49802 \text{ kW}.$$

Calculate the exergy destruction of the plant:

$$I_{in, \text{ plant}} = \phi_{in, \text{ plant}} - \dot{W}_t = 4980 - 10842.$$

Therefore, the exergy destruction of the plant is

$$I_{in, \text{ plant}} = 38960 \text{ kW}.$$

Calculate the efficiency of the plant:

$$\eta_{II} = \frac{W_t}{\phi_{in, \text{ plant}}} = \frac{10842}{49,802} = 0.2177$$
$$\eta_{II} = 21.77\%.$$

## 4.7 Biomass power plant

Biomass is green and does not increase the $CO_2$ level because the atmospheric $CO_2$ gas is captured in photosynthesis. Biomass is available worldwide and can be used to produce electricity on a wide scale (from 1MW to 15 MW). Commonly for large-scale biomass power generation, the most effective way to generate power is co-fired with coals, achieving 45% efficiency. Meanwhile, in a solid biomass-fired plant, the fuel is typically combusted to generate steam using a steam turbine (Rankine cycle). The low working pressures in the organic Rankine cycle power plants have overcome the problem of high costs for requirements such as steam boilers. A further advantage is the engine's long working life due to the working fluid's characteristics which are different from steam as it does not corrode the turbine blades. The organic Rankine cycle overcomes the issue of the small quantity of input fuel available in many regions because a small-sized power plant utilizes an efficient cycle.

**EXAMPLE 11** Biomass is burnt to generate water steam in an electric power plant. Assume the system runs at an ideal Rankine cycle as shown in figure 4.17. The inlet turbine condition is 12 MPa and the saturated vapor mass flow rate is 120 kg s$^{-1}$. The condenser operates at 4.0 kPa. Determine:
   a) The power output and the heat transfer in the boiler in kW,
   b) The thermal efficiency,
   c) The flow rate of the condenser cooling water in kg s$^{-1}$ if the cooling water is subject to a temperature increase of 20 °C with a constant pressure through the condenser.

*Properties* $T_2 = 300$ K $P_1 = 12$ MPa $P_2 = 4.0$ kPa $\dot{m}_{t,\,in} = 120$ kg s$^{-1}$
Specific heat of water ($C_p$) at 300 K is 4.179 kJ kg$^{-1}$.
*Analysis* The model used in this example is a steam power plant powered by biomass fuel. The diagram is shown in figure 4.15.
*Solution*
   State 1, using the steam table at $P_1 = 12$ MPa $s_{fg} = 5.4922$ kJ kg$^{-1}$ K$^{-1}$
   process 1–2 isentropic $h_{fg}$ at 12 MPa = 2684.6 kJ kg$^{-1}$
   Specific entropy of wet steam for state 2 at 0.004 MPa = $s_{2,\,wet} = s_g = 5.4922$ kJ kg$^{-1}$K$^{-1}$
   $s_{2g}$ = specific entropy of dry steam at 0.004 MPa = 8.4922 kJ kg$^{-1}$ K$^{-1}$
   (by interpolation between 2 states)
   $s_{2l}$ = specific entropy of saturation liquid at 0.004 MPa = 0.4102 kJ kg$^{-1}$ K$^{-1}$

$$s_{2,\,wet} = s_{2g}x_2 + (1 - x_2)s_{2,\,l}$$
$$x_2 = \frac{s_{2,\,wet} - s_{2,\,l}}{s_{2,\,g} - s_{2,\,l}} = \frac{5.4922 - 0.4102}{8.4922 - 0.4102} = 0.6288 = 0.63\%$$

$h_{2,g}$ at 0.004 MPa = 2551.94 kJ kg$^{-1}$ $h_{2,l}$ at 0.004 MPa = 117.946 kJ kg$^{-1}$

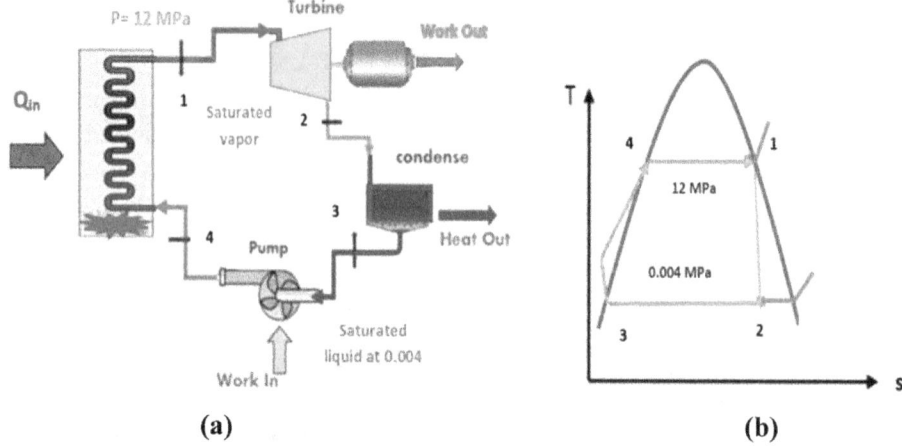

**Figure 4.17.** (a) Schematic representation of steam cycle. (b) $T$–$s$ diagram.

$$h_{2,\,wet} = h_{2g}x_2 + (1 - x_2)h_{2,\,l}$$
$$h_{2,\,wet} = (2551.94)(0.6288) + (1-0.6288) * 117.946$$
$$h_{2,\,wet} = 1648.44 \text{ kJ kg}^{-1}.$$

Work done per kg of steam entering turbine.

$$w_t = \Delta h = h_{1,\,g} - h_{2,\,wet}$$
$$w_t = 2684.6 - 1648.44 = 1036.16 \text{ kJ kg}^{-1}.$$

Steam flow rate $= 120$ kg s$^{-1}$

$$W_t = m * w_t = 120 * 1036.16 = 124339.2 \text{ kW}.$$

b) The rate of heat transfer to the steam passing through the boiler $q = h_{1,\,g} - h_{4,\,l}$ ignoring the work done by pump $h_{4,\,l} = h_{3,\,l}$.

$$Q_{in} = m * \left(h_{1,\,g} - h_{4,\,l}\right) = 120 * (2684.6-117.946)$$

$$Q_{in} = 307998.48 \text{ kW}.$$

a) Thermal efficiency $\eta_{th} = \dfrac{W_t}{Q_n} = \dfrac{124339.2}{307998.48} = 0.40$

$$\eta_{th} = 40\%$$

b) Mass flow rate absorbed by cooling water is given by

$$Q_{out} = mc\Delta T$$

where $m$ is the cooling water mass flow rate, $C_p = 4.179$ kJ kg$^{-1}$ K$^{-1}$, $\Delta T = 20$ K

$$Q_{out} = \dot{m} * (h_{2,\,wet} - h_{2,\,l}) = 120 * (1648.44 - 117.946)$$

$$Q_{out} = 183659.28 \text{ kW}$$

$$\dot{m} = \frac{183659.28}{4.179 * 20} = 2197.4 = 2.2 * 10^3 \text{ kg s}^{-1}.$$

## 4.8 How could the efficiency of the Rankine cycle be increased?

The majority of electric power on the globe is produced by steam power plants, with a small improvement in thermal efficiency yielding significant savings to the fuel requirements. For this reason, most efforts are made to improve the steam power plants thermal efficiency. The fundamental modifications to improve the power cycle thermal efficiency are either to increase the average temperature during heat addition to the working fluid in the boiler which should be as high as possible or decrease the average temperature as low as possible during heat rejection from the working fluid in the condenser.

## 4.9 Rankine cycle in ocean thermal energy conversion (OTEC)

A new clean technology known as ocean thermal energy conversion (OTEC) works between the ocean temperatures difference, cold deep water, and warm surface water in the tropics and subtropics to generate unbounded energy without fossil fuel fuels (figure 4.18).

Such a cycle boasts a competitive advantage over alternative electricity production sources; it is available, affordable, and clean [14].

The world's oceans surface absorbs about 80% of the Sun's solar energy and is replenished daily under any weather condition. Processes in a closed cycle ocean thermal energy conversion start with water flowing through a large pipe into the heat exchanger. A liquid such as ammonia is heated with a low boiling point, thus creating steam, it turns a turbine generator to produce electricity. A second pipe draws out cool water from the deep ocean. This cold water condenses the steam and back to liquid. The process repeats as the ammonia is recycled, creating unlimited clean energy 24 h a day, 365 days a year (figure 4.19).

The $T$–$s$ diagram of the ocean thermal energy conversion system is presented in figure 4.20.

The conventional ocean thermaleEnergy conversion cycle is a closed cycle with an evaporator, a turbine, a condenser, three designated pumps for working fluid, surface seawater, and deep seawater, as shown in figure 4.19. The warm ocean water evaporates the working fluid while the cold ocean water condenses the working fluid. At the ocean thermal energy conversion (OTEC) be made up of four main components site, the warm surface ocean water temperature varies seasonally at 24 °C–30 °C, while the cold deep ocean water remains 5 °C–9 °C. The temperature

**Figure 4.18.** Ocean thermal energy source.

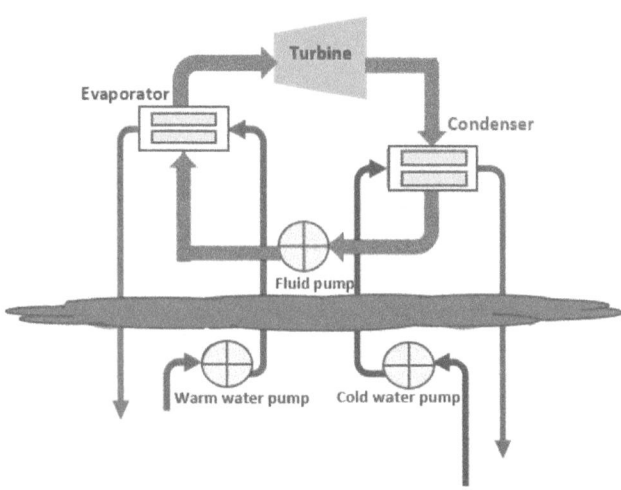

**Figure 4.19.** Closed cycle ocean thermal energy conversion.

difference in the ocean water needs approximately 20 °C to keep the ocean thermal energy conversion system operates at constant power [15].

Since the surface seawater provides abundant energy, the preheating is performed using an outside heat source. To deal with efficiency drop, a multi-stage turbine, which utilizes an external heat source to perform interstage superheating has been proposed. However, the multi-stage expansion will require a higher amount of cooling for the condenser, increasing the mass flow rate and power required to pump deep seawater. Therefore, a closed Rankine cycle engine is used with a closed condenser loop to minimize the possible negative effect on the underwater environment due to deep seawater upwelling. In a commercial plant, titanium is used as the material for the pipe to overcome biofouling. This adjustment will also minimize the

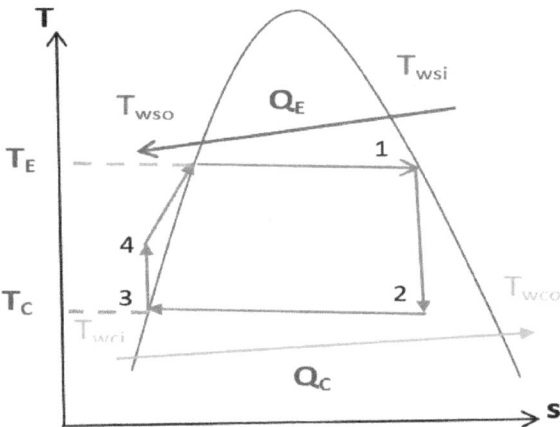

**Figure 4.20.** Ocean thermal energy conversion cycle $T$–$s$ diagram.

biofouling effect on the inside of the heat exchanger tube, allowing cheaper materials to manufacture the condenser.

### 4.9.1 Tyap of ocean thermal energy conversion (OTEC) cycles

The ocean thermal energy conversion (OTEC) can generate power continuously if the Sun keeps heating the surface seawater. The three forms of ocean thermal energy conversion (OTEC) plants are a closed cycle, open cycle, and hybrid cycle. The closed cycle uses a temperature difference to drive a turbine and subsequently generate electricity by vaporizing and condensing a working fluid. In the open cycle, warm surface water enters a vacuum chamber where it is flash-evaporated, the vapor drives a turbine to produce electricity. The remaining water vapor becomes desalinated water which is condensed using cold seawater, and this pure water can be collected and supplied for human use or returned to the ocean. The hybrid cycle combines features of both the closed and the open cycles with great potential to gain higher efficiencies to produce energy and fresh water.

#### 4.9.1.1 Closed cycle ocean thermal energy conversion
The ocean thermal energy conversion (OTEC) plant is a closed Rankine cycle as shown in figure 4.19, it uses similar working fluids to those used in refrigeration applications. Warm surface seawater is used to evaporate the working liquid at high pressure and then the working fluid is expanded in a turbine that drives a generator to produce power. After expansion, the vapor is condensed in a condenser using the cold deep seawater as the coolant. Finally, a liquid pump pressurizes the condensed ammonia from the low-pressure condenser to the high-pressure evaporator and then the process starts over.

#### 4.9.1.2 Open cycle ocean thermal energy conversion
An open cycle works in the same manner as the closed Rankine cycle except that warm seawater alone is employed as the working fluid. Warm seawater is

evaporated in the evaporator, and the resulting vapor is used to drive a low-pressure steam turbine and is condensed by the cold seawater in the condenser. In this version, the condensate, desalinated water, is a byproduct that can be used as a freshwater source. Open cycle ocean thermal energy conversion employs direct heat exchangers which has significant economic benefits to the closed Rankine cycle OTEC. Such a cycle needs to work under partial vacuum conditions and without non-condensable gas. This operation increases power consumption and consequently lowers efficiency. Furthermore, the low density of steam requires massive turbines for significant power output. For these reasons, most of the ocean thermal energy conversion (OTEC) cycles operate on the closed Rankine cycle (figure 4.21).

### 4.9.2 Working fluid of Rankine cycle

The working fluid performs a necessary function in the cycle. A favorable working fluid should have the essential thermophysical properties needed for the cycle operation and should stay chemically stable at the preferred temperature range. Working fluid choice affects the system efficiency, working conditions, environmental impact, and monetary viability. The essential factors influencing this system's thermodynamic and thermophysical requirements are boiling point, latent heat, specific heat, thermal conductivity, flash point, boiling temperature, chemical stability, and toxicity.

Ammonia is the best working fluid because it has a suitable boiling temperature (28 °C–32 °C) for the ocean thermal energy conversion (OTEC) cycle purpose, however, it is poisonous and can be hazardous to the environment. With the current improvement of working fluids, they are now less dangerous, with zero ozone depletion potential we can replace ammonia with such fluids, these include hydrocarbons (HCs) and hydrofluorocarbons (HFCs). The best performing working fluid is R-123, but it contributes to ozone depletion and global warming. Isopentane is considered an environmentally friendly working fluid for the system and the

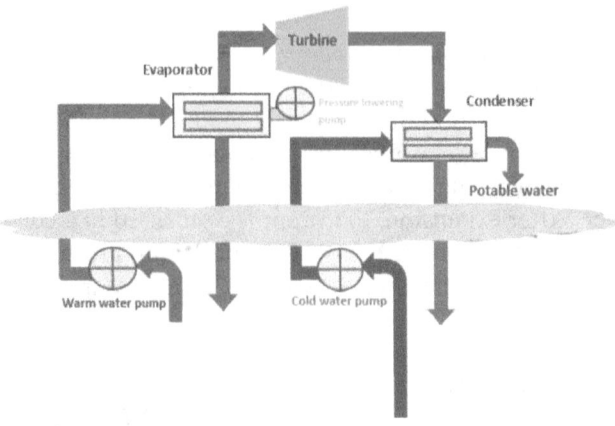

**Figure 4.21.** Open cycle ocean thermal energy conversion.

second-best in performance. The selection of working fluid becomes more manageable when its properties are investigated.

### 4.9.3 Thermodynamic analysis

The four main components of the ocean thermal energy conversion (OTEC) cycle are the evaporator, turbine, condenser, and coolant pump. The thermodynamic analysis of the performance of the ocean thermal energy conversion (OTEC) cycle made up of the four main components is given as follows [16]:

    i. Each of the components are in a steady state.
    ii. Pressure drops and heat losses are neglected.
    iii. The system is fully insulated.
    iv. Isentropic efficiency is given for all pumps and turbines.

The general mass and energy equilibrium equations are depicted below, respectively.

$$\sum \dot{m}_{\text{in}} = \sum \dot{m}_{\text{out}} \tag{4.62}$$

$$E_{\text{in}} = E_{\text{out}} \tag{4.63}$$

$$Q + W + \sum m_{\text{in}}\left(h_{\text{in}} + \frac{v_{\text{in}}^2}{2} + gz_{\text{in}}\right) = Q_{\text{out}} \sum m_{\text{out}}\left(h_{\text{out}} + \frac{v_{\text{out}}^2}{2} + gz_{\text{out}}\right). \tag{4.64}$$

All processes are steady state and potential and kinetic energy are neglected.

$$W_{\text{in}} + Q_{\text{in}} = W_{\text{out}} + Q_{\text{out}}. \tag{4.65}$$

The pump energy balance equation is written as

$$W_p = \dot{m}_{\text{in}}(h_{\text{out}} - h_{\text{in}}) = \dot{m}_{\text{in}}v(P_2 - P_1) \tag{4.66}$$

where, $\dot{m}_{\text{in}}$ is the mass flow rate into the working fluid pump or the seawater mass flow rate into the seawater pump.

Heat rate absorbed from the warm seawater, $Q_{e,\,\text{ws}}$ is,

$$Q_{e,\,\text{ws}} = m_{\text{ws}}c_p\Delta T_{\text{ws}} \tag{4.67}$$

where $m_{\text{ws}}$ is the mass flow rate of warm seawater.

Heat rate rejected into the cold seawater, $Q_{c,\,\text{cw}}$ is

$$Q_{c,\,\text{cw}} = m_{\text{cs}}c_p\Delta T_{\text{cs}} \tag{4.68}$$

where, $m_{\text{cs}}$ is the mass flow rate of cold seawater.

The heat rate supplied to the cycle (**evaporator**) is given as

$$Q_e = \dot{m}_{\text{wf}}\,(h_{\text{out}} - h_{\text{in}}) \tag{4.69}$$

where, $\dot{m}_{\text{wf}}$ is working fluid mass.

Heat rejection rate from the condenser of the cycle, $Q_c$ is written as

$$Q_c = \dot{m}_{wf} \ (h_{in} - h_{out}). \tag{4.70}$$

The energy balance in the turbine is written as

$$W_T = \dot{m}_{wf}(h_{in} - h_{out}). \tag{4.71}$$

The **net work done** in a **OTEC cycle** is given by

$$W_{net} = W_t - W_{ws} - W_{cs} - W_{wf}. \tag{4.72}$$

Thermodynamic cycle efficiency:

$$\eta_t = \frac{W_{net}}{Q_{in}}. \tag{4.73}$$

**EXAMPLE 12** Ocean thermal energy conversion (OTEC), shown below, is a renewable energy conversion technology that makes use of the temperature difference between surface seawater and deep seawater for power generation. OTEC is based on the Rankine cycle that uses a refrigerant as the working fluid to produce work with heat from low-temperature thermal energy. Let us consider using R-134a for OTEC. The pump increases the pressure of R-134a to 500 kPa, the turbine inlet temperature is at 25 °C, and the condenser pressure is 360 kPa. Calculate the working fluid mass flow rate needed if the cycle output is 1 MW of power.

*Properties* $T_2 = 25$ °C $= 298$ $K$ $P_1 = P_4 = 100$ kPa $P_2 = P_3 = 360$ kPa
$T_3 = 5$ °C power output $= 1$ MW.
*Analysis* The model used is ocean thermal energy conversion (OTEC). The diagram is shown in figure 4.22. Refrigerants R-134a
*Solution*
$P_2 = P_3 = 500$ kPa, $P_1 = P_4 = 360$ kPa, $T_3 = 25$ °C
Power $= 1$ MW
$h_1 = h_f$ at 360 kPa $= 57.82$ kJ kg$^{-1}$ $v_1 = v_f$ at 360 kPa $= 0.000\,7839$ m$^3$ kg$^{-1}$

$$w_p = v_1(P_2 - P)$$

$$w_p = 0.0007839(500-360) = 0.1097 \text{ kJ kg}^{-1}$$

$$h_2 = w_p + h_1 = 0.1097 + 57.82 = 57.93 \text{ kJ kg}^{-1}$$

$h_3 = h$ at 500 kPa, 25 °C $= 265.31$ kJ kg$^{-1}$ $s_3 = s$ at 500 kPa, 25 °C $= 0.9431$ kJ kg$^{-1}$ K$^{-1}$
$s_g$ at 360 kPa $= 0.9160 < 0.9431 \Longrightarrow$ state 4 is superheated
$h_4 = h$ at 360 kPa, 0.9431 kJ kg$^{-1}$ K$^{-1}$ $= 258$ kJ kg$^{-1}$

$$power = \dot{m} * w_{net} = \dot{m}[(h_3 - h_4) - (h_2 - h_1)]$$

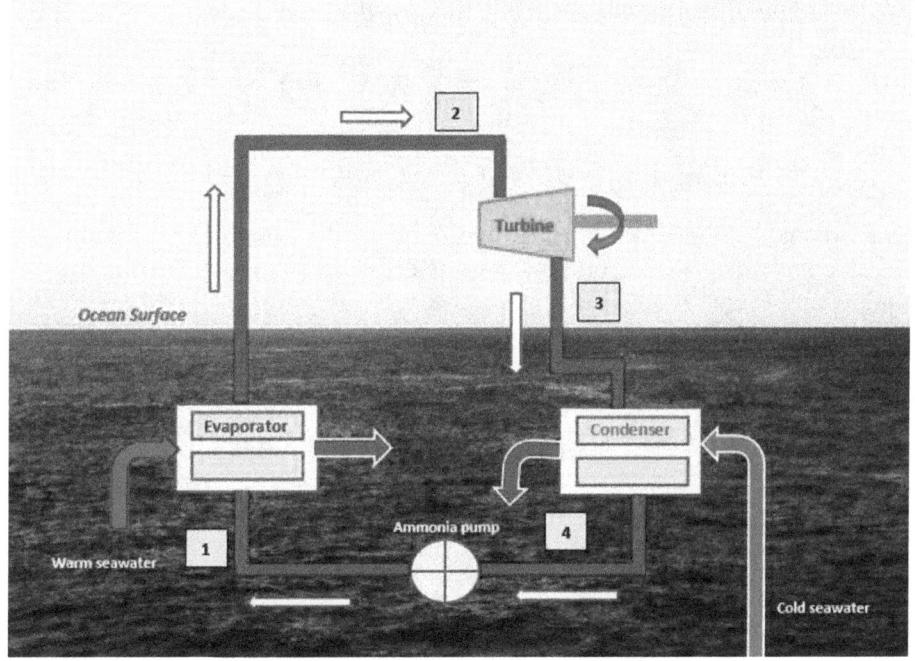

**Figure 4.22.** Ocean thermal energy conversion (OTEC).

$$1000 = \dot{m} * (h_3 - h_4 - h_2 + h_1)$$

$$\dot{m} = \frac{1000}{(265.31 - 258 - 57.93 + 57.83)} = \frac{1000}{7.2}$$

$$m = 138.89 \text{ kg s}^{-1}.$$

### 4.9.4 Exergy analysis

In this section, exergy thermodynamic laws apply to the ocean thermal energy conversion (OTEC) cycle. The irreversibility rate is defined as the cycle's loss values. According to each control volume's inlet and outlet, thermodynamic calculations are primarily dependent on the mass, energy, entropy, and exergy equilibrium equations, therefore, some assumptions of the thermodynamic process should be taken before calculation. The assumptions are:,

- No heat transfer from the cycle.
- The kinetic and potential energies are neglected for the thermodynamic calculation.
- The cycle runs at steady state process.
- Neglect any pressure losses in the pipes and system components.
- The working fluid is assumed to be a saturated liquid at the Ocean Thermal Energy Conversion (OTEC) cycle pump inlet.

The second law of thermodynamics equations is depicted below.

$$\sum \dot{m}_{in} s_{in} + S_{gen} + \sum \left(\frac{\dot{Q}_{in}}{T_s}\right)_{in} = \sum \dot{m}_{out} s_{out} + \sum \left(\frac{\dot{Q}_{out}}{T_s}\right)_{out} \qquad (4.74)$$

$$\sum \dot{m}_{in}\varphi_{in} + \phi_Q + \phi_W = \sum \dot{m}_{out}\varphi_{out} + \phi_Q + \phi_W + I \qquad (4.75)$$

where, 'in' and 'out' denotes the inlet and outlet of the component. In addition, $Q$, $W$, $S_{gen}$, $h$, and $I$ describe the rate of heat transfer, work, entropy generation, specific enthalpy, and exergy destruction rate, respectively.

$$\varphi = \varphi_{ph} + \varphi_{ch} + \varphi_{pt} + \varphi_{kn} \qquad (4.76)$$

where $\varphi_{ph}$, $\varphi_{ch}$, $\varphi_{pt}$, and $\varphi_{kn}$ denote the physical, chemical, potential, and kinetic exergy, and $\varphi_{ch}$, $\varphi_{pt}$, and $\varphi_{kn}$ are neglected. The physical exergy is

$$\varphi_{ph} = h - h_0 - T_0(s - s_0). \qquad (4.77)$$

The exergy value of heat transfer and work is

$$\phi_Q = \left(1 - \frac{T_0}{T}\right)Q \qquad (4.78)$$

$$\phi_W = W. \qquad (4.79)$$

The OTEC system is constructed according to the seawater temperature.

$$m_{in,\,ws} = m_{out,\,ws}, \; m_{in,\,cs} = m_{out,\,cs}, \; m_{in,\,wf} = m_{out,\,wf}. \qquad (4.80)$$

Pump:

$$\dot{m}_1 h_{in} + W_P = \dot{m}_2 h_2 \qquad (4.81)$$

$$\dot{m}_1 s_{in} + S_{gen,\,P} = \dot{m}_2 s_2 \qquad (4.82)$$

$$I_P = \dot{m}_1 \varphi_{in} - \dot{m}_2 \varphi_2 + W_P. \qquad (4.83)$$

Evaporator:

$$\dot{m}_{in,\,ws} h_{in,\,ws} + \dot{m}_{in,\,wf} h_{in,\,wf} = \dot{m}_{out,\,ws} h_{out,\,ws} + \dot{m}_{out,\,wf} h_{out,\,wf} \qquad (4.84)$$

$$\dot{m}_{in,\,ws} s_{in,\,ws} + \dot{m}_{in,\,wf} s_{in,\,wf} + S_{gen,\,P} = \dot{m}_{out,\,ws} s_{out,\,ws} + \dot{m}_{out,\,wf} s_{out,\,wf} \qquad (4.85)$$

$$I_e = \dot{m}_{in,\,ws}\varphi_{in,\,ws} + \dot{m}_{in,\,wf}\varphi_{in,\,wf} - \left(\dot{m}_{out,\,ws}\varphi_{out,\,ws} + \dot{m}_{out,\,wf}\varphi_{out,\,wf}\right). \qquad (4.86)$$

Turbine:

$$\dot{m}_{in,\,wf} s_{in,\,wf} + S_{gen,} = \dot{m}_{out,\,wf2} s_{out,\,wf} \qquad (4.87)$$

$$I_t = \dot{m}_{in,\,wf}\varphi_{in,\,wf} - \dot{m}_{out,\,wf}\varphi_{out,\,wf} + W_t. \qquad (4.88)$$

Condenser:

$$\dot{m}_{\text{in, cs}}h_{\text{in, cs}} + \dot{m}_{\text{in, wf}}h_{\text{in, wf}} = \dot{m}_{\text{out, cs}}h_{\text{out, cs}} + \dot{m}_{\text{out, wf}}h_{\text{out, wf}} \qquad (4.89)$$

$$\dot{m}_{\text{in, cs}}s_{\text{in, cs}} + \dot{m}_{\text{in, wf}}s_{\text{in, wf}} + S_{\text{gen, }P} = \dot{m}_{\text{out, cs}}s_{\text{out, cs}} + \dot{m}_{\text{out, wf}}s_{\text{out, wf}} \qquad (4.90)$$

$$I_c = \dot{m}_{\text{in, cs}}\varphi_{\text{in, cs}} + \dot{m}_{\text{in, wf}}\varphi_{\text{in, wf}} - \left(\dot{m}_{\text{out, cs}}\varphi_{\text{out, cs}} + \dot{m}_{\text{out, wf}}\varphi_{\text{out, wf}}\right). \qquad (4.91)$$

## 4.10 Brayton cycle

In 1872, an American engineer, George Bailey Brayton, made progress in the study of heat engines by patenting a constant pressure internal combustion engine, at first using vaporized gas but later using liquid fuels such as kerosene. The air-standard Brayton cycle is ideal. At the initial state, atmospheric pressure air enters a steady flow compressor inlet, which draws air into the engine and pressurizes in the compressor; then, compressed air feeds it to the combustion chamber at speeds of hundreds of miles per hour. The combustion system comprises a ring of fuel injectors that inject a steady flow of fuel into the combustion chamber, where it mixes with the air. The mixture is burned at very high temperatures. The combustion products at high temperature and high-pressure expand through the complicated array of stationary and rotating blades turbine. As hot combustion gas expands through the turbine, it spins the rotating blades to perform a dual function, drives the compressor, and turns a generator to produce electricity. The cycle shown in figure 4.26 is for constant specific heats on $P$–$v$ and $T$–$s$ diagrams.

### IDEAL BRAYTON CYCLE
1. Isentropic process, ambient air is drawn into the compressor and compressed to a higher temperature and pressure.
2. Isobaric process fuel and compressed air then run through a combustion chamber, where fuel is burned at constant pressure. The combustion products at high-temperature gases are sent to a turbine.
3. Isentropic process, the high-temperature gases expand through a turbine and produce power. Some of the work is used to drive the compressor.
4. Isobaric process, the exhaust gases leave the turbine.
5. The cycle's thermal efficiency is defined as the network delivered by the cycle divided by the heat added to the working fluid. Gas turbines are Brayton cycles with three components: a gas compressor to pressurize the air, a combustion chamber where heat is added to the fuel, and a turbine where the exhaust gases are expanded.

### Exergy flow diagram
As shown in the pressure-volume diagram, the constant pressure processes (isobaric processes) follow the isobaric lines for the gas that are shown in figure 4.23(b) as a horizontal line, and the adiabatic processes are represented by

paths connecting these horizontal lines, and all the paths illustrate the processes for a cycle. Figure 4.23 shows the temperature-entropy diagram ($T$–$s$ diagram). The diagram's points specify the thermodynamic state, the horizontal axis represents the specific entropy, and the vertical axis represents absolute temperature. $T$–$s$ diagrams are a useful tool because they help demonstrate the heat transfer during a process. For reversible processes, the area under the $T$–$s$ curve represents the heat transferred to the system during that process.

### IDEAL BRAYTON CYCLE
1. Compression via adiabatic process –
2. Heat addition via isobaric process –
3. Expansion via adiabatic process –
4. Heat rejection via isobaric process –

## THE FIRST LAW OF THERMODYNAMICS
The energy of the system can be neither be created nor destroyed; it just changes form. The first law of thermodynamics defines internal energy as a state function and provides a formal energy conservation statement. However, it gives no information about the direction in which processes can spontaneously occur.

$$Q_{C.V} + \sum \dot{m}_i\left(h_i + \frac{V_i^2}{2} + g. \, Z_i\right) = \sum \dot{m}_i\left(h_e + \frac{V_e^2}{2} + g. \, Z_e\right) + W_{C.V}. \qquad (4.92)$$

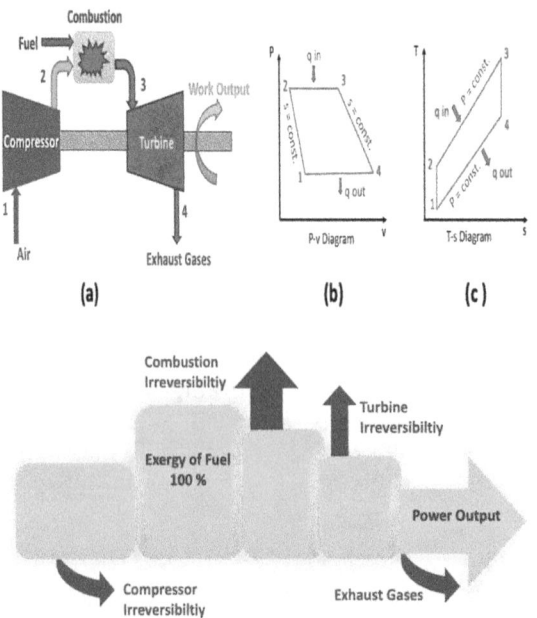

**Figure 4.23.** (a) Brayton cycle, (b) $P$–$v$ diagram, (c) $T$–$s$ diagram.

4-47

Thermal efficiency for the Brayton cycle is:

$$\eta_{th} = \frac{\text{Power output}}{\text{Heat input}} = \frac{Q_{add} - Q_{rej}}{Q_{add}} \tag{4.93}$$

$$\eta_{th} = 1 - \frac{Q_{reg}}{Q_{add}} = 1 - \frac{(T_4 - T_1)}{(T_3 - T_2)} \tag{4.94}$$

$$\frac{T_2}{T_1} = \left(\frac{P_2}{P_1}\right)^{\frac{(k-1)}{k}} = \left(\frac{P_3}{P_4}\right)^{\frac{(k-1)}{k}} = \frac{T_3}{T_4}. \tag{4.95}$$

Thus

$$\eta_{th} = 1 - \frac{1}{r_P^{\frac{(k-1)}{k}}} \tag{4.96}$$

$$r_P = \frac{P_2}{P_1} = \frac{P_3}{P_4} \tag{4.97}$$

where $r_P$ is called the pressure ratio and $k = cp/cv$ is the specific heat ratio.

**EXAMPLE 13** In a Brayton cycle, the air is used as a working fluid with inlet conditions of 27 °C, 0.1 MPa, and compressor pressure ratio 6.25. The cycle maximum temperature is 800 °C. Find
   (a) The compressor specific work,
   (b) The turbine specific work,
   (c) The specific heat supplied,
   (d) The cycle efficiency.

*Properties* $T_1 = 27$ °C $= 300$ K $P_1 = 100$ kPa $r_P = 6.25$ $T_3 = 800$ °C $= 1073$ K.
   *Analysis* The model used is an ideal gas with constant specific heat, at 300 K, and each process is steady state with no potential and kinetic energy changes. The diagram is shown in figure 4.24.
*Solution*
   Since process 1–2 is isentropic,

$$\frac{T_2}{T_1} = r_p^{(\gamma-1)/\gamma} = 6.25^{(1.4-1)/1.4} = 1.689$$

$$T_2 = 506.69 \text{ K}$$

a) The specific work of the compressor

$$w_c = c_p(T_2 - T_1) = 1.005 * (506.69 - 300)$$
$$w_c = 207.72 \text{ kJ kg}^{-1}$$

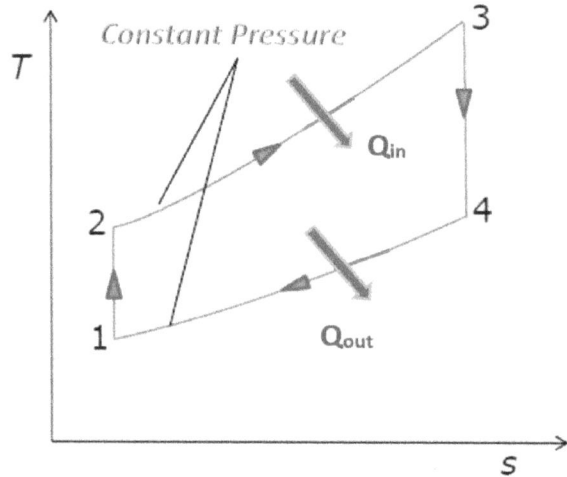

**Figure 4.24.** Brayton cycle T-s diagram.

Process 3–4 is also isentropic,

$$\frac{T_2}{T_4} = r_p^{(\gamma-1)/\gamma} = 6.25^{(1.4-1)/1.4} = 1.689$$

$$T_4 = 635.29 \text{ K}$$

b) The specific work of the turbine

$$w_t = c_p(T_3 - T_4) = 1.005 * (1073 - 635.29)$$

$$w_t = 439.89 \text{ kJ kg}^{-1}$$

c) The specific heat input

$$q_{in} = c_p(T_3 - T_2) = 1.005 * (1073-506.69)$$

$$q_{in} = 569.14 \text{ kJ kg}^{-1}$$

d) Cycle efficiency

$$\eta_{th} = \frac{(W_t - W_c)}{Q_{in}} = \frac{439.89-207.72}{569.14} = 0.408$$

$$\eta_{th} = 40.8\%.$$

**The second formula (Clausius):** It is not possible to transfer heat from a cold to a hot body, but it can happen if work is done at the same time, which could be explained throughout the working of an electric refrigerator. Entropy is a formula of the second law of thermodynamics that measures a physical property most associated with a state of disorder, randomness, or uncertainty. The entropy does not change for reversible processes.

$$\Delta S_{\text{rev}} = 0. \tag{4.98}$$

The entropy increases for irreversible processes.

$$\Delta S_{\text{irrev}} = +. \tag{4.99}$$

The entropy of the whole (closed) system (Sun, Earth, and space) always increases.

$$\Delta S \neq -. \tag{4.100}$$

The entropy as mentioned is a functioning state which depends on the initially and finally state system therefore the alteration will be as

$$\Delta S = S_2 - S_1. \tag{4.101}$$

For a steady state steady flow process, control volume entropy rate balance may be written as

$$\sum j \frac{Q_j}{T_j} + \sum i \dot{m}_i \cdot s_i - \sum e \dot{m}_e \cdot s_e + S_{\text{gen}} = 0. \tag{4.102}$$

### 4.10.1 Exergy analysis

Exergy analysis is a useful tool for power plant development and is based on thermodynamic quantities (entropy and enthalpy). The main target of the investigation of exergy in the power plants' applications is detecting and effectively reducing every part of the plant's irreversibility. Exergy analysis is a powerful tool for developing, evaluating and improving energy efficient systems. Generally, analysis of exergy is the appropriate method for any system that required work and heat transfer. Exergetic efficiency displays the real system effectiveness.

Chemical exergy standard tables have been published for a vast amount of types accounting for Gibbs free energy and an approximation of universal concentration exergy. If neglecting the nuclear, magnetic, electrical, and surface tension effects, a system's total exergy will be four components: physical exergy $\phi_{\text{PH}}$, kinetic exergy $\phi_{\text{KN}}$, potential exergy $\phi_{\text{PT}}$, and chemical exergy $\phi_{\text{CH}}$.

$$\phi = \phi_{\text{PH}} + \phi_{\text{KN}} + \phi_{\text{PT}} + \phi_{\text{CH}}. \tag{4.103}$$

Equation (4.92) can be expressed on unit-of-mass basis.

$$\varphi = \varphi_{\text{PH}} + \varphi_{\text{KN}} + \varphi_{\text{PT}} + \varphi_{\text{CH}} \tag{4.104}$$

Kinetic exergy

$$\varphi_{\text{KN}} = \frac{1}{2} \cdot v^2 \tag{4.105}$$

Potential exe

$$\varphi_{PT} = g \cdot z \qquad (4.106)$$

Chemical exergy

$$\varphi_{CH} = \sum_i x_i \varphi_{ch, i} + RT_0 \sum X_i \ln x_i \qquad (4.107)$$

Physical flow exergy for simple compressible pure substances is given as

$$\varphi_{PT} = (h - h_0) - T_0(s - s_0). \qquad (4.108)$$

Exergy analysis for the cycle processes, the control volumes of turbines, compressors, heat exchangers, pipes, and ducts operate at a steady state. Thus, there are no changes in their mass, energy, entropy, exergy, and volumes. The total exergy of a steady flow that entering into a system due to heat, work, and mass transfer is equal to the summation of exergy leaving and the exergy destroyed. Then the rating form of the general exergy balance for a steady-flow process is

$$\sum_j \left(1 - \frac{T_0}{T_j}\right) \cdot Q_j - W_{CV} + \sum \dot{m}_i \cdot \varphi_i - \sum \dot{m}_e \cdot \varphi_e - \dot{D}_D = 0. \qquad (4.109)$$

The exergy destruction is a measure of the thermodynamic imperfection of a process and is expressed in terms of lost work potential. In general, the expression for exergy destruction as articulated by Kotas (1995) is

$$I = T_0 \left[ (S_{ex} - S_{in}) - \sum_{i=1}^{n} \frac{Q_i}{T_i} \right]. \qquad (4.110)$$

### 4.10.2 Exergy analysis of the Brayton cycle

There are many definitions for exergy efficiency. The following equation represents the exergy balance of the whole cycle.

$$\phi_{fuel} = W_{net} + \phi_{fg} + \Delta\phi \qquad (4.111)$$

where $\phi_{fuel}$ is exergy of fuel, $\phi_{fg}$ exergy of flue gases, and $\Delta\phi$ is exergy destruction.

#### EXERGY OF FUEL

The fuel is in thermal and mechanical equilibrium with the environment ($T = T_0$, $P = P_0$), but it is not in chemical equilibrium. The fuel at the ambient state has exergy because it is not in global equilibrium with the environment. Chemical reactions are possible. Chemical energy is released in the chemical reaction with oxygen from the ambient air, and this energy is converted into work or another energy form. The expression for the exergy of fuel at environment temperature and pressure is the following,

$$\phi_{\text{fuel}}(P_2,\ T_0) = \phi_{\text{fuel}}(P_0,\ T_0) + R_m T_0 \ln\frac{P_2}{P_0} \qquad (4.112)$$

where $R_m$ is the universal gas constant.

Since the pressure $P_2$ is variable ($r_p$ is variable), the exergy of fuel is variable.

$$\eta_{\text{II}} = \frac{W_{\text{net}}}{\phi_{\text{fuel}}}. \qquad (4.113)$$

EXERGY EFFICIENCY

Exergy efficiency, defined as exergy output divided by the total exergy input, is often considered as exergy efficiency of the whole cycle.

Exergy flow rate of air at state 1

$$\phi_1 = \dot{m}_a[(h_1 - h_0) - T_0(s_0 - s_1)] \qquad (4.114)$$

$T_0,\ h_0,\ s_0$ are the properties referring to dead state.

$$\dot{I} = T_0 S_{\text{gem}}. \qquad (4.115)$$

Compressor: Entropy balance

$$s_{\text{in}} - s_{\text{out}} + s_{\text{gen}} = (\Delta s)_{\text{ch}}(\Delta s)_{\text{ch}} = 0 \qquad (4.116)$$

$$s_{\text{gen}} = s_{\text{out}} - s_{\text{in}} = s_2 - s_1 \qquad (4.117)$$

$$\phi_1 = \dot{m}_a[(h_{\text{out}} - h_{\text{in}}) - T_0(s_{\text{out}} - s_{\text{in}})]. \qquad (4.118)$$

*Exergy destruction (Irreversibility) of compressor*

$$I_c = T_0 \dot{m}_a(s_{s,\ \text{out}} - s_{\text{in}}). \qquad (4.119)$$

If the process of compression is not reversible, then $s_{s,\ \text{out}} - s_{\text{in}} \neq 0 \Rightarrow$ exergy destruction ($I_c$) $\neq 0$

*Exergy destruction ratio of compressor %=exergy destruction in compressor/exergy of fuel*

Similarly, for the turbine,

$$s_{\text{in}} - s_{\text{out}} + s_{\text{gen}} = (\Delta s)_{\text{ch}}(\Delta s)_{\text{ch}} = 0 \qquad (4.120)$$

$$s_{\text{gen}} = s_{\text{out}} - s_{\text{in}} = s_{s,\ \text{out}} - s_{\text{in}} \qquad (4.121)$$

$$\phi_t = (\dot{m}_a + \dot{m}_f)[(h_{s,\ \text{out}} - h_{\text{in}}) - T_0(s_{s,\ \text{out}} - s_{\text{in}})]. \qquad (4.122)$$

Exergy destruction (Irreversibility) of turbine

$$I_t = T_0(\dot{m}_a + \dot{m}_f)(s_{s,\ \text{out}} - s_{\text{in}}). \qquad (4.123)$$

If the expansion in the turbine is not reversible process, then $s_{s,\ \text{in}} - s_{\text{out}} \neq 0 \Rightarrow$ exergy destruction ($I_t$) will take place.

*Exergy destruction ratio of turbine %= exergy destruction in turbine/exergy of fuel*

Exergy of fuel

$$I_{fuel} = \left[1 - \frac{T_0}{T_{fuel}}\right] * \dot{m}_{fuel} * (HCV) \qquad (4.124)$$

Combustion Chamber: Exergy destruction

$$I_{c.c} = T_0 \dot{s}_{gem} \qquad (4.125)$$

$$I_c = T_0(\dot{m}s_{out} - \dot{m}s_{in}) \qquad (4.126)$$

$$I_c = T_0\left[\dot{m}_a s_{a,\,in} + \dot{m}_f s_{f,\,in} - (\dot{m}_a + \dot{m}_f)s_{out}\right]. \qquad (4.127)$$

Exergy destruction ratio of combustor %= exergy destruction in combustion chamber/exergy of fuel

The second-law efficiency of the cycle is

$$\eta_{II} = \frac{W_{net}}{Exergy\ \ of\ \ fuel}. \qquad (4.128)$$

Also

$$\eta_{II} = 1 - \frac{\sum \phi_{destroyed}}{Exergy\ \ of\ \ fuel} = 1 - \frac{I_c + I_{c.c} + I_t}{Exergy\ \ of\ \ fuel}. \qquad (4.129)$$

**EXAMPLE 14** A regeneration Brayton cycle with a working fluid of air as shown in the figure below, has a cycle pressure ratio of 7 and the compressor isentropic efficiency is 75%. The turbine exit temperature is 1150K and its isentropic efficiency is 82%. The cycle minimum temperature is 310 K and the generator effectiveness is 65%. Assuming the source and sink temperatures are 1800 K and 310 K respectively. Determine

(a) The temperature at the turbine exit.
(b) The work output.
(c) The thermal efficiency of the cycle.
(d) The total exergy loss that corresponds to the Brayton cycle.
(e) The exergy of the exhaust gases at the exit of the regenerator.

**Properties** Given, $r_p = 7$, $T_1 = 310$ K, $T_4 = 1150$ K

$$\eta_c = 75\% \ \ \eta_t = 82\% \ \ \epsilon = 70\%.$$

The gas turbine discussed in this example is equipped with a regenerator. For specified effectiveness, the thermal efficiency is to be determined. Analysis of the *T-s* diagram of the cycle is shown in figure 4.25. We first determine the enthalpy of the air at the exit of the compressor, using the definition of compressor efficiency.

***Analysis*** An ideal Brayton cycle is considered. Steady operation conditions, with no potential and kinetic energy change (figure 4.26).

***Solution***

Process 1–2 isentropic compression of ideal gas.

At $T_1 = 310$ K, from ideal gas properties of air

$$h_1 = 310.24 \text{ kJ kg}^{-1}, \; P_2 = 1.5546 \text{ kPa}$$
$$P_2 = r_p * P_1 = 7 * 1.23 = 10.88 \text{ kPa}.$$

At $P_2 = 10.88$ kPa, from ideal gas properties of air

$$h_{2.s} = 533.98 + (544.35 - 533.98)\left(\frac{10.88 - 10.37}{11.10 - 10.37}\right)$$
$$h_{2.s} = 541.256 \text{ kJ kg}^{-1}.$$

Isentropic efficiency of the compression is given by

$$\eta_{s,c} = \frac{h_{2,s} - h_1}{h_2 - h_1}$$
$$0.75 = \frac{541.256 - 310.24}{h_2 - 310.24}$$
$$h_2 = 618.26 \text{ kJ kg}^{-1}.$$

At $4 = 1150$ K, from ideal gas properties of air

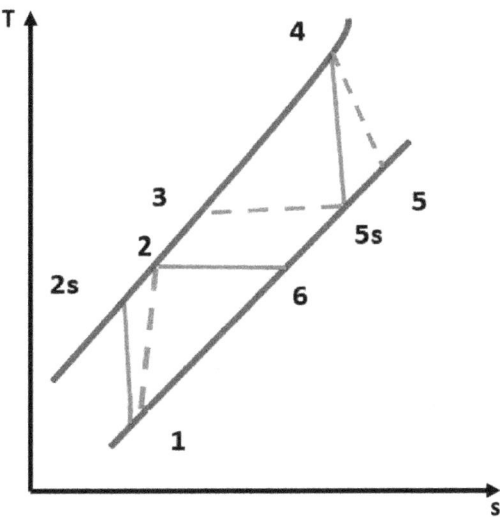

**Figure 4.25.** T-s diagram.

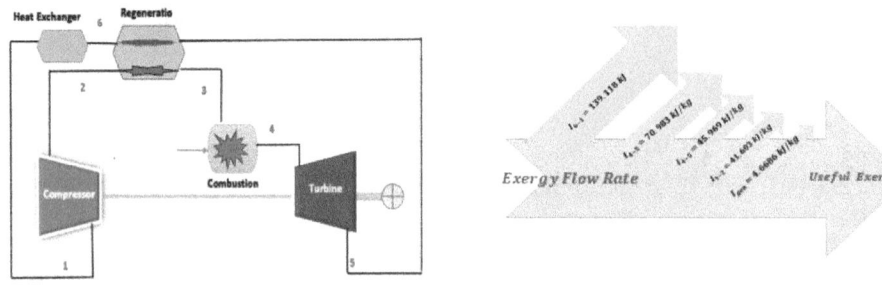

**Figure 4.26.** Regeneration gas turbine and irreversibility diagram.

$$h_4 = 1207.57 + (1230.92-1207.57)\left(\tfrac{1150-1140}{1160-1140}\right)$$

$$h_4 = 1219.245 \text{ kJ kg}^{-1}$$

$$P_4 = 193.1 + (207.2-193.1)\left(\tfrac{1150-1140}{1160-1140}\right)$$

$$P_4 = 200.15 kP.$$

Process 4–5 isentropic expansion of ideal gas.

$$\frac{P_4}{P_5} = r_p = \frac{200.15}{P_5} = P_5 = 28.592.$$

At $P_5 = 28.592$ from ideal gas properties of air

$$h_{5,s} = 702.52 + (713.27-702.52)\left(\tfrac{28.592-27.29}{28.80-27.29}\right)$$

$$h_{5,s} = 711.789 \text{ kJ kg}^{-1},$$

turbine Isentropic efficiency is given by

$$\eta_t = \frac{h_4 - h_5}{h_4 - h_{5,s}}$$

$$0.82 = \frac{1219.245 - h_5}{1219.245-711.789}$$

$$h_5 = 803.131 \text{ kJ kg}^{-1}.$$

(a) The exit temperature of air from turbine
At $h_5 = 803.131$ kJ kg$^{-1}$ from ideal gas properties of air

$$T_5 = 780 + (800-780)\left(\tfrac{803.13-800.03}{821.95-800.03}\right)$$

$$T_5 = 782.82 \text{ K}.$$

(b) Calculate the net work

$$\varepsilon = \frac{h_3 - h_2}{h_5 - h_2}$$

$$0.65 = \frac{h_3 - 618.26}{803.13 - 618.26}$$

$$h_3 = 738.425 \text{ kJ kg}^{-1}$$

$$w_{net} = (w_t - w_c) = (h_4 - h_5) - (h_2 - h_1)$$

$$w_{net} = (1219.245 - 803.131) - (618.26 - 310.24)$$

$$w_{net} = 108.094 \text{ kJ kg}^{-1}.$$

(c) The heat supplied is

$$q_{in} = (h_4 - h_3) = (1219.245 - 738.425)$$

$$q_{in} = 480.82 \text{ kJ kg}^{-1}.$$

The thermal efficiency is defined as

$$\eta_{th} = \frac{w_{net}}{q_{in}} = \frac{108.094}{480.82}$$

$$\eta_{th} = 0.2248 = 22.48\%.$$

(d) Using the ideal gas properties air table to find the entropies at 1 and 4

At $T = 310$ K $\dot{s}_1 = 1.73495$ kg kJ K$^{-1}$
At $T = 1150$ K $\dot{s}_4 = 3.02355$ kg kJ K$^{-1}$
Find $\dot{s}_2$ at $h_2 = 618.26$ kJ kg$^{-1}$

$$\dot{s}_2 = 2.42644 + (2.44356 - 2.42644)\left(\frac{618.26 - 617.53}{628.07 - 617.53}\right)$$

$$\dot{s}_2 = 2.42763 \text{ kJ kg}^{-1} \text{ K}^{-1}.$$

Find $\dot{s}_5$ at $h_5 = 803.13$ kJ kg$^{-1}$

$$\dot{s}_5 = 2.69013 + (2.70400 - 2.60319)\left(\frac{738.425 - 734.82}{745.62 - 734.82}\right)$$

$$\dot{s}_5 = 2.69405 \text{ kJ kg}^{-1} \text{ K}^{-1}.$$

Find $\dot{s}_3$ at $h_3 = 738.425$ kJ kg$^{-1}$

$$\dot{s}_3 = 2.60319 + (2.61803 - 2.60319)\left(\frac{738.425 - 734.82}{745.62 - 734.82}\right)$$

$$\dot{s}_3 = 2.60814 \text{ kJ kg}^{-1} \text{ K}^{-1}.$$

Calculate the heat rejection by the regenerative

$$q_{out} = q_{in} - w_{net} = 480.82 - 108.094$$

$$q_{out} = 372.726 \text{ kJ kg}^{-1}.$$

Calculate the specific heat at state 6

$$h_6 = h_1 + q_{out} = 310.24 + 372.726$$

$$h_6 = 682.966.$$

Find $\dot{s}_6$ at $h_6 = 682.966$ kJ kg$^{-1}$

$$\dot{s}_6 = 2.5286 + (2.54175 - 2.52589)\left(\frac{682.966 - 681.14}{691.82 - 681.14}\right)$$

$$\dot{s}_6 = 2.5286 \text{ kJ kg}^{-1} \text{ K}^{-1}.$$

Calculate the exergy destruction associated with process 1–2

$$I_{1-2} = T_0(s_2 - s_1) = T_0\left(\dot{s}_2 - \dot{s}_1 - R\ln\frac{P_2}{P_1}\right)$$

$$I_{1-2} = 310(2.42763 - 1.73495 - 0.287\ln 7)$$

$$I_{1-2} = 41.603 \text{ kJ kg}^{-1}.$$

Calculate the exergy destruction associated with process 4–5

$$I_{4-5} = T_0(s_5 - s_4) = T_0\left(\dot{s}_5 - \dot{s}_4 - R\ln\frac{P_5}{P_4}\right)$$

$$I_{4-5} = 310\left(2.69405 - 3.02355 - 0.287\ln\frac{1}{7}\right)$$

$$I_{4-5} = 70.983 \text{ kJ kg}^{-1}.$$

Calculate the exergy destruction associated with regeneration process

$$I_{gen} = T_0[(s_3 - s_2) + (s_6 - s_5)]$$

$$I_{gen} = 310[(2.60814 - 2.42763) + (2.5286 - 2.69405)]$$

$$I_{gen} = 4.6686 \text{ kJ kg}^{-1} \text{ K}^{-1}.$$

Calculate the exergy destruction associated with process 3–4

$$I_{3-4} = T_0\left(s_4 - s_3 + \frac{-q_{in}}{T_H}\right) = T_0\left(\dot{s}_4 - \dot{s}_3 - R\ln\frac{P_4}{P_3} - \frac{q_{in}}{T_H}\right)$$

$$I_{3-4} = 310\left(3.02355 - 2.60814 - 0.287\ln1 - \frac{480.82}{1800}\right)$$

$$I_{4-5} = 45.969 \text{ kJ kg}^{-1}.$$

Calculate the exergy destruction associated with process 6–1

$$I_{6-1} = T_0\left(s_1 - s_6 + \frac{q_{\text{out}}}{T_L}\right) = T_0\left(\dot{s}_1 - \dot{s}_6 - R\ln\frac{P_1}{P_6} - \frac{q_{\text{out}}}{T_L}\right)$$

$$I_{6-1} = 310\left(1.73495 - 2.5286 - 0.287\ln1 - \frac{372.726}{300}\right)$$

$$I_{6-1} = 139.118 \text{ kJ kg}^{-1}.$$

Calculate the change of entropy for the exit of the regenerator.

$$(s_6 - s_0) = \left(\dot{s}_6 - \dot{s}_0 - R\ln\frac{P_6}{P_1}\right)$$

$$(s_6 - s_0) = (2.5286 - 1.73498 - 0.287\ln1)$$

$$(s_6 - s_0) = 0.79362 \text{ kJ kg}^{-1}\text{K}^{-1}.$$

Calculate the stream exergy at the exit of the regenerator.

$$\phi_6 = (h_6 - h_0) - T_0(s_6 - s_0)$$

$$\phi_6 = (682.966 - 310.24) - 310(0.79362).$$

Thus, the exergy of the exhaust gases is

$$\phi_6 = 126.704 \text{ kJ kg}^{-1}.$$

### 4.10.3 Solar Brayton cycle

The heat from solar radiation can be used in different cycles to produce power as shown in figure 4.27. An effective hybrid open solar Brayton cycle was run for the first time in 2002. The air was heated up to 1000 K in the combustion chamber. More hybridization was attained during the EU Solhyco project, running a hybridized Brayton cycle with solar energy and biodiesel. Brayton configurations have the potential to attain high normalized efficiency. Solar Brayton power plant efficiency depends on various parameters and depends on the solar receiver and heat absorption types [17]. There are two ways to heat the working fluid, either directly, using volumetric or tubular receivers, or indirectly, utilizing molten salt. Atmospheric conditions like wind cause heat losses in the receiver, affecting efficiency, meaning all these parameters must be considered in the receivers'

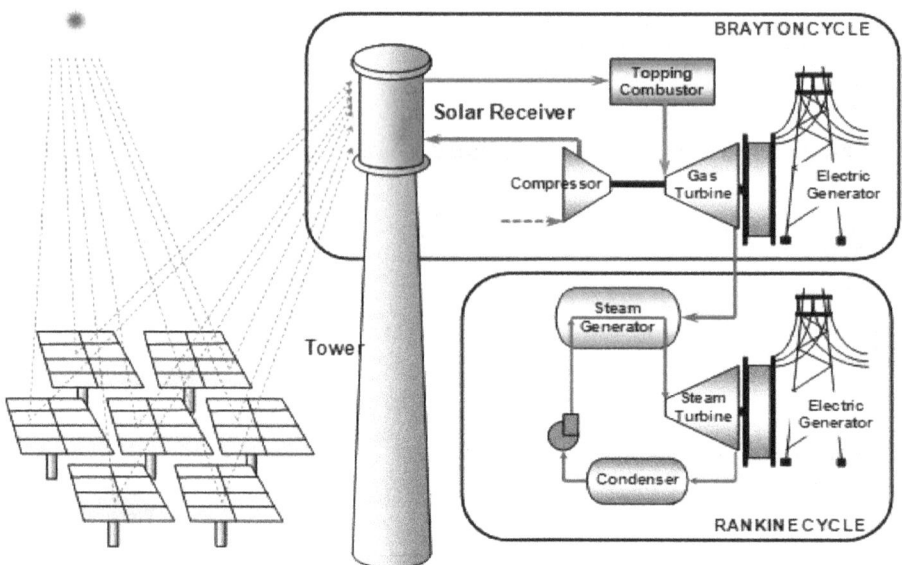

**Figure 4.27.** Layout of a solar combined-cycle power generation, comprising the solar tower concentrating system. Source: https://prec.ethz.ch/research/solar-power/solargasturbine.html.

modeling. There are various types of turbines that are applicable in Brayton cycles coupled with solar energy [18].

**EXAMPLE 15** A Brayton cycle with a working fluid of supercritical $CO_2$ at 7 MPa, 40 °C and a mass flow rate of 10 kg s$^{-1}$ enters the inlet compressor. The compressor pressure ratio is 10. The maximum temperature in the cycle is 600 °C. The isentropic efficiencies are 85% and 90% for the compressor and turbine, respectively. The effectiveness of the regenerator is 80%. Compare the performance of two configurations of a solar-driven Brayton cycle for electrical power generation with: (i) no regenerator and (ii) with a regenerator (figure 4.28).

*Properties* Considering the Brayton cycle with $CO_2$ as the working fluid. The properties values can be obtained using the properties table for $CO_2$.

So, $c_p$ for $CO_2$ at 600 K$c_{p_{CO_2}}$ = 1.07 kJ kg$^{-1}$ K$^{-1}$

$c_p$ for $CO_2$ at 900 K$c_{p_{CO_2}}$ = 1.18 kJ kg$^{-1}$ K$^{-1}$

$\gamma$ for carbon dioxide = 1.28

*Analysis* (1) Steady operating environment. (2) The carbon dioxide assumptions are appropriate. (3) No changes in kinetic and potential energy. (4) We consider the temperature and specific heats variation.

*Solution*

The $CO_2$ temperatures at the compressor and turbine exits are computed using isentropic relations:

Inlet temperature $T_i$ = 40 °C = 313 K

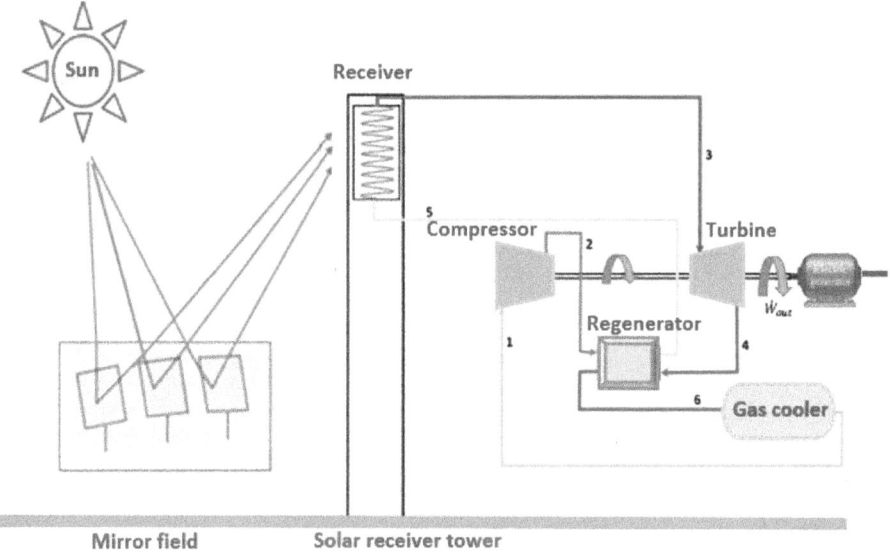

**Figure 4.28.** Solar Brayton cycle.

Inlet pressure $P_1 = P_1 = 7$ MPa compression ratio $r_p = \frac{P_2}{P_1} = \frac{P_3}{P_4}$

$$\therefore P_2 = P_3 = 10 * 7 = 70 \text{ MPa}.$$

Now, for the isentropic process $1-2_s$

$$\frac{T_{2,s}}{T_1} = r_p{}^{\Upsilon-1/\Upsilon}$$

$$T_{2,s} = 313 * 10^{1.28-1/1.28} = 517 \text{ K}$$

$$\eta_{s,c} = \frac{T_{s,2} - T_1}{T_2 - T_1}0.85 = \frac{517-313}{T_2 - 313}$$

$$T_2 = 553 \text{ K} \quad h_2 = 630 \text{ kJ kg}^{-1} \quad s_2 = 1.8 \text{ kJ kg}^{-1} \text{ K}^{-1}.$$

Constant pressure process $2-3$ $T_3 = 600\ °C = 873$ K
   Isentropic process $3-4_s$

$$\frac{T_3}{T_{4,s}} = r_p{}^{\Upsilon-1/\Upsilon} \quad T_{4,s} = \frac{873}{10^{1.28-1/1.28}} = 527.55 \text{ K}$$

$$\eta_{s,t} = \frac{T_3 - T_4}{T_3 - T_{4,s}}0.9 = \frac{873 - T_4}{873-527.55}$$

$$T_4 = 562.095 \text{ K}.$$

The values of enthalpy and entropy-based on pressures and temperatures that are calculated above can be obtained from the properties table for $CO_2$.

Therefore, the work per kilogram of $CO_2$ for the compressor and turbine may be found using the equation for the first law:

Compressor work $W_c = \dot{m}c_p(T_2 - T_1) = 10 * 1.07 * (553-313)$

$$W_c = 2568 \text{ kW}$$

Turbine output

$$W_t = \dot{m}c_p(T_3 - T_4) = 10 * 1.18 * (873-562.095)$$

$$W_t = 3668.67 \text{ kW}$$

The network

$$W_{net} = W_t - W_c 3 = 3668.67 - 2568$$

$$W_{net} = 1100.67 \text{ kW}$$

Rate of heat input and heat rejection can be expressed as

$$Q_{in} = \dot{m}c_p(T_3 - T_2) = 10 * 1.18 * (873-553)$$

$$Q_{in} = 3776 \text{ kW}$$

$$Q_{out} = \dot{m}c_p(T_4 - T_1) = 10 * 1.07 * (562.095-313)$$

$$Q_{out} = 2665.31 \text{ kW}.$$

Then the Brayton cycle thermal efficiency becomes

$$\eta_{th} = \frac{W_{net}}{\dot{Q}_{in}} = \frac{1100.67}{3776} = 0.2914$$

$$\eta_{th} = 29.149\%.$$

**Supercritical $CO_2$ as the working fluid with a regenerator**

Constant pressure process 2–3 $T_3 = 600 \,°C = 873 \text{ K}$ $P_3 = 70 \text{ MPa}$

Isentropic process 3–4$_s$ $s_{4,s} = s_3$ $P_3 = P_{4,s} = 70 \text{ MPa}$ $h_{4,s} = 750 \text{ kJ kg}^{-1}$

$$\eta_{s,t} = \frac{h_3 - h_4}{h_3 - h_{4,s}} \quad 0.9 = \frac{1080 - h_4}{1080-750}$$

$$h_4 = 783 \text{ kJ kg}^{-1} \quad s_4 = 2.58 \text{ kJ kg}^{-1} \text{ K}^{-1}$$

$$\epsilon = 0.6 = \frac{h_5 - h_2}{h_4 - h_2} = \frac{h_5 - 630}{783-630}$$

$$h_5 = 721.8 \text{ kJ kg}^{-1} \quad s_5 = 2 \text{ kJ kg}^{-1} \text{ K}^{-1}$$

By energy balance in the regenerator

$$q_{2-5} = q_{4-6}$$

$$h_5 - h_2 = h_4 - h_6$$

$$721.8 - 630 = 783 - h_6$$

$$h_6 = 691.2 \text{ kJ kg}^{-1} \quad s_6 = 2.4 \text{ kJ kg}^{-1} \text{ K}^{-1}$$

The values of enthalpy and entropy-based on pressures and temperatures that are calculated above can be obtained from the properties table for $CO_2$.

Therefore, the specific work of $CO_2$ for the compressor and turbine can be calculated using the equation for the first law:

Compressor work $W_c = \dot{m}(h_2 - h_1) = 10 * (630-430)$

$$W_c = 2000 \text{ kW}.$$

Turbine output

$$W_t = \dot{m}(h_3 - h_4) = 10 * (1080-783)$$

$$W_t = 2970 \text{ kW}.$$

The network

$$W_{\text{net}} = W_t - W_c = 2970-2000$$

$$W_{\text{net}} = 970 \text{ kW}.$$

The rate of heat input and heat rejection can be expressed as

$$Q_{\text{in}} = \dot{m}(h_3 - h_5) = 10 * (1080-721.8)$$

$$Q_{\text{in}} = 3582 \text{ kW}$$

$$Q_{\text{out}} = \dot{m}(h_6 - h_1) = 10 * (691.2-430)$$

$$Q_{\text{out}} = 2612 \text{ kW}.$$

Then the Brayton cycle thermal efficiency becomes

$$\eta_{\text{th}} = \frac{W_{\text{net}}}{\dot{Q}_{\text{in}}} = \frac{970}{3582} = 0.27074$$

$$\eta_{\text{th}} = 27.09\%.$$

## 4.11 Stirling cycle

A Stirling engine is a closed thermodynamic cycle using air or alternative gases as the working fluid, it is operated by the compression and expansion at different temperatures to convert heat energy to mechanical work and uses an external heat source (figure 4.29).

The Stirling engine is distinguished from other closed cycle hot air engines as it is a regenerative heat engine that always uses gaseous working fluids, it needs a specific type of internal heat exchanger and thermal store, called regenerators. A Stirling engine does not have exhaust valves for high-pressure outlet gasses. No combustion occurs inside the engine's cylinders; therefore, it is a quiet engine, and it is considered one of the most efficient machines. The Stirling cycle involves a series of proceedings that change the gas's pressure inside the engine to produce work. Several gases' properties could affect the operation of Stirling engines: if there is a fixed amount of gas in a fixed volume and the temperature of that gas is raised, then the pressure will increase, or if there is a fixed amount of gas and it is compressed (decrease the volume of its space), then the temperature of that gas will increase.

Initially, a simple engine consists of two cylinders. The expansion cylinder is heated using an external heat source, while the compression cylinder is cooled using an external cooling source. The two cylinders' gas chambers are connected to control the movement, and there are pistons connected mechanically by a linkage to determine how they will move relative to one another. The Schmidt theory is based on the isothermal expansion and compression of an ideal gas, and it is the simplest method and very useful during cycle development.

**Exergy flow diagram**

The Stirling engine ultimately operates by a sequence of four steps, and each sequence step is reversible, and the four steps together form the Stirling cycle. Figure 4.30 shows the schematic diagram for each step in the Stirling cycle and a pressure-volume diagram of the Stirling cycle [19].

The four steps of the ideal Stirling cycle are:

(a) Stroke 1–2: in this stage, heat is transferred to the gas at $T_H$ from the reservoir to maintain a constant temperature. The power piston moves upwards towards the displacer piston thus the working gas is compressed (Isothermal compression).

(b) Stroke 2–3: At this stage, the working gas is heated which raises its temperature at constant volume (Constant volume heating).

**Figure 4.29.** The Stirling engine.

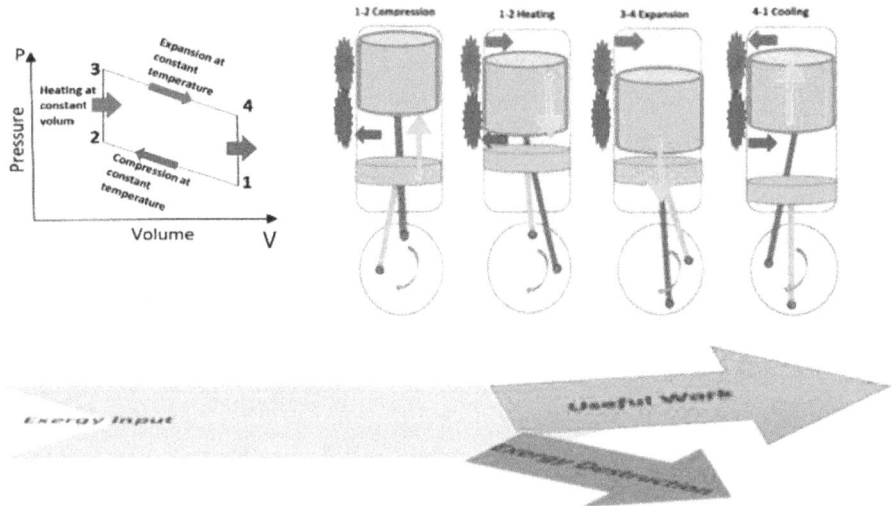

**Figure 4.30.** Stirling cycle and *P–v* diagram.

(c) Stroke 3–4: in this stage, the gas expands at a constant high temperature as the power piston moves down and work is being done by the expansion proces. (Isothermal compression).

(d) Stroke 4–1: At this stage, working gas returns to its starting temperature at a constant volume, and the cycle is completed (Constant volume cooling).

### 4.11.1 Stirling working fluid

A Stirling-cycle engine operates on a closed regenerative thermodynamic cycle, with cyclic compression and expansion of a gaseous working fluid at a different temperature, where the flow is controlled by volume changes to produce work. The working fluid is a fixed mass of gas, usually air, hydrogen, or helium. The air was the first and only working fluid in the early 19th century, whereas hydrogen and helium became the preferred working fluids for modern machines. These working fluids possess a more significant dynamic heat transfer coefficient and are both chemically inert and easily liquefied.

### 4.11.2 Energy analysis

The Stirling engine uses a compressible fluid as the working fluid in a closed system. A Stirling cycle consists of four processes:

- An isothermal expansion with heat addition from the reservoir.
- Isothermal compression with heat rejection to the reservoir.
- A rejection with internal heat transfer from the working fluid to the regenerator.
- Heat addition with internal heat transfer from the regenerator back to the working fluid at constant volume (figure 4.31).

We consider the heat transferred during all four processes, which will allow us to evaluate the ideal Stirling engine's thermal efficiency. The Stirling engine has the advantage of working with two constant volume processes instead of two isentropic processes. It has the maximum mechanical performance compared with another reciprocating thermal engine when operating in the same working conditions. An analysis of the four ideal Stirling cycle processes using the first law of thermo-dynamics, the network is done by the cycle, is given as the following equations.

$$W_{net} = W_{1-2} + W_{3-4}. \tag{4.130}$$

Apply the first law analysis of an ideal gas to determine the heat transferred.

$$W_{net} = W_{1-2} + W_{3-4} = Q_{in} - Q_{out}. \tag{4.131}$$

Thermal efficiency.

$$\eta_{th} = \frac{W_{net}}{Q_{in}}. \tag{4.132}$$

**EXAMPLE 16** Consider an engine operating on an air-standard Stirling cycle, it operates with maximum pressure at 3600 kPa and the minimum pressure is 50 kPa. The maximum volume is 12 times the minimum volume, and the low-temperature source is at 20 °C. The temperature difference between the external reservoirs and the air is 5 °C. Determine the specific heat added to the cycle and the net specific work of the cycle.

*Properties* Obtain the properties of air from the air table.
At room temperature
Gas constant $R = 0.287$ kJ kg$^{-1}$ K$^{-1}$ $k = 1.4$
Specific heats at constant pressure $C_p = 1.005$ kJ kg$^{-1}$ K$^{-1}$

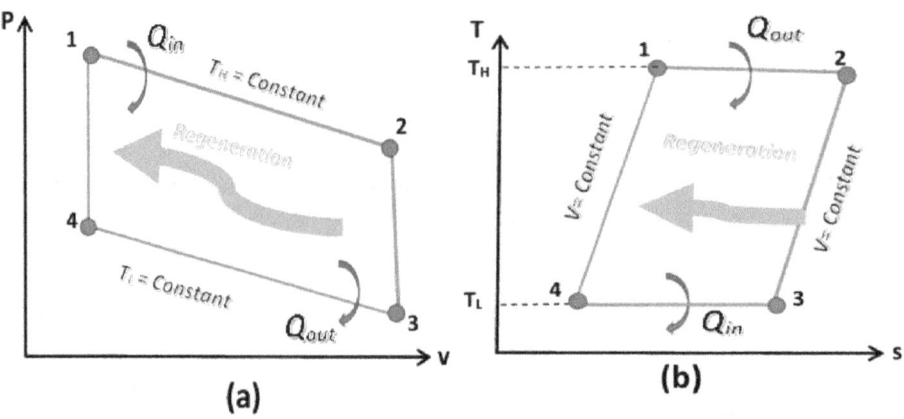

Figure 4.31. Thermodynamic Stirling cycle: (a) $P$–$v$ diagram. (b) $T$–$s$ diagram.

Specific heats at constant volume $C_v = 0.718$ kJ kg$^{-1}$ K$^{-1}$.

*Analysis* Consider that this process is a steady-state, and work, heat transfer, and changes in potential energy can be neglected. Air is an ideal gas with constant specific heats.

The ideal gas equation is applied to the isothermal process 3–4 giving

$$\frac{P_4}{P_3} = \frac{v_3}{v_4}.$$

Here, $p_4$ and $p_3$ are the pressure at state 3 and 4 respectively and $\frac{v_3}{v_4}$ is the volume ratio.

$$P_4 = P_3\frac{v_3}{v_4} = 50 * 12 = 600 \text{ kPa}$$

Since process 4–1 is constant volume process,

$$T_1 = T_4\left(\frac{P_1}{P_4}\right) = 298 * \left(\frac{3600}{600}\right) = 1788 \text{ K}.$$

Calculate the rate of the heat addition process 1–2 by using the relation of work done in an isothermal process.

$$q_{\text{in}} = w_{1-2} = RT_1\ln\frac{v_2}{v_1} = 0.287 * 1788 * \ln 12$$

then, the rate of heat transfer during process 1–2 is

$$q_{\text{in}} = 1275 \text{ kJ kg}^{-1}.$$

To calculate the rate of heat rejection during process 3–4 use the relation of work done in an isothermal process.

$$q_{\text{out}} = w_{3-4} = RT_3\ln\frac{v_4}{v_3} = 0.287 * 298 * \ln\frac{1}{12} = 212.5 \text{ kJ kg}^{-1}.$$

The rate of heat transfer during process 3–4 is

$$q_{\text{out}} = -212.5 \text{ kJ kg}^{-1}.$$

Calculate the net work done

$$w_{\text{net}} = q_{\text{in}} - q_{\text{out}} = 1275 - 212.5.$$

Therefore, the net work done is

$$w_{\text{net}} = 1063 \text{ kJ kg}^{-1}.$$

### 4.11.3 Exergy analysis

Exergy translates as the maximum available useful work that is obtainable from the system at a given state in a specified reference environment, ($P_0 = 1$ bar, $T_0 = 298.15$ K) as the dead state. The objective of exergy analysis is to examine the locations, reasons, and exergy destruction (Irreversibility) in the system. To analyze the exergy value of a flowing stream of mass in a particular system, there are two common practices to the exergy balance equation for each component in this system. The first way is the physical exergy associated with the stream flow in the system, exergy analysis applies to the inlet and exit of each component of the system, which can be synthetically described as a network of energy flows, where different kinds of irreversible energy conversion processes may occur. The way is to differentiate between the fuel, products, and losses of the system based on their economic value, which is called a functional method. In this chapter, the first way is evaluated, and an exergy balance equation for each component has been applied. Generally, the exergy equation for a control volume of a component in a steady-state system can be expressed as following equations. The exergy analysis methodology can help select optimal operating parameters and comparing different process configurations. Exergy balance leads to the concept of exergetic efficiency, which takes into consideration the quality of energy. The exergy of the amount of heat supplied varies due to the temperature variation.

$$\varphi_q = q(1 - (T_0/T)) \qquad (4.133)$$

$$\varphi_n = (h - h_0) - T_0(s - s_0). \qquad (4.134)$$

Thus, it is obvious that exergy depends on the parameters of the state. Equation 4.2 can be rearranged as

$$\varphi_n = (h_1 - T_0 s_1) - (h_0 - T_0 s_0) \qquad (4.135)$$

$$= h_1 - T_0 s_1 + C_1 \qquad (4.136)$$

where $C_1$ is a constant for given ambient conditions.

Any value for $C_1$ can be chosen, as one is usually only interested in the difference between the exergy values and not in the absolute value. So, the equation for the exergy function can be expressed as

$$\varphi = h - T_0 s + C_a \qquad (4.137)$$

where $C_a$ is an arbitrary constant.

**EXAMPLE 17** An engine operates on an air standard Stirling cycle using a low-temperature reservoir at 20 °C. permitting a 5 °C temperature difference between the external reservoirs and the air when appropriate. It is designed such that its maximum pressure is 3600 kPa and minimum pressure is 50 kPa and the maximum volume is 12 times the minimum volume. Determine the exergy destruction for each process in the Stirling cycle in kJ kg$^{-1}$.

*Properties* Obtain the properties of air from the air table.

At room temperature

Gas constant $R = 0.287$ kJ kg$^{-1}$ K$^{-1}$ $k = 1.4$

Specific heats at constant pressure $C_p = 1.005$ kJ kg$^{-1}$ K$^{-1}$.

Specific heats at constant volume $C_v = 0.718$ kJ kg$^{-1}$ K$^{-1}$.

*Analysis* This is a steady state process, there is no work or heat transfer, potential and kinetic energy changes are neglected. Air is an ideal gas, and the specific heat is constant.

**Solution**

First, we determined the state 4 pressure and state1 temperature using the ideal gas law.

First Process 3–4 (Isothermal):

Calculate the pressure at state 4 by using the equation,

$$\frac{P_4}{P_3} = \frac{V_3}{V_4} \quad P_4 = P_3\left(\frac{V_3}{V_4}\right).$$

Substitute 50 kPa for $P_3$ and 12 for $\left(\frac{V_3}{V_4}\right)$

$$P_4 = 50(12) = 600 \text{ kPa}.$$

Process 4–1 (Constant volume process)

Calculate the temperature at state 1 by using the equation,

$$\frac{T_1}{T_4} = \frac{P_1}{P_4} \quad T_1 = T_4\left(\frac{P_1}{P_4}\right).$$

Substitute 298 K for $T_4$, 600 kPa for $P_4$, and 2600 kPa for $P_1$

$$T_1 = 298\left(\frac{3600}{600}\right) = 1788 \text{ K}.$$

Apply the first law of thermodynamic to process 1–2 to calculate the heat input to the system.

$$\dot{q}_{in} = u_{1-2} + w_{1-2}u_{1-2} = 0$$

$$q_{in} = RT_1 \ln\frac{v_2}{v_1}.$$

Substitute 0.287 kJ kg$^{-1}$ K$^{-1}$ for $R$, 1788 K for $T_1$, and 12 for $\frac{v_2}{v_1}$

$$q_{in} = 0.287 * 1788 * \ln 12 = 1275 \text{ kJ kg}^{-1}.$$

Apply the first law of thermodynamic to process 3–4 to calculate the heat rejection to the system.

$$\dot{q}_{out} = u_{3-4} + w_{3-4}u_{3-4} = 0$$

$$q_{out} = RT_3 \ln\frac{v_4}{v_3}.$$

Substitute 0.287 kJ kg$^{-1}$ K$^{-1}$ for $R$, 298 K for $T_3$, and 12 for $\frac{v_3}{v_4}$

$$q_{\text{out}} = 0.287 * 298 * \ln\frac{1}{12} = -212.5 \text{ kJ kg}^{-1}.$$

Negative sign indicates heat rejection in process 3–4

Calculate the values of exergy destruction associated the cycle processes by using the equation.

$$I = T_0 s_{\text{gen}} = T_0\left(\Delta s - \frac{q_{\text{in}}}{T_{\text{source}}} + \frac{q_{\text{out}}}{T_{\text{sink}}}\right).$$

Exergy destruction during process 1–2.

$$I_{1-2} = T_0\left((s_2 - s_1) - \frac{q_{\text{in}}}{T_{\text{source}}}\right)$$

$$= T_0\left(c_v\ln\left(\frac{T_2}{T_1}\right) + R\ln\left(\frac{v_2}{v_1}\right) - \frac{q_{\text{in}}}{T_{\text{source}}}\right)$$

$$= T_0\left(c_v\ln(1) + R\ln\left(\frac{v_2}{v_1}\right) - \frac{q_{\text{in}}}{T_{\text{source}}}\right)$$

$$= T_0\left(0 + R\ln\left(\frac{v_2}{v_1}\right) - \frac{q_{\text{in}}}{T_{\text{source}}}\right)$$

$$= T_0\left(0 + R\ln\left(\frac{v_2}{v_1}\right) - \frac{q_{\text{in}}}{T_{\text{source}}}\right)$$

$$= 298 * \left(0.287 * \ln(12) - \frac{1275}{1788}\right) = 298 * (0.7132 - 0.71308)$$

$$I_{1-2} = 0.035 \text{ kJ kg}^{-1}.$$

Exergy destruction during process 2–3.

During this process, there is no interaction with the surroundings to cause a exergy destruction.

$$I_{2-3} = 0.$$

Therefore, there is no exergy destruction during process 2–3

Exergy destruction during process 3–4

$$I_{3-4} = T_0\left((s_4 - s_3) - \frac{q_{\text{out}}}{T_{\text{sink}}}\right)$$

$$= T_0 \left( c_v \ln\left(\frac{T_4}{T_3}\right) + R\ln\left(\frac{v_4}{v_3}\right) - \frac{q_{out}}{T_{sink}} \right)$$

$$= T_0 \left( c_v \ln(1) + R\ln\left(\frac{v_4}{v_3}\right) - \frac{q_{out}}{T_{sink}} \right)$$

$$= T_0 \left( 0 + R\ln\left(\frac{v_4}{v_3}\right) - \frac{q_{out}}{T_{sink}} \right)$$

$$= 298 * \left( 0.287 * \ln\left(\frac{1}{12}\right) - \frac{212.5}{298} \right) = 298 * (-0.7132 - 0.71308)$$

$$I_{3-4} = -0.035 \text{ kJ } kg^{-1}.$$

Exergy destruction during process 4–1.

During this process, there is no interaction with the surroundings to cause a exergy destruction.

Therefore, there is no exergy destruction during process 4–1

$$I_{4-1} = 0.$$

### 4.11.4 Solar Stirling engine

Renewable energy sources have become increasingly significant as there is an increasing energy demand globally, and the price of conventional fuel has risen. We want to utilize sustainable energy and solar energy to power the Stirling engine. This method of power production is not accompanied by pollution and noise. A solar-powered Stirling engine is a heat engine powered by a temperature gradient generated by the Sun as shown in figure 4.32. This system needs a massive dish that concentrates solar energy to the main point at the dish center. The concentrated solar energy operates the Stirling cycle where the heat flows from a hot supply to a cold sink to produce work that drives a generator to produce electric power [20].

## 4.12 The value of Kalina cycle in engineering

The Kalina cycle is a result of the development of the Rankine cycle. Dr. Alexander Kalina made advancements to the Rankine cycle at the end of the 1970s and early 1980s. The Kalina cycle is an unconventional thermodynamic cycle used to convert thermal energy from a relatively low temperature heat source to mechanical power. The Kalina cycle is a modified closed Rankine cycle that includes a separator, evaporator, absorber, and a reducing valve as shown in figure 4.33. The Kalina cycle uses two working fluids at different boiling temperatures, and it is unlike the Rankine cycle, where it uses one fluid of a pure substance. This cycle differs from conventional Rankine steam cycles and has a high heat recovery because of the close temperature match between the cycle heat receiving and the source heat availability and the high recuperation level [21]. The Kalina power plant components consist of

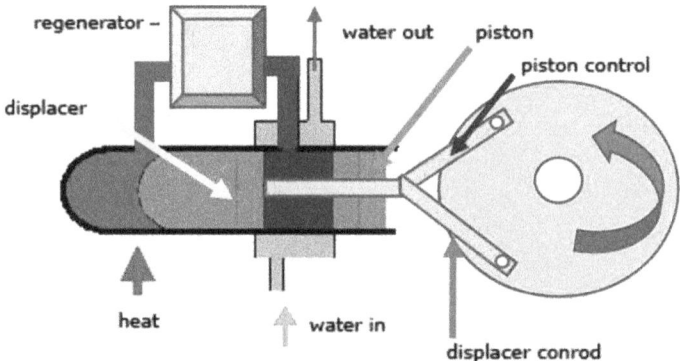

**Figure 4.32.** Solar Stirling cycle.

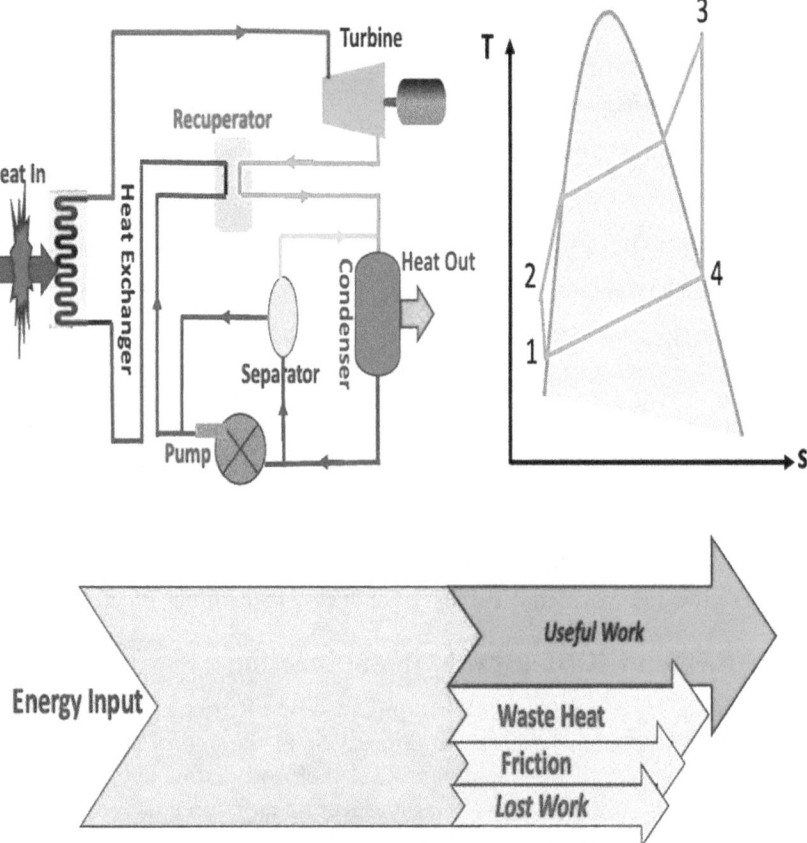

**Figure 4.33.** Kalina cycle and T-s diagram.

the same devices (turbine, pumps, valves, etc.) as a conventional steam power plant. The molecular weight of the mixture ammonia–water working fluid differs slightly from pure water. The significant difference between the two cycles is Rankine heat addition and rejection happen at uniform temperature during phase change; in contrast, the heat addition and heat rejection in the Kalina cycle happens at varying temperatures, even during phase change since the fluid is a mixture. The Kalina cycle has a lower average heat rejection temperature and higher average heat addition temperature, leading to high thermal efficiency. The Kalina cycle is recognized as one of the most thermodynamically efficient power cycle technologies in the world.

The five steps in the thermodynamic cycle include:
1. Compression in the pump.
2. Heat addition in the heat exchanger (boiler).
3. Expansion in the turbine.
4. Recuperation in the recuperators.
5. Heat rejection in the condenser.

Many modifications can be applied to the Kalina cycle, depending on the required use. Another Kalina cycle is illustrated in figure 4.34.

The mixture of the ammonia–water working fluid enters the heat recovery component before entering into the generator at high pressure and high temperature in the form the concentrated vapor of ammonia. The concentrated vapor then enters the superheater to increase the ammonia vapor temperature, subsequently, the vapor exits from the superheater and passes through the turbine. As the vapor expands in the turbine and decreases in pressure and temperature, converting the thermal energy to mechanical energy. After leaving the turbine the vapor passes across the absorber where the weak solution mixes with the strong solution at a low temperature. At this stage, the concentrated solution is formed at low temperature, then is pumped once more the generator via the heat recovery unit, and the cycle is repeated.

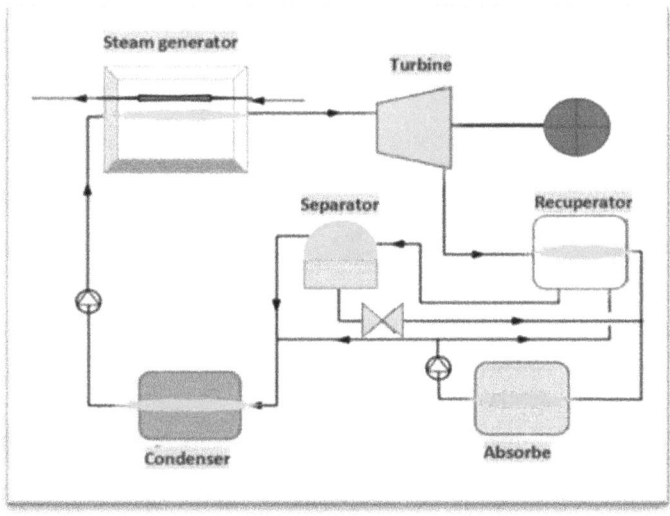

**Figure 4.34.** Basic configuration of a Kalina cycle.

### 4.12.1 Energy and exergy analysis of Kalina cycle

The thermodynamic analysis of the Kalina cycle components is straightforward since mass and energy are conserved. All the above components are assumed to be a steady-state flow; therefore, the energy equation is usable with the assumption that the kinetic and potential energy changes are negligible [22], the equations are reduced to

$$\sum m_{\text{in}} = \sum m_{\text{out}} \tag{4.138}$$

$$\sum Q + \sum \dot{m}_{\text{in}} h_{\text{in}} = \sum W + \sum \dot{m}_{\text{out}} h_{\text{out}} \tag{4.139}$$

$$\text{Steam generator } Q_{\text{SG}} = \dot{m}_{\text{SG, in}}(h_{\text{SG, out}} - h_{\text{SG, in}}). \tag{4.140}$$

Separator Mass balance of the ammonia in the separator is written as follows.

$$\dot{m}_{s,\text{ in}} x_{s,\text{ in}} = \dot{m}_{s,\text{ out1}} x_{s,\text{ out1}} + \dot{m}_{s,\text{ out2}} x_{s,\text{ out2}} \tag{4.141}$$

$$\dot{m}_{s,\text{ in}} = \dot{m}_{s,\text{ out1}} + \dot{m}_{s,\text{ out2}} \tag{4.142}$$

$$\dot{m}_{s,\text{ in}} * h_{s,\text{ in}} = \dot{m}_{s,\text{ out1}} * h_{s,\text{ out1}} + m_{s,\text{ out2}} * h_{s,\text{ out2}}. \tag{4.143}$$

Turbine

$$\dot{W}_t = \dot{m}_{t,\text{ in}}(h_{t,\text{ out}} - h_{t,\text{ in}}) \tag{4.144}$$

Absorber

$$Q_{\text{abs}} = m_{\text{abs, in}}(h_{\text{abs, in}} - h_{\text{abs, out}}) \tag{4.145}$$

Recuperator

$$\dot{m}_{R1,\text{ in}}(h_{1,\text{ out}} - h_{1,\text{ in}})_R = \dot{m}_{2,\text{ in}}(h_{2,\text{ out}} - h_{2,\text{ in}})_R \tag{4.146}$$

Valve

$$\dot{m}_{V,\text{ in}} * h_{V,\text{ in}} = \dot{m}_{V,\text{ out}} * h_{V,\text{ out}} \tag{4.147}$$

Condenser

$$Q_c = m_{c,\text{ in}}(h_{c,\text{ in}} - h_{c,\text{ out}}) \tag{4.148}$$

Pump

$$W_p = v_{p,\text{ in}}(P_{t,\text{ out}} - P_{t,\text{ in}}) \tag{4.149}$$

$$\eta = \frac{W_t - W_p}{Q_{\text{in}}}. \tag{4.150}$$

## 4.12.2 Exergy analysis

Exergy, unlike energy, represents the quality of a stream of energy rather than the quantity. As such, the analysis of exergy flow and their destruction offers a better baseline for the cycle's performance and its components. The exergy analysis is applied to every process to calculate the maximum available work input to the cycle utilized and how much energy has been lost. It makes a clear distinction between energy losses to the environment and internal irreversibilities in a process.

$$S_{\text{gen}} = \frac{dS}{dt} + \sum S_{\text{out}} \sum S_{\text{in}} \sum \frac{Q_i}{T_i} \geqslant 0 \qquad (4.151)$$

where,

$S_{\text{gen}}$ represents the generated entropy in the control volume,

$\frac{dS}{dt}$ the accumulation of entropy inside the control volume,

$\sum S_{\text{out}}$ and $\sum S_{\text{in}}$ the flow of entropy in and out of the control volume respectively,

$\sum \frac{Q_i}{T_i}$ the entropy generation associated with heat transfer.

$$I = T_0 S_{\text{gen}} \qquad (4.152)$$

The exergy definition is the maximum reversible work achievable when a system is in equilibrium with the environment. The total exergy of a system when the system is at rest with the surroundings is equal to physical exergy plus chemical exergy; no magnetic, electric, nuclear, and surface tension effects are included [23].

$$\phi = \phi_{Ph} + \phi_{Ch}. \qquad (4.153)$$

The physical exergy component is calculated using the following relation:

$$\phi_{Ph} = \dot{m}((h - h_0) - T_0(s - s_0)). \qquad (4.154)$$

Chemical exergy is a maximum work when the system is brought into reaction with reference substances present in the environment. The chemical exergy of the flow is calculated using the following equation:

$$\phi_{Ch} = \dot{m}_i \left( \left( \frac{\overline{\varphi}_{Ch,\,\text{NH}_3}^0}{M_{\text{NH}_3}} \right) Y + \left( \frac{\overline{\varphi}_{Ch,\,\text{H}_2\text{O}}^0}{M_{\text{H}_2\text{O}}} \right) (1 - Y) \right) \qquad (4.155)$$

where

$\overline{\varphi}_{Ch,\,\text{NH}_3}^0$ and $\overline{\varphi}_{Ch,\,\text{H}_2\text{O}}^0$ are the standard molar specific chemical exergies of ammonia and water respectively,

$Y$ is the mass fraction of ammonia.

At a known temperature of $T$, the $\phi_Q$ can be calculated from:

$$\phi_Q = \sum \left( 1 - \frac{T_0}{T} \right) Q. \qquad (4.156)$$

The exergy analysis equation is applied to every component of the Kalina cycle in the following equations.

Exergy balance in the turbine:

$$\triangle \phi_t = \sum_{in} \varphi_i \dot{m}_{in} - \sum_{exit} \varphi_e \dot{m}_{out} = W_{act} \qquad (4.157)$$

$$I_{Turbine} = T_0 m_{in}(s_{in} - s_{out}). \qquad (4.158)$$

Exergy balance in a pump:

$$\triangle \phi_{Pump} = \sum_{in} \varphi_{in} \dot{m}_{in} - \sum_{out} \varphi_{out} \dot{m}_{out} - W_{in} \qquad (4.159)$$

$$I_{pump} = T_0 m_{in}(s_{out} - s_{in}). \qquad (4.160)$$

Exergy balance in the separator

$$\dot{m}_{in} = \dot{m}_{1,\,out} + \dot{m}_{2,\,out} \qquad (4.161)$$

$$\triangle \phi_{Sep} = \varphi_{in} \dot{m}_{in} - (\varphi_{out} \dot{m}_{out})_1 - (\varphi_{out} \dot{m}_{out})_2 \qquad (4.162)$$

$$I_{Sep} = T_0((m_1 s_1 + m_2 s_2)_{out} - m_{in} s_{in}). \qquad (4.163)$$

Exergy balance in the steam generator

$$\triangle \phi_{SG} = \sum_{in} \varphi_{in} \dot{m}_{in} - \sum_{out} \varphi_{out} \dot{m}_{out} \qquad (4.164)$$

$$I_{SG} = T_0 m_{in}(s_{in} - s_{out}). \qquad (4.165)$$

Exergy balance in the absorber

$$Q_{abs} = m_{abs,\,in}(\varphi_{abs,\,in} - \varphi_{abs,\,out}) \qquad (4.166)$$

$$I_{abs} = T_0 m_{in}(s_{in} - s_{out}) - m_{wa} cp_{wa} \ln \frac{T_0}{T_{s,\,abs}}. \qquad (4.167)$$

Exergy balance in the condenser

$$\triangle \phi_{Cond} = \sum_{in} \varphi_{in} \dot{m}_{in} - \sum_{out} \varphi_{out} \dot{m}_{out} = \qquad (4.168)$$

$$I_{Cond} = T_0 m_{in}(s_{in} - s_{out}) - m_{wa} cp_{wa} \ln \frac{T_0}{T_{s,\,Cond}}. \qquad (4.169)$$

The second law efficiency can be defined as the ratio of exergy output to exergy input. The exergy output depends on the degree of irreversibility of the cycle. Therefore,

$$\eta_{II} = \frac{\text{useful exergy out}}{\text{exergy input}} = 1 \frac{\text{exergy destruction}}{\text{exergy input}}. \qquad (4.170)$$

### 4.12.3 Kalina cycle working fluid

The Kalina cycle is a new cycle in heat recovery and power generation that uses an ammonia–water mixture as a working fluid. The Kalina cycle's unique feature is in the working fluid. The working fluid is composed of two different fluids with different boiling points. Since a two-fluids mixture will boil over a range of temperatures, the ratio of the two fluids can be varied in other parts of the system, this overall effect increases the process's thermodynamic efficiency. The Kalina cycle has been studied for different applications for low-temperature heat sources. Furthermore, there are different types of Kalina cycle families, which are known by their unique names. For example, the KCS5 cycle applies to direct fuel-fired power plants. The KCS6 type applies to gas turbines in the form of combined cycles, although there are types of Kalina cycles that are designed for utilizing low-temperature heat sources such as KCS11 and KCS34. Although the ammonia–water working pair has zero ozone depletion potential and low global warming potential, toxic special safety procedures against the leak should be considered [24]. Therefore, there is a need to use other working pairs to replace water–ammonia in the Kalina cycle. There are alternative working fluids that are used in the domain of refrigeration and air conditioning such as a mixture of refrigerants including CFC (chlorofluorocarbon), HCFC (hydrochlorofluorocarbons), HFC (hydrofluorocarbons), and commercial products like R407C, $CO_2$–hydrocarbon blends, $CO_2$–dimethyl ether (DME), and R32–hydrocarbons. These refrigerants are chosen because of their appropriate ecological characteristics like zero ozone depletion, low global warming, and non-toxicity. A mixture of hydrocarbon refrigerants with $CO_2$ reduces their flammability and provides a reasonable control of the carbon dioxide level depending on the concentration of the mixture. R32 is a good alternative working fluid, this refrigerant is energy-efficient due to its relatively high pressure and density, thus, R32 mixtures. The zeotropic mixtures of HFC could be used in the Kalina cycle, such as R22–R134a and R32–R134a. The principle of forming the zeotropic mix is to mix fluids with different boiling points so that the evaporation or condensation process occurs over a temperature range.

### KALINA CYCLE ADVANTAGES
- Generates more power than conventional steam power plants by 10% to 50%.
- Has lower initial capital costs since it has a smaller heat exchanger and no heat transfer oil loop (compared to ORC systems).
- Less supervision and has lower plant loads.
- Uses standard, easily available, and to a large degree proven plant components.
- Superior heat transfer which decreases demand for cooling water and infrastructure.
- The least maintenance downtime.

RENEWABLE ENERGY

### 4.12.4 Kalina cycle with heat recovery energy

Dr. Alexander Kalina proposed a new basic cycle in 1984 to be used as a bottoming cycle to utilize the waste heat provided from the exhausts of gas turbines as a heat source [25] as shown in figure 4.35. This cycle arrangement is called Kalina cycle system 1 (KCS 1). The working fluid in this cycle is an ammonia–water mixture, the concentration of ammonia in this cycle is not steady and varies over the process, and any modification of the cycle arrangement depends on the applications. For example, low-temperature geothermal applications can use KCS 2, and KCS 5 is suitable for direct fuel-fired plants.

### 4.12.5 Kalina cycle waste heat recovery

Using the Kalina cycle to recover industrial waste heat has a high energy conversion efficiency because the ammonia–water mixture working fluid evaporates and condenses at varying temperatures, leading to a suitable temperature match with the heat source and cooling water in the vapor generator and condenser, which reduces irreversible heat loss in the heat exchangers. Figure 4.36 illustrates an example of the Kalina cycle recovering waste heat [26].

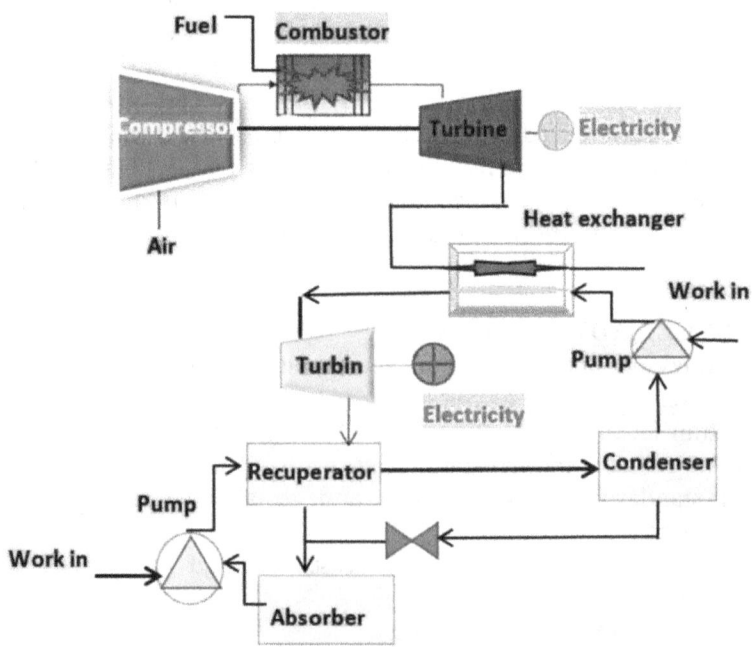

**Figure 4.35.** Simple schematic diagram of the combined Brayton–Kalina cycle.

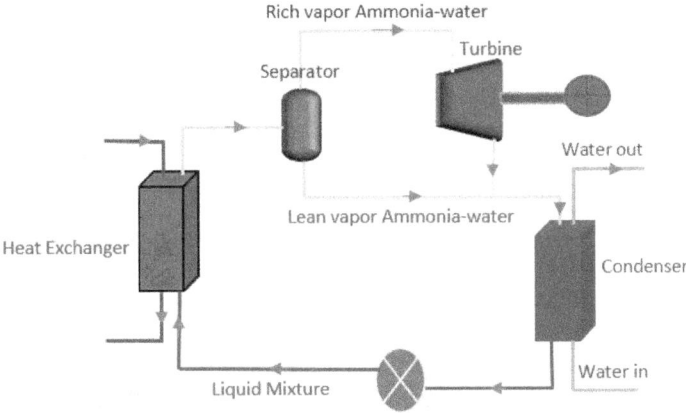

**Figure 4.36.** A Kalina cycle recovering waste heat recovery.

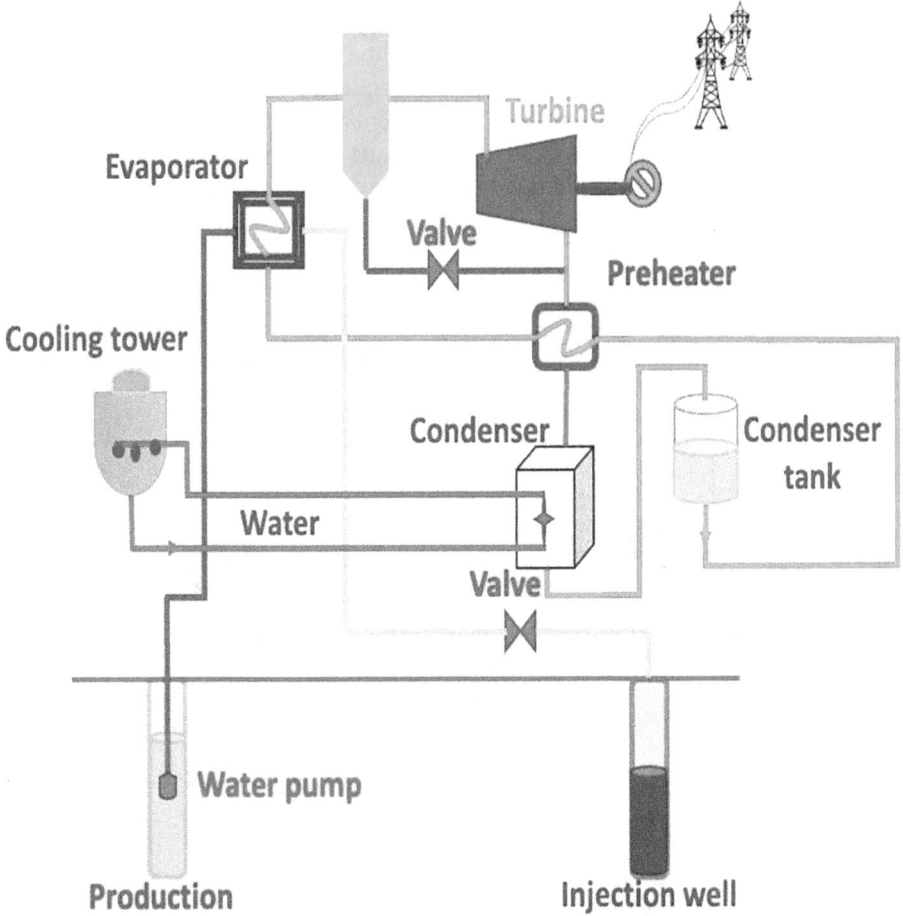

**Figure 4.37.** Kalina cycle geothermal power plant.

### 4.12.6 Geothermal Kalina cycle

Geothermal energy is considered clean and renewable, and geothermal energy can power the Kalina cycle, as shown in figure 4.37. The closed Kalina cycle uses low-temperature steam, water, or brine from the geothermal source to heat the working fluid and uses cooling water or air to cool the working fluid down in closed heat exchangers. Because the Kalina cycle's working fluid consists of binary fluids and has different boiling temperatures, the solution boils and condenses over a range of temperatures gives more heat extracted from the geothermal fluid than with a pure working fluid. The working fluid's boiling temperatures depend on the heat input temperature, which is adjusted by modifying the ratio between the working fluid components; as a result, the average temperature of the working fluid in the heating process is higher, and the cooling process's average temperature is lower, thus giving the cycle a higher efficiency, also to increase efficiency, a more complex design with additional separators and recuperators is needed.

### 4.12.7 Solar Kalina cycle

The solar Kalina cycle is promising, a feasible way to generate power from low- and medium-temperature solar heat. The solar Kalina cycle is driven by solar heat of approximately 100–500 °C. Figure 4.38 shows the schematic diagram of solar-driven Kalina cycle. The system contains three subsystems: a solar collector with a variable aperture area, the oil–ammonia exchanger and the Kalina cycle. The heat collected by the heat transfer fluid is exchanged with the ammonia–water mixture in the oil–ammonia exchanger [27].

## 4.13 Thermal energy storage

Thermal energy storage (TES) is one form of energy storage. This technology is stored thermal energy by heating or cooling a storage medium to use hours, days, or

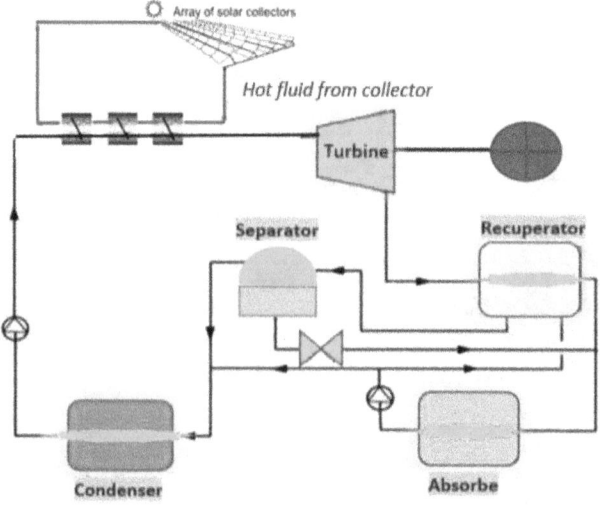

**Figure 4.38.** Solar Kalina cycle.

months later for heating and cooling applications and power generation. Such technology has become an important part of renewable energy technology systems since it can reduce peak demand, energy consumption, $CO_2$ emissions, and costs while also increasing energy systems' overall efficiency. Thermal energy storage systems are divided into three types: sensible heat, latent heat, and thermochemical.

Thermal energy storage is cheapest than electricity storage, and it has a high potential of integrating renewable energy sources such as wind and solar into the heating or cooling application. It also provides several benefits to heating and cooling output, increasing its efficiency, and integrating other heat sources as industrial waste heat or seawater. The three steps that have described the process of storing thermal energy are sensible, latent, and chemical storage. In sensible storage systems, the heat exchange takes place with a change of temperature without changing the storage medium phase. The heat is transferred in this system to the liquid or solid storage medium by increasing its temperature or cooling the storage medium. Some of those systems such as water, molten salt, thermal oil, and concrete. A phase change material is used as the storage medium in the latent storage systems and stores the energy by changing its phase. The heat is transferred to the storage medium by melting it. At the same time, it is extracted through solidification, the mediums that can be used are sodium nitrate, potassium nitrate, calcium nitrate, sodium sulfate, potassium carbonate, etc. The method of storing heat in the thermochemical energy storage is using a reversible chemical reaction that involves high energy reactions. An intricate compound is broken down into simpler compounds by providing heat for an endothermic reaction and storing the separated compounds until the stored energy is required once more. When the stored energy needs to be extracted, an exothermic reaction can occur in the isolated compounds, releasing this energy as heat. The most common application for thermal energy storage is in solar thermal systems.

### 4.13.1 Integrating thermal energy storage with concentrated solar power

Thermal energy storage provides a workable solution to the challenge of using solar energy because of reduced energy production when the Sun sets or is blocked by clouds. In a concentrating solar power system, the Sun's rays are reflected onto a receiver, which creates heat used to generate electricity that can be used after a sunny period.

Primarily based on this structure, the thermal energy storage systems are divided into active and passive systems [28].

**Active System**

The active storage systems are systems wherein the storage medium 'flows' within the plant. There are two functional systems:

1. Active direct systems, in this system, the solar receiver heat transfer fluid is also used as the storage medium; a model of this system is a molten salt central receiver solar field with two-tank molten salt storage and figure 4.39 shows an example.
2. Active indirect systems. In this system, the heat transfer fluid within the solar receiver is unlike the fluid in the storage medium. An example of this system

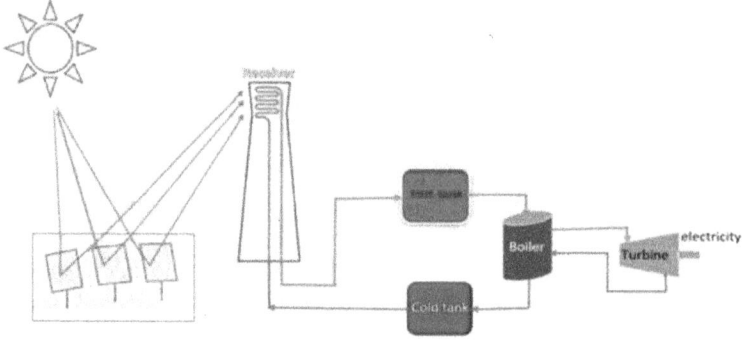

**Figure 4.39.** Flowchart of an active direct TES integration system.

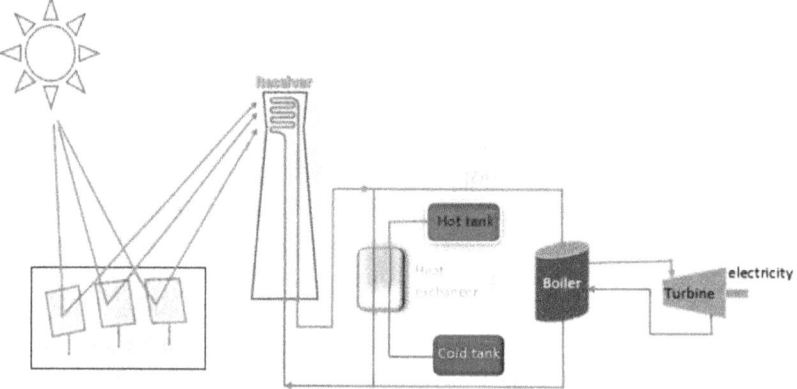

**Figure 4.40.** Flowchart of an active indirect TES integration system.

is a parabolic trough solar field with thermal oil as the heat transfer fluid as shown in figure 4.40.

**Passive Systems**

In contrast to the active systems, the storage is a solid material and stationery, and the heat is transferred between the fluid the storage material. A fluid with high thermal conductivity is a requirement.

**PROBLEMS**

1) What does exergy represent? What does the exergy balance tell us? How can we use exergy analysis?
2) What factors result in exergy destruction? How can exergy destruction be reduced in a process?
3) An ideal Rankine cycle using water as its working fluid operates with supercritical pressure at the high-pressure side of the cycle. Calculate the thermal efficiency of the cycle if the state entering the turbine is at 25 MPa, 500 °C, and the condenser pressure is 5 kPa. Determine

    a) The steam quality at the turbine exit,

    b) The cycle efficiency.

    (Ans. $x_e = 0.717$, $\eta = 44\%$)

4) Consider a power plant operating on an ideal Rankine cycle using water as a working fluid, operating with the high pressure of 4 MPa in the boiler and the low pressure of 20 kPa in the condenser and a turbine inlet temperature of 700 °C. determine the value of the exergy destruction in each of the components of the cycle when heat is being rejected to the atmosphere at 15 °C and heat is supplied from an energy reservoir at 750 °C

    (Ans. $I_b = 926$ kJ kg$^{-1}$ K$^{-1}$, $I_c = 304$ kJ kg$^{-1}$ K$^{-1}$)

5) A solar thermal plant is using an ideal Rankine cycle to generate power. The thermal fluid is heated in the solar troughs and then the steam is heated in the evaporator. Power is generated by the turbine and the steam is condensed by transferring heat to a cooled water stream. The mass flow rate in the cycle is 20 kg s$^{-1}$. The evaporator operates at 10 MPa and saturated liquid water leaves the condenser at 60 °C. Determine:

    a) Turbine power output (in MW) if steam enters the turbine at 700 °C and 10 MPa,

    b) Rankine cycle thermal efficiency if the pump consumes 202 kW of power.

    (Ans. $P = 26.85$ MW, $\eta_{th} = 40.58\%$)

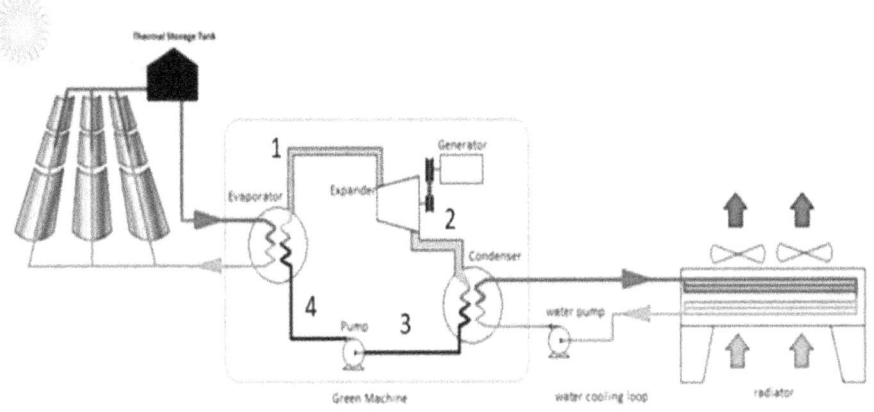

6) An ideal Brayton refrigeration cycle with a compressor pressure ratio of 3, uses air as a working fluid which enters the compressor at 100 kPa, 270 K. The temperature at the turbine inlet is 310 K. For the compressor and turbine with isentropic efficiencies of 80 and 88% respectively, determine:

    a) The coefficient of performance,

    b) The exergy destruction in the turbine,

    c) The exergy destruction in the compressor.

    (Ans. $COP = 65.7\%$, $I_c = 25$ kJ kg$^{-1}$ K$^{-1}$, $I_t = 10.07$ kJ kg$^{-1}$ K$^{-1}$)

7) Consider an engine operating on an ideal Stirling cycle using helium as the working fluid. It has an operating temperature range of 300 K to 2000 K and a pressure range of 150 kPa to 3 MPa. If the mass of helium used in the cycle is 0.12 kg, determine:
   a) The thermal efficiency of the cycle,
   b) The amount of heat transfer in the regenerator,
   c) The amount of work output per cycle.
   (Ans. $\eta_{th}$ = 85%, $Q_{reg}$ = 35.6 kJ, $W_t$ = 465.5 kJ)
8) Evaluate the rate of entropy change of air and the rate of entropy generation during the steady process. Compression of air by a 5 kW compressor from a from 100 kPa and 17 °C to 600 kPa and 167 °C. Mass flow rate is 1.6 kg min$^{-1}$.
   Assumptions:
   (1) Air is an ideal gas with variable specific heat. (2) No change in kinetic and potential energy. (3) Surrounding temperature is 17 °C
   (Ans. $\Delta S_{air}$ = 0.00248 kW K$^{-1}$, $\dot{S}_{gen}$ = 0.00083 kW K$^{-1}$)

# 167 °C

# 600 kPa

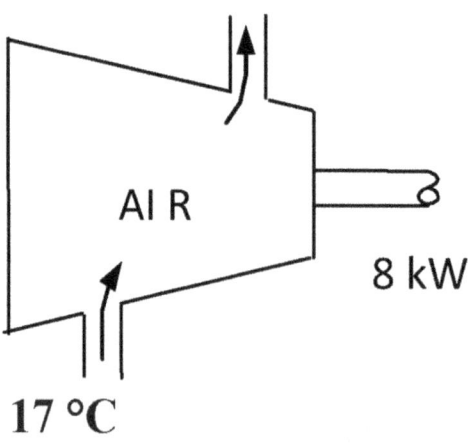

# 17 °C

# 100 kPa

9) Steam expands adiabatically from state 1 to state 2 in a two stage turbine. Calculate the reversible power output and the second-law efficiency and exergy destruction rate.

- Assumption:

(1) Steady-flow process. (2) No changes in kinetic and potential energy. (3) Ambient temperature ($T_0$) is 25 °C.

(Ans. $W_{rev}$ = 5808 kW, $\eta_{II}$ = 86.1%)

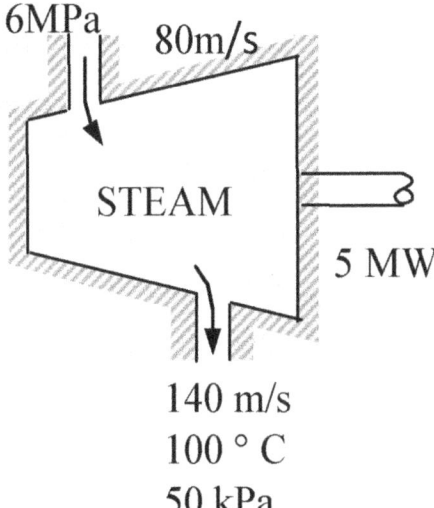

80 m/s,

600 °C

6MPa   80m/s

STEAM   5 MW

140 m/s
100 ° C
50 kPa

10) Steam expands adiabatically from state 1 to state 2 in a two-stage turbine (similar to problem 9). Calculate the exergy destruction rate.

Assumption: (1) Steady-flow process. (2) No changes in kinetic and potential energy. (3) Ambient temperature ($T_0$) is 25 °C.

(Ans.   $I_t$ = 457 kW)

11) A mixture of saturated liquid-vapor water is in an insulated cylinder. The cylinder is attached to a supply line, and the steam is permitted to go into the cylinder until all of the liquid is vaporized. Determine the amount of steam that entered the cylinder and the exergy destroyed.

Assumptions:

(1) This is an unsteady process as conditions inside the device are changing throughout the process, however, the process can be analyzed as a uniform-flow process for the reason that the state of fluid stays constant at the inlet. (2) A quasi-equilibrium expansion process. (3) Kinetic energy and potential energy can be ignored. (4) The device is insulated and consequently heat transfer can be ignored.

(Ans. $m_{in}$ = 8.27 kg, $I_{cy}$ = 2832 kJ)

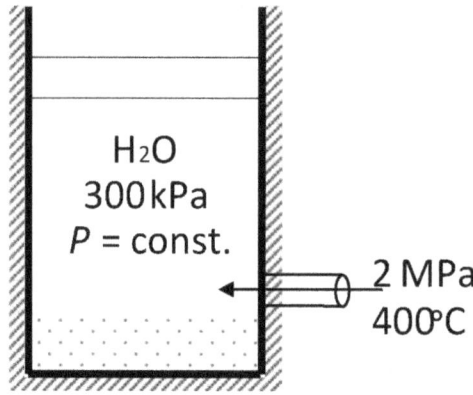

12) Water is heated using hot oil in a heat exchanger. The oil temperature at the outlet and the exergy destruction rate in the heat exchanger are to be determined.

Assumptions:
(1) Exits at a steady state. (2) The heat exchanger is insulated in order so we can say the heat loss to the environment is negligible and as a result heat transfer from the hot fluid is the same as the heat transfer to the cold fluid. (3) No changes in the kinetic and potential energies of the fluid streams. (4) The properties of the fluid are constant.

(Ans. $T_{\text{out}}$ = 129.1 °C, $I_{HE}$ = 219 kW)

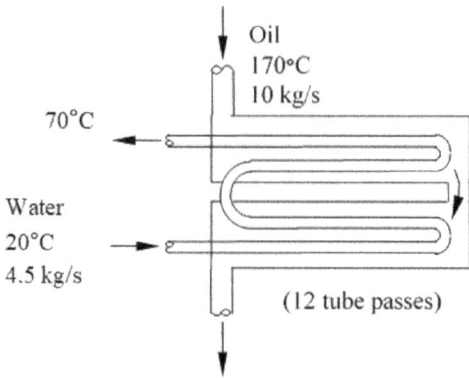

# References

[1] Borgnakke C and Sonntag R E 2019 *Fundamentals of Thermodynamics* 7th edn (New York: Wiley)

[2] Yadav U S *et al* June 2020 Theoretical study on solar integrated waste heat recovery systems for power industries *Int. Res. J. Eng. Technol. (IRJET)* **07**

[3] Sharma M and Dev R 2015 Working fluid selection for low temperature organic Rankine cycle *Conference: ISES Solar World Congress, At: EXCO, Daegu Korea*

[4] Darvish K and Ehyaei M A 2015 Selection of optimum working fluid for organic rankine cycles by exergy and exergy-economic analyses *Sustainability* **7** 15362–83

[5] Long R *et al* 2014 Exergy analysis and working fluid selection of organic Rankine cycle for low grade waste heat recovery *Energy* **73** 475–83

[6] Wang R, Jiang L and Ma Z *et al* 2019 Comparative analysis of small-scale organicrankine cycle systems for solar energy utilisation *Energies* **12** 829

[7] Mondejar M E *et al* 2017 Prospects of the use of nanofluids as working fluids for organic Rankine cycle power systems *Energy Procedia* **129** 160–7

[8] Saadatfar B, Fakhrai R and Fransson T 2014 Conceptual modeling of nano fluid ORC for solar thermal polygeneration *Energy Procedia* **57** 2696–705

[9] Sharma K V, Akilu S, Hassan S and Hegde G 2017 Considerations on the thermophysical properties of nanofluids *Engineering Applications of Nanotechnology, Topics in Mining, Metallurgy and Materials Engineering* (Berlin: Springer)

[10] Sami S 2019 Impact of magnetic field on the enhancement of performance of thermal solar collectors using nanofluids *Int. J. Ambient Energy* **40** 1–10

[11] Marin A and Dobrovicescu A *et al* 2014 Energy and exergy analysis of an organic Rankine Cycle *U.P.B. Sci. Bull., Series D* **76**

[12] Baral S 2019 Experimental and techno-economic analysis of solar-geothermal organic rankine cycle technology for power generation in Nepal *Int. J. Photoenergy* 5814265

[13] Norazreen Binti Samsuri 2017 A simulation study on performance enhancement of technoeconomic efficiency of ocean thermal energy conversion cycle using different working fluids *Degree of Master of Philosophy, UniversitiTeknologi Malaysia*

[14] Yilmaz F 2019 Energy, exergy and economic analyses of a novel hybrid ocean thermal energy conversion system for clean power production *Energy Convers. Manag.* **196** 557–66

[15] Chen F *et al* Thermodynamic analysis of rankine cycle in ocean thermal energy conversion *Int. J. Simul.: Syst. Sci Technol.* DOI:10.5013/IJSSST.a.17.13.07

[16] Fontaine K, Yasunaga T and Ikegami Y 2019 OTEC maximum net power output using carnot cycle and application to simplify heat exchanger selection *Entropy* **21** 1143

[17] Le Roux W G, Bello-Ochende T and Meyer J P 2010 Minimization and optimum distribution of entropy generation for maximum net power output of the small-scale open and direct solar thermal Brayton cycle *Conference: First Postgraduate Renewable Energy Symposium (Lynedoch, Cape Town)*

[18] Ahmadi M H *et al* 2018 A review on solar-assisted gas turbines *Energy Sci Eng.* **6** 658–74

[19] YANMAR Technical Review January 2017 Development of zero emission generating system *Stirling Engine*

[20] He M M 2016 Stirling engine for solar thermal electric generation, *PhD Thesis Electrical Engineering and Computer Sciences University of California at Berkeley*

[21] Online course and simulator for engineering thermodynamics, Kalina cycle.

[22] Ahmad M and Karimi M N March 2016 Thermodynamic analysis of Kalina cycle *Int. J. Sci. Res. (IJSR)* **5**

[23] Nasruddin and Usvika R *et al* 2009 Energy and exergy analysis of Kalina cycle system (KCS) 34 with mass fraction ammonia-water mixture variation *J. Mech. Sci. Technol.* **23** 1871–6

[24] Elsayed A *et al* 2013 Thermodynamic performance of Kalina cycle system 11 (KCS11): feasibility of using alternative zeotropic mixtures *Int. J. Low-Carbon Technol.* DOI:10.1093/ijlct/ctt020

[25] Dincer I and Bicer Y 2020 Chapter 4—integration of conventional energy systems for multigeneration *Integrated Energy Systems for Multigeneration* (New York: Academic) pp 143–221

[26] Jouhara H and Khordehgah N *et al* 2018 Waste heat recovery technologies and applications *Therm. Sci. Eng. Prog.* **6** 268–89

[27] Hong H, Gao J and Qu W *et al* November 2017 Thermodynamic analyses of the solar-driven Kalina cycle having a variable concentration ratio *Appl. Therm. Eng.* **126** 997–1005

[28] Alnaimat A and Rashid Y 2019 Advances in concentrated solar power: a perspective of heat transfer Heat and Mass Transfer - Advances in Science and Technology Applications (London: InTech Open) DOI:10.5772/intechopen.84575

**IOP** Publishing

Thermodynamic Cycles for Renewable Energy Technologies

K R V Subramanian and Raji George

# Chapter 5

## Waste heat recovery

**Prakriti Gupta and G M Madhu**

A large amount of waste heat is generated in many industrial processes which is not effectively utilized and discarded to environment. Many of the by-products from the industrial process can be effectively utilized for heat generation. Recovering the waste energy and effectively using the same within the process results in the reduction in product cost and improves efficiency of the process and reduces the greenhouse effect. In this work a comprehensive review has been made on pinch analysis for effective heat integration and an attempt is made to explain the methodology of recovering the waste heat and minimize use of heating and cooling utilities with examples. Optimization of heat integration process using the LINGO program has been demonstrated with suitable examples. The possibility of generation of power with the waste heat generated and use waste heat for heating purposes (cogeneration) is also explained. The various systems for thermal energy storage and their efficacy compare to one another is dealt in brief.

## 5.1 Introduction

Many developed and developing countries like India have seen a sharp increase in the demand for fuel and energy in the past few decades with the boom in manufacturing industries. This has resulted in the spike of fuel prices and global warming, amongst other consequences. The increase in energy demand and emission of greenhouse gases to boggling extents has forced us to think of effective utilization of waste heat from domestic households and industries thereby reducing the consumption of the energy and to enhance the productivity. Heat energy accounts for 50% of global energy consumption and contributes 40% of the greenhouse gas emissions [1]. Out of this, nearly 43% of the heat is used in manufacturing/industrial organisations. A large amount of heat is lost through exhaust gases, liquid streams (heat exchangers) and to the environment (operating losses). Heat losses are classified based on the temperature range of the sources, namely, high-, medium- and low-temperature heat losses. Recovery of the waste heat for all ranges will allow the system to work at very high efficiencies and productivity.

Based on the type of process and its operating conditions, waste heat is rejected from below atmospheric temperatures (chilled condition) to very high temperatures such as that lost by flue gases from industrial furnaces. Heat recovered at high temperatures will be of higher quality with greater efficiency as opposed to heat recovered from lower temperatures, by virtue of the flexibility of its reuse on sinks with a wider temperature range. This is a consequence of the unidirectional nature of heat flow from a higher temperature source to a lower temperature sink, as proposed by the second law of thermodynamics. Recovered heat must be effectively used for heat and mass transfer operations. It can be used in heat exchangers for heating and cooling, steam generators, the product of which can be used for steam blanketing and absorption operations. Along with heat recovery, if cogeneration (heating and power generation) is be employed, utilization of maximum amount of heat recovered to its highest potential may be possible. For example, heat recovered from the flue gases of blast furnaces can be used to produce very high-pressure (VHP) and high-pressure steams (HP). Very high-pressure steam can then, in turn, be employed for generation of power and other grades of steam for heating purposes.

It is estimated that the waste heat potential from iron and steel industries alone, is more than 1000 °C. The heat potential in the temperature range of 500 °C–1000 °C is available, with its potential being restricted in the cement, iron and steel sectors [2]. The heat recovery potential increases in the 200 °C–500 °C range for pulp and paper industries. For majority of the industrial sectors, a large amount of waste heat lies in the temperature range of 100 °C–200 °C. The potential of heat quality obtained below 100 °C lies in drying operations, food and beverage sectors. The waste heat from the industries is tantamount to 6%–10% of their own consumption [3]. The typical temperatures of industrial processes are listed in table 5.1.

Many industrial processes offer great potential for waste heat recycle, heat recovery and cogeneration. Cogeneration refers to the amalgamation of heat recycling and power generation. For instance, high calorific flues gases from furnaces can be effectively utilized for heat recovery as well as power generation, essentially by generating high-quality steam. Industrial processes not only generate unused heat; many combustible by-products are also laid waste due to lack of integration. It is imperative to consider burning these wastes in existing boilers and furnaces to produce steam, a viable source of heat energy. Depending on the quality of the steam produced at a specific pressure, heat utilization demands can be determined by heat integration, whereas the non-heating demands (stripping and absorption processes) can be determined by mass integration strategies [5]. There are many strategies, tools and techniques available to identify waste heat recovery and recycle; which would ensure optimal heat integration with high efficiency for industrial processes. One of the most widely used techniques for both mass and heat integration is the pinch analysis.

**Table 5.1.** Various industrial processes and waste heat temperatures [4].

| Process | Temperature range (°C) |
|---|---|
| Exhaust gases temperatures from industrial manufacturing | |
| Catalytic crackers | 800–1000 |
| Cement kilns | 1000–1250 |
| Aluminum refining furnaces | 1000–1200 |
| Copper refining furnaces | 1200–1300 |
| Solid waste incinerators | 1200–1600 |
| Steel heating furnaces | 1500–1700 |
| Glass melting furnaces | 1600–2800 |
| Petrochemical and chemical process industries | |
| Boiling | 100–250 |
| Distillation | 100–250 |
| Evaporation | 100–200 |
| Heating in food processing | 50–200 |
| Drying | 50–200 |
| Food and beverage industries | |
| Food processing, breweries, dairy | 100–200 |
| Clean-in-place washing, washing bottles | 50–150 |
| Solvent extraction and distillation of vegetable oil | 50–150 |

## 5.2 Process integration

Process integration is a holistic approach to process design, retrofitting, and operation which underscores the unity of industrial processes [6], emphasis being laid on the integration of various unit operations. Process integration is useful in designing and operating industrial processes to work faster, better, cheaper, safer, and greener [5] as one single unit. To minimize the heat loss and maximize the use of heating and cooling utilities, it is necessary to maximize the heat exchange among the various process streams.

## 5.3 Targeting

One of the most crucial steps of process integration is the identification of a target utilization or generation, depending on the resource that needs to be minimized or maximized respectively. Targeting signifies the identification of a standard to which the process would be efficiently optimized. Consider the following example shown in figure 5.1: A teacher living in a busy and developed city has to travel every day from her house to her school all across the city. Between these points A and B there are

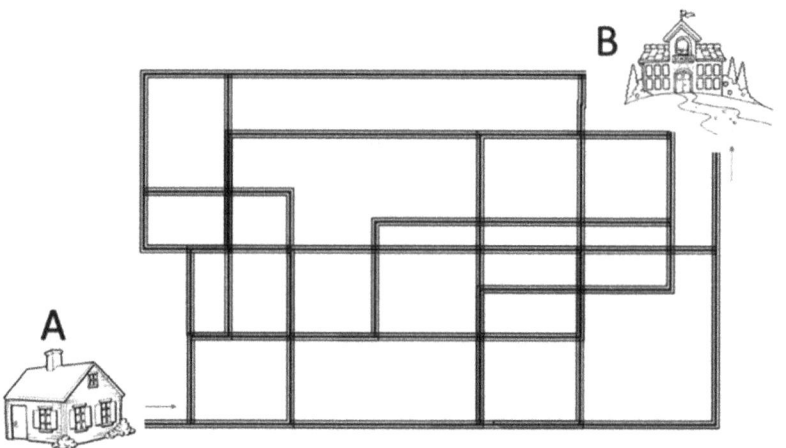

**Figure 5.1.** Target to find least distance between point A and point B.

multiple routes that can be traversed. She uses her GPS to show her the best route, which gives her multiple possibilities, all of which show different distances. The *target* here would signify the least distance that needs to be travelled from her house to the school without explicitly elucidating the details such as the by-lanes that need to be taken, how long she would have to wait in a traffic jam, whether there are any road blocks on the way and what the individual speed limits are for each road.

The *target* of a process is the yardstick to which the process components must be designed and most often are the parameter that is reported in layman terms. For the same process, there can be multiple targets defined, such as for waste generation, fresh resource consumption, heat generation and heating utilities. Once a target is set, process synthesis can be carried out, wherein various unit operations can be schemed out and recycle strategies laid forth.

## 5.4 Pinch analysis

Pinch analysis is an effective technique used to determine and design a process that minimizes waste generation, raw material consumption and the usage of heating and cooling utilities; and maximizes heat recovery. Pinch technique provides thermodynamically attainable energy targets (heat pinch) without delineating the explicit analysis of individual processes. This technique also helps us in calculating the minimum cooling and heating utilities required after direct integration. Various researchers explained heat pinch analysis in a lucid manner [7–9]. Heat pinch analysis is also applied to combined heat and power integration, which helps us determine the quality and quantity of heating requirements.

### 5.4.1 Heating and cooling utilities estimation based on heat pinch analysis

Consider a chemical processing facility that has been illustrated in figure 5.2. The feed mixture (C1) is first heated to 450K, then fed to a catalytic reactor where an

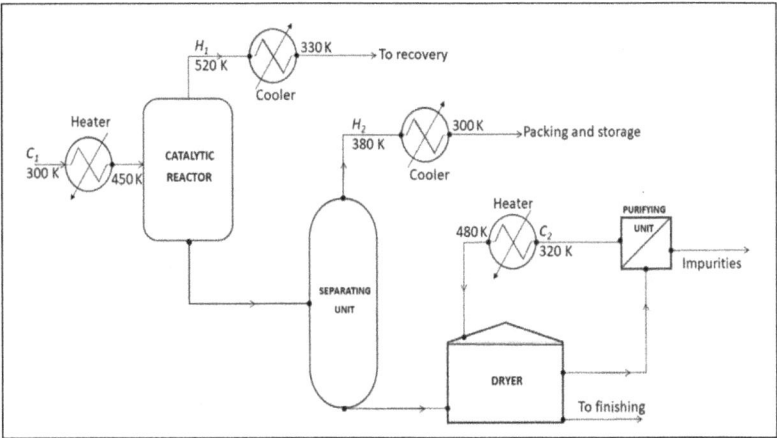

**Figure 5.2.** A typical chemical processing facility.

exothermic reaction takes place. The off-gases leaving the reactor (H1) at 520K are cooled to 330K prior to being forwarded to the recovery unit. The mixture leaving the bottom of the reactor is separated into vapour fraction and slurry fraction. The vapour fraction (H2) exits the separation unit at 380K and is to be cooled to 300 prior to storage. The slurry fraction is dried with a hot gas at 480K. The gas is purified to get rid of volatile impurities and recycled to the washing unit. During purification, the temperature drops to 320K. To address the heating and cooling requirements we need to employ heat exchangers and external utilities. If separate heating and cooling utilities were employed, then the operating cost increases along with large amount of heat losses. In the present process there is no heat integration. It is always advantageous to attempt to transfer heat from process hot streams to process cold streams by integration, which results in simultaneous reduction in heating and cooling utilities as well as in operating costs. Integrated processes will always result in conserving natural resources, which is required to produce heating and cooling utilities, reducing the product cost.

Pinch analysis is employed to find the minimum heating and cooling requirements in the integrated process. This provides requirements of overall heating and cooling utilities after integration without going into the details of the process. Pinch analysis is of two types, graphical pinch analysis and algebraic pinch analysis [5].

*5.4.1.1 Graphical pinch analysis*
The procedure employed to find the heating and cooling utilities required for a multi-stream process is as follows. Only counter flow arrangement of fluids is employed in heat exchangers to achieve better efficiency.

1. From the process flow diagram find the number of hot streams (stream which is to be cooled) and cold streams (stream which has to be heated).
2. Let hot stream temperatures be denoted by 'T' and cold stream temperatures be denoted by 't'.

3. Assume the heat capacity is constant over the entire operating range. Let the heat load for hot fluid be $HH = F_h C_{ph}(t_s - t_t)$. where $F_h$ is flow rate of hot fluid, $C_{ph}$ be the specific heat, $T_s$ be the supply temperature and $T_t$ be the target temperature of hot fluid.

4. Let the heat load for cold fluid be $HC = F_c C_{pc}(T_t - T_s)$. where $F_c$ is flow rate of cold fluid, $C_{pc}$ be specific heat, $t_s$ be the supply temperature and $t_t$, the target temperature of cold fluid.

5. In the counter flow heat exchanger, the difference between hot and cold fluid temperature along the heat exchanger is assumed to be constant. Minimum temperature difference ($\Delta T_{min}$) is assumed to be 10 °C. Minimum temperature difference must be estimated by plotting a graph of total annual cost versus temperature difference [5].

6. Rank the hot and cold streams based on the target temperature for hot streams and based on the supply temperature for cold streams.

7. Calculate the heat loads for hot streams. Find cumulative heat loads based on the ranking of streams and plot against the temperature as shown in figure 5.2. Superimpose the hot streams as shown in figure 5.3(a). After superimposition the generate hot composite stream curve is shown in figure 5.3(b).

8. Find cold stream temperature using equation ($t = T - \Delta T^{min}$). Calculate the heat loads for hot streams. Find cumulative heat loads based on the ranking of streams and plot against the temperature as shown in figure 5.4(a). After superimposition the generated cold stream composite curve as shown in figure 5.4(b) and generate cold composite stream curve.

9. Plot both the lines in a same graph as shown in figure 5.5.

10. Hot composite curve is stationery while the cold stream composite curve can either be moved up or down until it touches the hot stream composite curve. Overlapping of the curves is not permitted. The point where the hot and cold stream meet is known as the thermal pinch point, as shown in figure 5.6. Minimum hot and cold utility loads can be calculated as shown in the figure 5.6. More than one pinch point is possible without overlapping of the curves. The pinch point represents that zone of zero heat exchange.

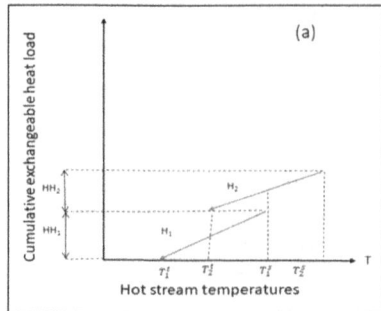 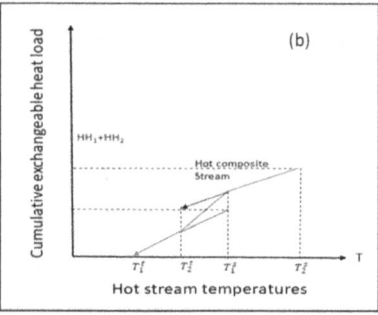

**Figure 5.3.** Plot of cummulative exchangable heat load versus hot stream temperatures (a) before super-imposition (b) after super imposition [5].

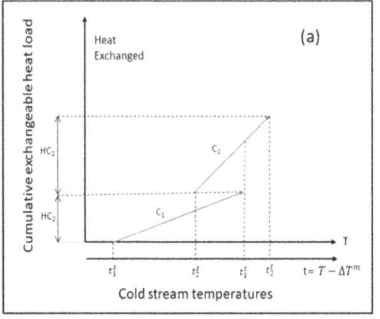 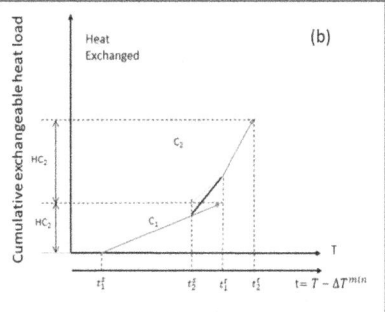

**Figure 5.4.** Plot of cummulative exchangable heat load versus cold stream temperatures (a) before superimposition (b) after super imposition [5].

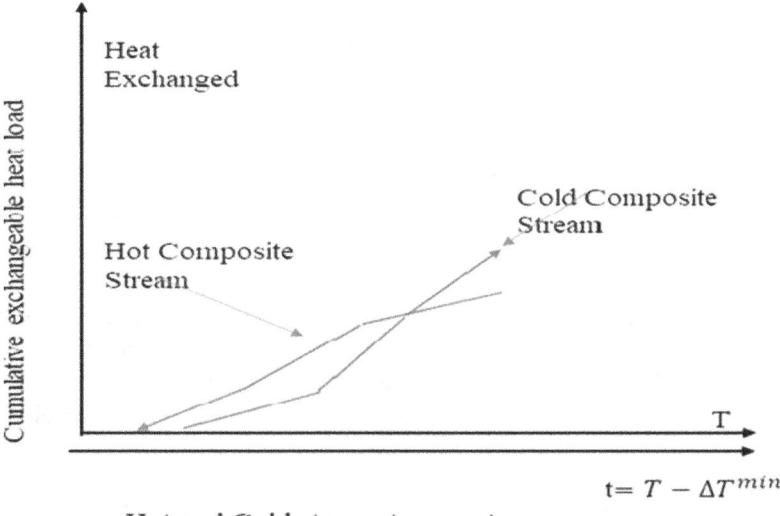

**Figure 5.5.** Plot of cummulative exchangable heat load versus cold and stream temperatures.

### 5.4.1.2 Algebraic pinch analysis

1. Based on minimum temperature difference prepare the temperature interval diagram, which must cover all supply temperatures and target temperatures of hot fluid and cold fluid streams. Find the number of temperature intervals.
2. Represent the supply temperatures and target temperatures for all hot and cold streams in the temperature interval diagram.
3. Find the heat loads for all hot and cold streams within the supply and target temperature ranges.
4. Prepare exchangeable load interval diagram for hot and cold streams for all intervals.
5. Prepare a cascade diagram by taking exchangeable heat of hot stream as input and cold stream as output for all the intervals.

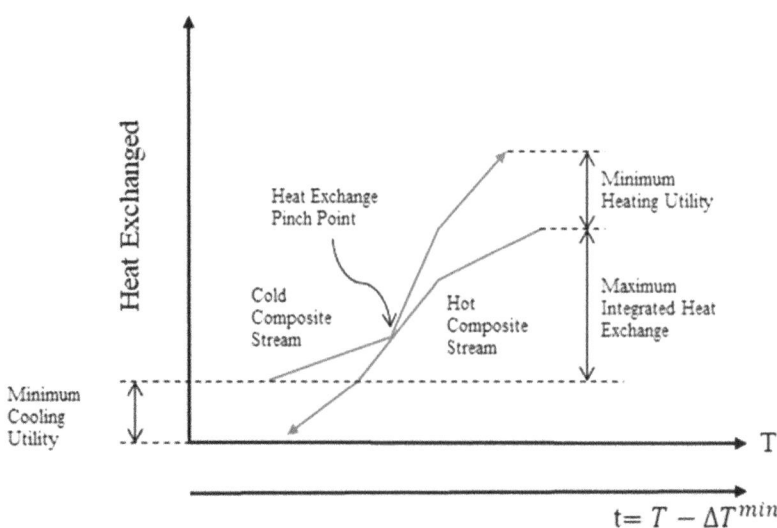

**Figure 5.6.** Plot of cummulative exchangable heat load versus cold and stream temperatures [5].

6. Find the residual heat by carrying out heat balance across each interval. Assume residual heat entering the first cascade is zero. Find the maximum negative residual heat. Maximum negative value is minimum deficit heat load.
7. Prepare a revised cascade diagram by adding maximum negative deficit heat to first cascade and find the residual heat across each cascade by heat balance. The residual heat leaving the last cascade is minimum cooling required. The point where residual heat is zero is known as pinch point.

## 5.5 Heat recycling

In a typical processing facility, there are often multiple heating and cooling requirements for individual process units. An exothermic reactor, for instance, may liberate high quantities of heat which needs to be reduced to ensure the safety of the plant and equipment, and at the other end, certain finished goods may require drying to ensure extended shelf life. It would benefit the operation in terms of heating and cooling utilities if the heat were extracted from the reactor and redirected to the drying unit. If such a scenario were to be extended to multiple streams, certain questions need to be answered, such as which heat generating stream ought to be redirected to which heat consuming unit or how much additional heat must be supplied after integrating the network of streams. This forms the basis of heat targeting and thus the synthesis of heat exchange networks (HEN). The objective of formulating an HEN is to determine the minimized external heating and cooling utilities after an optimized heat integration has been carried out amongst most, if not all, of the streams that need temperature control.

In the following section we shall discuss two methods of carrying out heat exchange with the help of an example. Consider a chemical process plant as

**Table 5.2.** Stream data for the chemical process.

| Stream | Flow rate × specific heat (kW °C$^{-1}$) | Supply temperature (K) | Target temperature (K) | Enthalpy change (kW) |
|---|---|---|---|---|
| H1 | 10 | 520 | 330 | 1900 |
| H2 | 5 | 380 | 300 | 400 |
| C1 | 19 | 300 | 450 | 2850 |
| C2 | 2 | 320 | 480 | 320 |

represented in figure 5.2. It is necessary to calculate the heating and cooling utilities required for the service of the process industry. The data for the hot and cold streams of the chemical process plant is shown in table 5.2.

The cost of heating and cooling utilities of the plant needs to be minimized. This can be done using two methods, namely, by constructing a heat cascade diagram algebraically and by graphical thermal pinch analysis. We first discuss the algebraic method to solve this problem.

### 5.5.1 Algebraic approach for heat integration

A cascade diagram for heat analysis is carried out by the construction of the following:

- Temperature interval diagram (TID).
- Individual tables of exchangeable heat load (TEHL) for the hot and cold streams.
- Heat cascade diagram.

A temperature interval diagram represents the exchangeability of heat from each of the stream between their supply and target temperatures. First we identify the range of operation which includes the lowest and highest temperatures of both cold and hot streams. The hot and cold stream at any interval follows

$$T = t + \Delta T$$

where $T$ is the temperature of the hot stream and $t$ is that of the cold stream. For any combination of streams, $T$ must be greater or equal to $t$ for thermodynamic feasibility in accordance with the second law of thermodynamics. In an ideal configuration of streams, $\Delta T$ is minimized to an optimum as a trade-off between operating costs and fixed establishment costs. Let $\Delta T^{\min}$ be the optimum temperature difference of the system, minimized at the thermal pinch point.

For the problem statement, consider $\Delta T^{\min} = 10$ K. To construct the TID, we first identify the range of all the streams and all target and supply temperatures. We must then specify at what interval each of the hot and cold streams are operating.

Next, the heat loads of each interval is calculated for the hot and cold streams. Let the number of hot and cold stream be $N_H$ and $N_C$ respectively. If $H_i$ is the hot stream considered for recycling in the $j$th interval, where $i = 1, 2, 3 .. N_H$ and $j$ is the number of intervals, then the heat load of the hot stream in the $j$th interval is

$$Heat\ load = F_i \times Cp_i \times \Delta T_j$$

where $\Delta T_j$ is the temperature difference in the interval and $F_i$ and $Cp_i$ are the flow rate and specific heat of the stream.

Similarly, if $C_i$ is the cold stream considered for recycling in the $j$th interval, where $i = 1, 2, 3 .. N_C$ and $j$ is the number of intervals, then heat load of the cold stream in the $j$th interval is

$$Heat\ load = f_i \times C'p_i \times \Delta T_j$$

where $\Delta T_j$ is the temperature difference in the interval and $f_i$ and $C'p_i$ are the flow rate and specific heat of the stream respectively. Temperature interval diagram for the chemical process plant based on hot and cold stream temperatures is represented in table 5.3. The entry and exit temperatures of each stream is shown with the arrows. The decreasing trend indicates the hot streams, which is losing heat and the increasing trend of the cold stream temperature implies heat gain.

Based on the temperature interval diagram exchangeable heat load table 5.4(a) and table 5.4(b) is prepared for hot and cold streams based on sensible heat transfer. The total heat load for hot and cold streams is the sum of the exchangeable heat load in each interval for all the streams for both cold and hot stream curves.

Using these values, we can now construct our cascade diagram in the following format (figure 5.7) where initial residual load is zero as no heat exchange takes place at the entry:

Considering each interval in the cascade, heat is supplied by the hot steam as input and heat is removed by the cold stream as output. The residual heat load entering the first interval is taken as zero. Heat balance for each interval is

**Table 5.3.** Temperature interval diagram for the problem statement.

| Interval | Hot Streams | T (K) | t (K) | Cold streams |
|---|---|---|---|---|
| | | 520 | 510 | |
| 1 | | 490 | 480 | |
| 2 | H1 | 460 | 450 | |
| 3 | | 380 | 370 | C2 |
| 4 | | 340 | 330 | |
| 5 | | 330 | 320 | |
| 6 | H2 | 320 | 310 | C1 |
| 7 | | 310 | 300 | |
| 8 | | 300 | 290 | |

**Table 5.4.** (a) and 4(b): Exchangeable heat loads of hot and cold streams.

| Interval | Load of H1 (kW) | Load of H2 (kW) | Total load (kW) | Interval | Load of C1 (kW) | Load of C2 (kW) | Total load (kW) |
|---|---|---|---|---|---|---|---|
| 1 | 300 | | 300 | 1 | | | |
| 2 | 300 | | 300 | 2 | 60 | | 60 |
| 3 | 800 | | 800 | 3 | 160 | 1520 | 1680 |
| 4 | 400 | 200 | 600 | 4 | 80 | 760 | 840 |
| 5 | 100 | 50 | 150 | 5 | 20 | 190 | 210 |
| 6 | | 50 | 50 | 6 | | 190 | 190 |
| 7 | | 50 | 50 | 7 | | 190 | 190 |
| 8 | | 50 | 50 | 8 | | | |

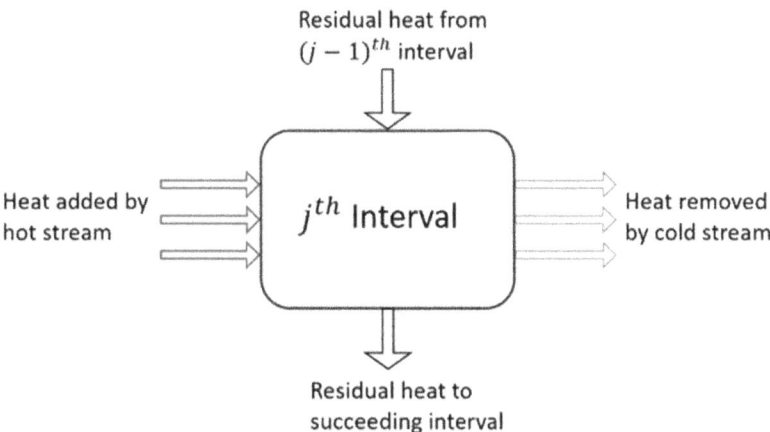

**Figure 5.7.** Heat exchange in the $j$th interval.

performed to find the residual load entering the next interval. After the cascade has been sketched out, we identify the residual heat with the least value, in this case it is *residual* $(j = 7) = -920$ kW as shown in figure 5.8(a). A revised cascade is constructed, but this time, with an addition of 920 kW so as to eliminate all negative residuals as shown in figure 5.8(b). *A negative residual heat implies heat must move from a cold process stream (heat sink) to a hot process stream (heat source) which is a thermodynamic impossibility.* This is the target minimum for the external heating utilities.

The interval where residual becomes zero represents the zone of no heat exchange, which implies $T = t$ for that particular interval of heat exchange. We can see that the *Thermal Pinch Point* occurs at the 7th interval, between hot stream H2 at $T = 310$ K and cold stream C1 at $t = 300$ K.

Residual heat at the bottom of the cascade represents the heat required by external cooling utilities. $CU1_{min} = 50$ kW and $HU1_{min} = 920$ kW.

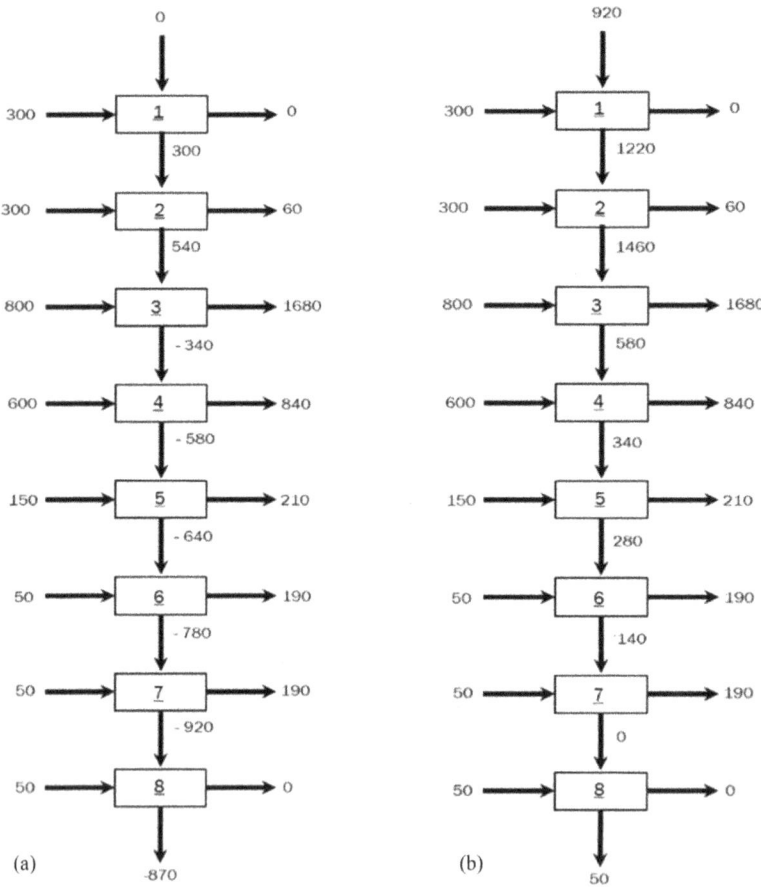

**Figure 5.8.** (a) Heat cascade before thermal integration. (b) Thermal pinch diagram after heat integration.

### 5.5.2 Graphical thermal pinch diagram for heat integration

Thermal pinch diagram is constructed as explained in the previous section. Firstly, the hot streams are rearranged in ascending order of the target temperature. The cold streams are rearranged in ascending order of their supply temperature. The cumulative load is then found out (depicted in table 5.5) and composite curves for heat rich and heat deficit curves are made separately, superimposing individual lines as shown in figures 5.9 and 5.10.

The head of the arrows represents the target temperature of the stream, whereas the tail represents the supply temperature. Next, we combine the two curves into one thermal pinch diagram, making sure that the cold composite curve does not, at any point lie on the right of the hot stream curve, thus creating a region of infeasibility as shown in figure 5.11.

It can be seen that the cold composite curve intersects the hot composite curve at $T = 310$, which represents the thermal pinch point. From the graph, the minimum cooling requirement CU1 is 50 kW and the minimum heating requirement is 920 kW, which is known as the thermal pinch diagram (figure 5.12).

**Table 5.5.** Rearranged stream data of the problem.

| Stream | Flow rate × specific heat (kW °C⁻¹) | Supply temperature (K) | Target temperature (K) | Enthalpy change (kW) | Cumulative enthalpy (kW) |
|---|---|---|---|---|---|
| H2 | 5 | 380 | 300 | 400 | 400 |
| H1 | 10 | 520 | 330 | 1900 | 2300 |
| | | | | | |
| C1 | 19 | 300 | 450 | 2850 | 2850 |
| C2 | 2 | 320 | 480 | 320 | 3170 |

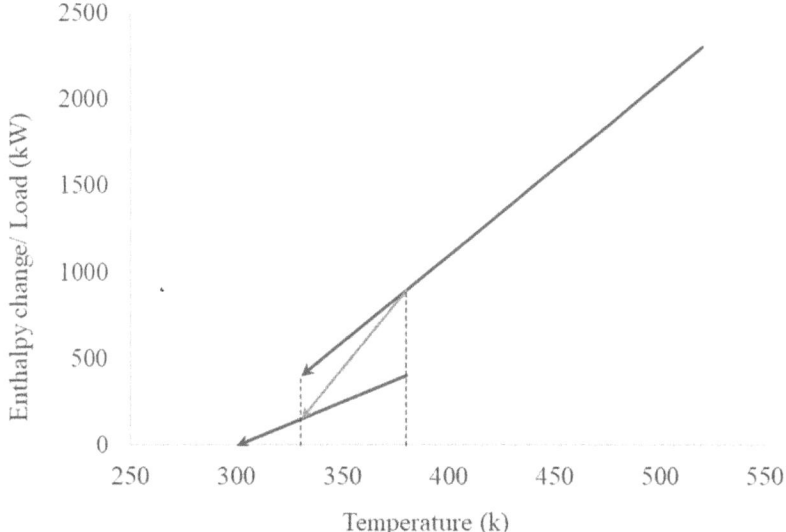

**Figure 5.9.** Hot stream composite curve.

It is important to note that, although the graphical pinch analysis is an elegant yet simple method to visualise the heat flow dynamics between streams, it comes at a cost of significant errors when there are multiple streams with wide range of temperature. It becomes hard to identify the pinch point in such scenarios. The practice of constructing HENs via the algebraic approach takes precedence as it provides an accurate target and can be iterated.

### 5.5.3 Mathematical approach for synthesis of heat exchange network to predict minimum heating and cooling utilities

Based on the transshipment formulation of Papoulias and Grossmann (1983) [8, 9], mathematical approach for synthesis of heat exchange networks was designed by El-Halwagi (2006) [5]. To ensure the thermodynamic feasibility of heat exchanger heat must go through the ware houses and represented by temperature intervals as explained in algebraic method. To optimize (minimize) the cost of operation it is necessary to adopt mathematical approach. The cost of the processes depends on the cost of heating and cooling utilities. The heat balance across the interval can be represented by the hat balance diagram across the interval.

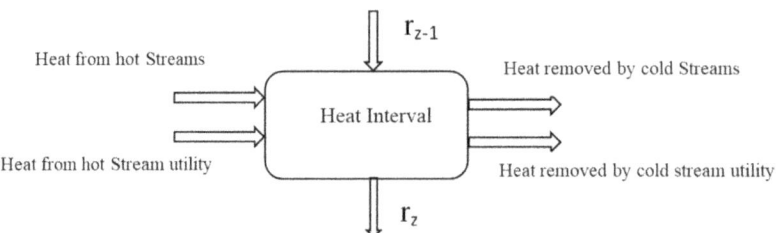

Writing a heat balance across the heat interval
Heat from hot stream-heat removed by cold stream
= heat form hot utility($Q_{HU}$) − heat removed by cold utility($Q_{HC}$) + $r_z - r_{z-1}$.

The objective function is to minimize cost of utility usage. The constraints are heating utility, cooling utility and residual heats, which can be greater than or equal to zero. The optimization of heat exchange networks can be carried out using LINGO programming. For heat integration problem which discussed for optimization of cost. Let us consider the heating utility is available at a cost of Rs 2.5 per kJ and cooling utility is available at Rs 3.5 per kJ. The plant is operating for 8000 h in a year.

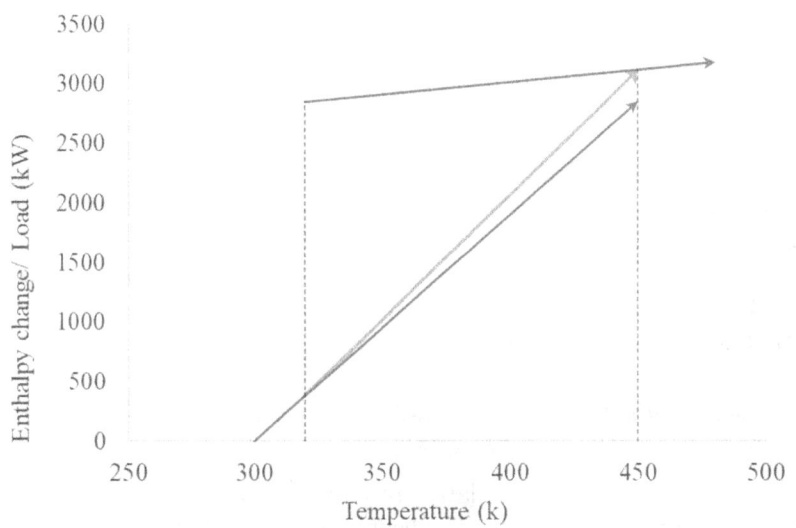

**Figure 5.10.** Cold stream composite curve.

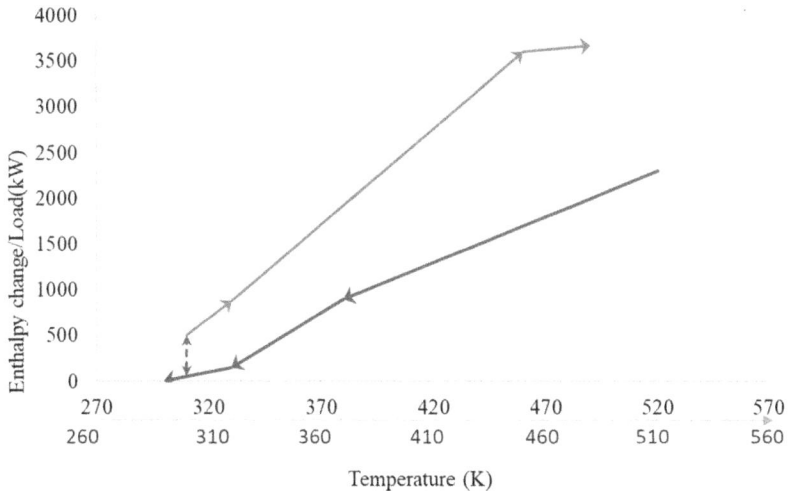

**Figure 5.11.** Intermediate thermal pinch diagram.

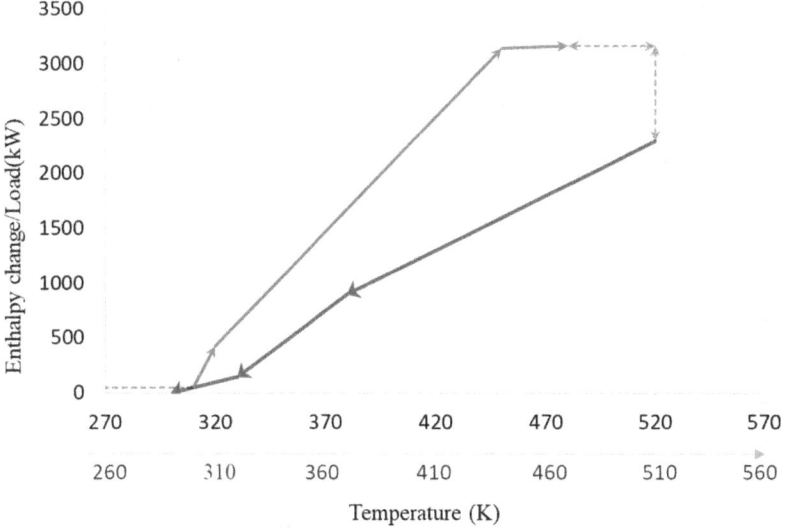

**Figure 5.12.** Thermal pinch analysis.

The objective function can be written as

Minimize = (2.5 × QHUmin + 3.5 × QCmin) × 3600 × 8000.

Based on the heat intervals as explained in algebraic method, the heat interval is considered to solved the problem using LINGO as explained by El-Halwagi [5].

```
min=(72000*QHmin)+(1000800*QCmin);
r1-QHmin=300;
r2-r1=240;
```

```
r3-r2=-880;
r4-r3=-240;
r5-r4=-60;
r6-r5=-140;
r7-r6=-140;
-r7+QCmin=50;
QHmin>0;
QCmin>0;
r1>=0;
r2>=0;
r3>=0;
r4>=0;
r5>=0;
r6>=0;
r7>=0;
end
```

```
Global optimal solution found.
  Objective value:                          0.1162800E+09
  Infeasibilities:                              0.000000
  Total solver iterations:                             0
  Elapsed runtime seconds:                          0.21

  Model Class:                                        LP

  Total variables:                   9
  Nonlinear variables:               0
  Integer variables:                 0

  Total constraints:                18
  Nonlinear constraints:             0

  Total nonzeros:                   27
  Nonlinear nonzeros:                0

Variable                    Value
QHMIN                    920.0000
QCMIN       50.00000
R1          1220.000
R2          1460.000
R3          580.0000
R4          340.0000
R5          280.0000
R6          140.0000
R7          0.000000
```

The results obtained for residual heat loads and pinch point by LINGO matches with the results of algebraic method and the total cost of the operation is Rs. 11 628 000.

## 5.6 Combined heat power integration

### 5.6.1 Heat recovery from high temperature equipment

The temperature regulation is imperative for equipment operating at very high conditions in industries like petrochemicals, metallurgy, glass, cement, ore extraction and fertilizers. Temperature of blast furnaces can range from 180 °C–1600 °C depending on the distance from the hearth [10] and that of electric arc furnaces can reach 1800 °C [11] as shown in the figure 5.13. Based on the temperature zones along the height of the furnace, different grades of steam can be generated. The region closest to the hearth of furnace liberates the highest amount of heat. The heat losses from this zone are heat rich sources and can be considered for VHP steam generation. The subsequent zones are then used for high pressure and medium pressure steam generation that find use as heating utilities. Whereas the zone furthest from the bottom, although does not produce quality heat for recovery, can nevertheless be used for boiling water. A schematic diagram below illustrates the potential of heat recovery from blast furnaces.

### 5.6.2 Combined heat and power integration

Heat, like work, is a form of a transient energy, which means it exists only when it is transferred from one body to the other. Its quantification is thus based on its ability to bring about changes in other forms of energy that are stored within a body [12–14] such as in terms of raising the temperature of fluid of known mass and specific heat (thermal energy) or setting a stationary object into motion (kinetic energy). When excess heat is liberated as a result of industrial processes, it serves as a potential source, not only for recycling as a heating utility for cold streams, but also to

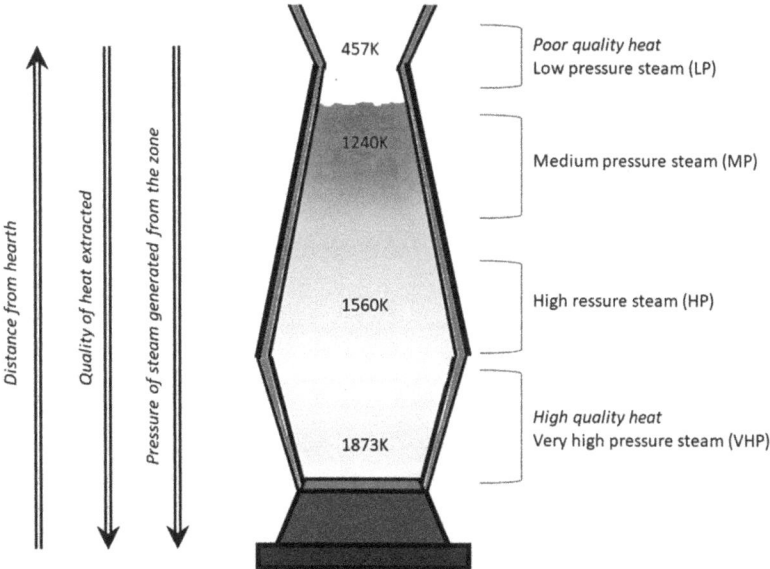

**Figure 5.13.** Temperature zones of a typical blast furnace.

generate electrical power to serve other processing units. This process is called cogeneration or combined heat and power (CHP).

Conventionally, power is generated through steam turbines, nuclear fission, wind turbines, geothermal turbines, solar cells, to name a few. These power generating units convert various forms of energy directly or indirectly to electrical energy. As of 2019, 62.7% of the total electricity generated globally is derived from the combustion of oil, natural gas and coal, all of which are exhaustible fuel sources [15]. These fossil fuels are essentially burnt in boilers and furnaces to produce various grades of steam. Very high-pressure (VHP) and high-pressure (HP) grade steam then deliver rotational shaft work to generate electrical power. This is the basic principle of steam turbine operation. The combination of mass and heat integration is essential for effective heat recovery and utilization, which is represented by a block diagram as shown in figure 5.14.

For the practice of heat integration, stream redirection plays a pivotal role in reducing the direct utility costs for heating and cooling. Additionally, the combination of heat sources and sinks can be used to generate power by installation of heat engines. Heat engine is a power generating equipment that converts heat energy to work as a result of energy transfer from a high-temperature source to a low-temperature sink.

Another means of heat recovery is from the flaring of combustible by-product streams to produce VHP steam for power generation. Large volumes of gases are flared in industries such as petroleum processing plants or refineries, oil-gas extraction and chemical plants to vent excessive pressure build-up and ease the load of the equipment and as a means of disposal of by-products that are otherwise deleterious to the environment. Emam A. Emam [16] reported that 150–170 billion $m^3$ gases are annually flared or vented. It is not only essential to minimize the waste generation, thereby reducing flaring of these combustible streams, it would benefit the operation of the plant if the heat generated by flaring of flue gases are be tapped for power generation. This establishes the holistic integration of the process as it conflates both mass-integration strategies with that of CHP fundamentals. Based on

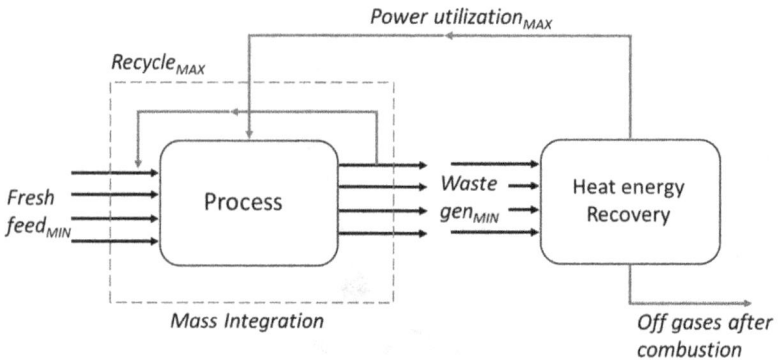

**Figure 5.14.** Combined mass and power integration strategy.

criteria such as cost of power, raw material price and environmental impact of the off-gases, one can prioritize amongst the following targets:

- Maximum recycle of combustible streams to reduce fresh feed procurement.
- Maximum power generation for optimal plant output.
- Minimum gaseous waste generation.

## 5.7 Thermal energy storage(TES)

So far we have discussed energy recycling and combined power generation from waste heat as effective and low cost means of sustainable heat utilization. Yet, it is still very far from a zero energy system [17]; a process that generates nearly as much energy as it consumes. A strategy that has a potential to achieve a net zero emission energy system in terms of heat employs the storage of excessive heat and inviable renewable energy sources. Thermal energy storage or *TES* is the method of retaining or storing heat for varied periods of time for future consumption. TES systems provide both environmental and economic benefits by reducing the consumption of non-renewable energy resources and prevent the loss of thermal energy by storing excess heat until it is consumed. [17] As heat energy is a transient form of energy, thermal energy needs to be converted to other forms for conducive future utilization. TES can be carried out by three popular mechanisms based on the type of the material selected for storing heat and cold:

- I. Sensible heat storage.
- II. Latent heat storage by the use of phase change materials (PCMs).
- III. Thermo-chemical storage (TCS).

### 5.7.1 Sensible heat storage

Sensible heat storage is a simple method wherein excess heat generated from a high-temperature energy source is used to change the temperature or 'sensible heat' of the storage medium without causing a change in phase, thereby generating a hot or cold potential for future use. Water is a versatile fluid with favourable attributes such as high thermal capacity, easy handling and storage, low volatility, non corrosivity and low cost. It is thus the most common medium used for this purpose in both domestic and industrial settings [18]. Sensible heat storage has paved way for renewable energy consumption such as solar and geothermal energy. The other materials used for storing the sensible heat is thermal oil, molten salts, liquid metals concrete and naturally available earth materials. Thermal oils are advantageous over water, as they remain in the liquid phase up to 250 °C to 400 °C under atmospheric conditions. Molten salts are preferred as heat storage medium for high temperature (>400 °C), which is highly preferred in solar thermal plants because of its thermophysical properties. Solar water heaters work in the following manner; heat from the Sun is tapped by the highly absorptive collectors or evacuated tubes depending on the design, that heat up the water in a tank. This water is then consumed through a centralised or distributed system. Borehole and aquifer storage also make use of the principle of sensible heat. Borehole storage refers to heat extraction of underground earth wherein vertical heat exchangers enable bidirectional heat transfer to and from the ground layers for surface use[19]. Aquifers, on the other hand, extract or dump heat into naturally occurring underground water springs depending on the heating or cooling requirements of the season [20]. During hot seasons, heat is discharged to the medium, and during the colder weathers, heat is extracted. In case of aquifers, low heat density of underwater springs necessitates that heat dumping during cold weather be coupled with the installation of heat pumps [21, 22]. It is important to understand the working mechanism of a generalised sensible heat storage system regardless of the medium used. The excess energy from a heat source is utilised to raise the thermal heat of the medium without bringing about its phase change. The heat sources could be waste heat generated by process systems or energy inherently present in renewable sources such as hot water springs, solar energy or the heat contained in earth by virtue of its heated core. This charged heat within the medium is then stored in a highly insulated vessel based on the design, periodicity of requirement and size of the vessel. The period of time for which the high temperature of the medium is retained is known as the storage time. At the time of utilization, heat is derived from the medium in the *discharging phase* of the storage cycle. This principle is illustrated in the schematic diagram (figure 5.15). It is important to note that the insulation may be synthetic by design as with industrial hot water tanks and solar water tanks, or exist in nature, such as the insulation provided by rock beds for underground heat sources.

### 5.7.2 Phase change materials (PCMs)

Sensible heat storage it is limited in its application due to certain setbacks. Sensible heat storing systems have low efficiency, low energy density and physical size that pose an impediment in the large-scale implementation of the system [23]. They also inevitably have a variable discharging temperature [24]. It is therefore used for daily and weekly plant requirements as opposed to long-term TES. A more superior alternative like the use of PCMs, therefore becomes significant. PCM-based TES

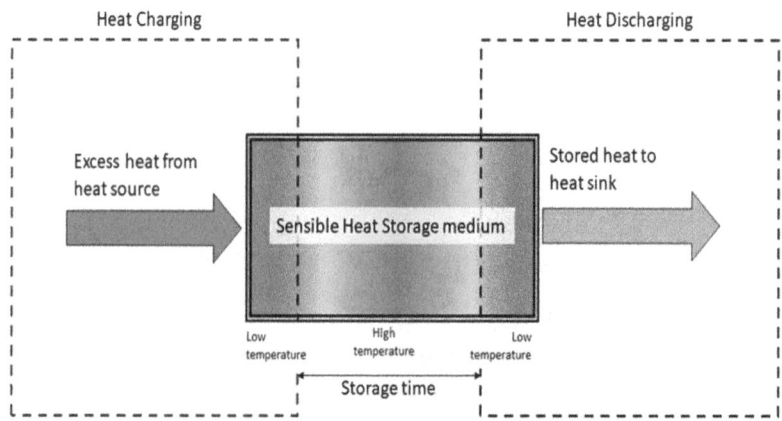

**Figure 5.15.** Principle of sensible heat storage systems.

**Table 5.6.** Thermal storage properties of common PCMs [18].

| PCM | Phase change temperature (°C) | Enthalpy of phase change (kJ kg$^{-1}$) | Density range (kg m$^{-3}$) |
|---|---|---|---|
| Ice (Solid to Liquid) | 0 | 333 | 920 |
| Water (Liquid to gas) | 100 | 2250 | 1000 |
| Organic PCM | | | |
| Paraffin's | −10 °C – 53 °C | 150–250 | 770–785 |
| Fatty acids | 32–69 | 150–200 | 940–980 |
| Esters | 11–43 | 122–216 | 800–880 |
| Alcohols | 23.3–165 | 110–280 | 790–900 |
| Glycols | 12.5–70 | 160–174 | 7000–7128 |
| Inorganic PCM | | | |
| Salt hydrates | 29–89 | 115–280 | 1450–1800 |
| Nitrates | 306–560 | 145–172 | 2113–2261 |
| Carbonate salts | 732–1330 | 142–509 | 1972–2930 |
| Chloride salts | 192–772 | 45–714 | 1502–2907 |
| Sulfate salts | 884–1460 | 84–203 | 2003–2680 |
| Fluoride salts | 850–1418 | 39–1044 | 1984–2640 |
| Alloys | 340–996 | 185–754 | 2380–7000 |

have higher heat capacities, can discharge heat energy at constant temperatures and use various materials based on the range of temperature and power requirements. The phase change materials can be either organic or inorganic materials. The organic materials include paraffins, fatty acids, esters, alcohols and glycols. Inorganic materials such as salt hydrates, nitrates, chlorides, carbonates, sulphates, fluorides, hydroxides, metals and metal alloys can be used as phase change materials. Table 5.6 shows certain PCMs with their phase change temperatures and heat capacities. [18]

TES systems employ the method of latent heat storage of materials with high thermal capacities. Heat is charged when the material undergoes change in phase and released when the material transitions from one phase to the other (Gas to liquid, liquid to solid or solid to liquid). These transitions occur at sharp and nearly isothermal conditions. PCM-based TES systems can effectively store 50–100 times more heat per unit volume than sensible heat storage systems [18]. Latent heat storage (LHS) using PCMs can be achieved through various techniques and methods. For example, encapsulated PCM technology employs an approach of incorporating PCMs (paraffin-like n-Dodecanal and n-Tridecane) into the target system with the objective of cooling. PCMs that solidify over nightfall, melt during the hot day, thereby passively cooling the space. This can be applied to building spaces and industrial cooling purposes.

### 5.7.3 Thermo-chemical storage (TCS)

Another promising strategy to achieve highly efficient thermal storage systems is via chemical reactions. Most reactions that require an initial activation energy are exothermic, but cannot necessarily be used to store heat. For a storage system, a medium must facilitate the transfer of waste heat at one instant of time to a consumption unit at a later time, thereby 'storing' the excess heat for the duration of inactivity without considerable losses. Chemical adsorption is reversible in nature and can be effectively employed for the storage of heat for long durations merely by adjusting the external conditions to favour either direction of reaction. It therefore makes for a viable pathway for TCS systems. Most commonly applied TCS systems benefit from the adsorption of water vapour on crystalline alumino-silicates, otherwise known as zeolites. Zeolites and its composites have a cage-like structure with cavities containing cations that house water molecules within it as a result of ion exchange. Water can be adsorbed or desorbed from the zeolite composite sites by what is called reversible dehydration [25]. The ease of movement of water depends on a lot of factors. One of which is the size, structure and chemical composition of the cavities. The stronger the water is adsorbed on the zeolite, higher will be the temperature of dehydration. For zeolites with larger pores, some water is released at lower temperatures. The sorption of water on the zeolites is controlled by the ambient humidity for TCS. As the humidity is decreased, the water molecules detach from the zeolite surface (desorption). This constitutes the charging phase of the TES. As the humidity is increased, water gets adsorbed and discharges heat. Therefore, heat energy can be adeptly stored by this mechanism which is illustrated in figure 5.16. High heat capacities of TCS systems make thermal energy transportation a reality. In theory, heat generated from a blast furnace unit could be transported to a paper mill for drying operations. The zeolite is charged at the heat source wherein water is lost, it is then transported to the drying plant where water is adsorbed and heat is liberated from the zeolite. Some of the sorptive materials that are being studied are alumino-phosphates, functional adsorbents, selective water adsorbents and porous salt hydrates [26].

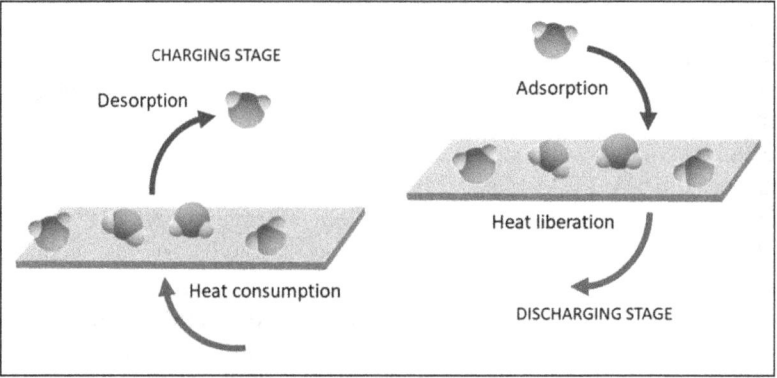

**Figure 5.16.** Cycle of charging and discharging of heat by sorptive activity of water molecules on the zeolite surface.

### 5.7.4 Comparative study of TES strategies

Although thermal energy storage is the future of the renewable energy sector and has the potential to enable zero energy systems, it still requires a tedious process of research to make the approach economically feasible. Sensible heat storage is the most common TES practice adopted by countries of all economic statuses to varied degrees. For countries that can disburse high capital for the purpose of designing environmentally friendly processes, PCM and TCS techniques can be implemented. Sensible heat storage materials have high thermal stability at elevated temperatures, ease of handling and low cost, they are used in high-temperature applications. The main disadvantage is low thermal stability during discharge. PCM latent heat is 50 to 100 times better than sensible heat storage. The major drawback is poor thermal conductivity and enormous change in volume. TCS has benefits like highest thermal storage density (per unit mass and unit volume), low heat losses and longer duration of heat storage. The main problem associated with TCS is sintering and grain growth resulting reduction in the porosity. PCM and TCS systems have a colossal advantage over conventional sensible heat storage in terms of energy density, discharge temperature control, efficiency of heat storage and independence from seasonal variations in weather. Table 5.7 illustrates the comparative parameters of each technology [27]

However they come with the drawback of material design and mechanism complexity which requires further research and development before commercial application. For high-temperature TES application PCMs and TCS require the design of stable material medium and reactors. This is the main reason that these technologies are not yet widespread in the industries. Sensible heat storage requires the design of super insulators to prevent energy leaks which may make the approach feasible. The choice of TES is therefore invariable depending on the application and requirement of the industry implementing it. It is needless to say that TES technology is a major competitor in the green energy sector and the future of zero energy systems.

## 5.8 Conclusion

Process integration plays a vital role in heat integration. Mass integration of material streams within a plant operating as a network of various streams containing the targeted species reduces the fresh feed requirement and losses at disposal. This significantly reduces the operating costs in terms of raw material expenditure. Minimizing waste generation simultaneously brings about de-bottlenecking at the

**Table 5.7.** Typical parameters of TES technologies.

| Parameter | Sensible (hot water) | PCM | Chemical reactions |
| --- | --- | --- | --- |
| Capacity (kWh t$^{-1}$) | 10–50 | 50–150 | 120–250 |
| Power (MW) | 0.001–10 | 0.001–1 | 0.01–1 |
| Efficiency (%) | 50–90 | 75–90 | 75–100 |
| Storage period (h,d,m) | d/m | h/m | h/d |
| Cost (€ kWh$^{-1}$) | 0.1–10 | 10–50 | 8–100 |

waste treatment plants as well as fresh feed consumption of the process plant. Heat integration refers to the maximum heat recovery from hot process streams to fulfil the requirement of the cold process streams, thereby reducing external heating and cooling requirements. Mass integration and heat integration can be done by graphical and algebraic pinch analyses without getting into intricate details of the processes. It provides an idea of possible heat and material recycle within the process. Heat recovery combined with power generation is referred to as combined heat and power or cogeneration. This strategy enables process plants to cut down on the power costs of operation by extracting usable power from waste heat rich streams. The scope of cogeneration is extensive as it is a low-cost strategy with definite high impact. With natural resources at the brink of exhaustion and staggering global emission rates, the need for reduction in consumption, recycle and integration of virgin resources without bringing about any substantial changes to the output of the plant has become inevitable. Mass and heat integration can be carried out for various targets simultaneously, for instance heat integration can be done without affecting the mass integration of a target species. It is necessary that all process plants adopt the integration approach to achieve highly efficient, cheap and clean processes.

A brief review is presented on thermal energy storage, which is the future of the renewable energy sector and has the potential to enable zero energy systems. Various thermal energy systems such as sensible heat storage, phase change materials and thermo chemical storage are discussed in detail. PCM latent heat is much higher than sensible heat storage. The major drawback is poor thermal conductivity and an enormous change in volume. TCS has benefits like highest thermal storage density (per unit mass and unit volume), low heat losses and longer duration of heat storage. PCM and TCS systems have a huge advantage over conventional sensible heat storage.

## References

[1] Kusch-Brandt S 2019 Urban renewable energy on the upswing: a spotlight on renewable energy in cities in REN21's *Renewables 2019 Global Status Report* 139

[2] Papapetrou M, Kosmadakis G, Cipollina A, Commare U L and Micale G 2018 Industrial waste heat: estimation of the technically available resource in the EU per industrial sector, temperature level and country *Appl. Therm. Eng.* **138** 207–16

[3] Holzleitner M, Moser S and Puschnigg S 2020 Evaluation of the impact of the new Renewable Energy Directive 2018/2001 on third-party access to district heating networks to enforce the feed-in of industrial waste heat *Util. Policy* **66** 101088

[4] Arzbaecher C and Parmenter K E F 2007 Industrial waste-heat recovery: benefits and recent advancements in technology and applications *Proc. of the ACEEE* pp 2–1

[5] El-Halwagi M M 2006 *Process Integration.* (Amsterdam: Elsevier)

[6] El-Halwagi M M 1997 *Pollution Prevention through Process Integration: Systematic design Tools.* (Amsterdam: Elsevier)

[7] Kemp I C 2011 *Pinch Analysis and Process Integration: A User Guide on Process Integration for the Efficient Use of Energy.* (Amsterdam: Elsevier)

[8] Papoulias S A and Grossmann I E 1983 A structural optimization approach in process synthesis—II: heat recovery networks *Comput. Chem. Eng.* **7** 707–21

[9] Papoulias S A and Grossmann I E 1983 A structural optimization approach in process synthesis—I: utility systems *Comput. Chem. Eng.* **7** 695–706

[10] Shenoy U V 1995 Heat exchanger network synthesis: process optimization by energy and resource analysis *Includes Two Computer Disks.* (Houston, TX: Gulf Publishing Company)

[11] Kemp I C 2019 *Pinch Analysis and Process Integration: A User Guide on Process Integration for the Efficient Use of Energy* Elsevier
Ubale D and Ubale P 2011 Numerical investigation of temperature distribution in blast furnace hearth *IOP Conference Series: Materials Science and Engineering* **691**

[12] Ubale D and Ubale P 2019 Numerical investigation of temperature distribution in blast furnace hearth *IOP Conf. Series: Materials Science and Engineering* 691

[13] Trejo E, Martell F, Micheloud O, Teng L, Llamas A and Montesinos-Castellanos A 2012 A novel estimation of electrical and cooling losses in electric arc furnaces *Energy* **42** 446–56

[14] Smith J M and Van Ness H C 1959 *Introduction to Chemical Engineering Thermodynamics.* (New York: McGraw-Hill)

[15] British Petroleum 2020 Statistical Review of World Energy 2020 (London: British Petroleum)

[16] Emam E A 2015 *GAS Flaring in Industry: An Overview.* **57** 532–55

[17] Davis S J, Lewis N S, Shaner M, Aggarwal S, Arent D, Azevedo I L and Benson S M *et al* 2018 Net-zero emissions energy systems *Science* **360**

[18] Alva G, Lin Y and Fang G 2018 An overview of thermal energy storage systems *Energy* **144** 341–78

[19] Kizilkan O and Dincer I 2015 Borehole thermal energy storage system for heating applications: Thermodynamic performance assessment *Energy Convers. Manage.* **90** 53–61

[20] Lee K S 2013 Underground thermal energy storage *Underground Thermal Energy Storage* pp 15–26 (London: Springer)

[21] Cruickshank C A and Baldwin C 2016 Sensible thermal energy storage: diurnal and seasonal *Storing Energy* pp 291–311 (Amsterdam: Elsevier)

[22] Zeghici R M, Oude Essink G H P, Hartog N and Sommer W 2015 Integrated assessment of variable density–viscosity groundwater flow for a high temperature mono-well aquifer thermal energy storage (HT-ATES) system in a geothermal reservoir *Geothermics* **55** 58–68

[23] Hauer A, Quinnell J and Lävemann E 2013 Energy storage technologies-characteristics, comparison, and synergies *Transition to Renewable Energy Systems* 1st ed. (*Wiley-VCH*)

[24] Khan Z, Khan Z and Ghafoor A 2016 A review of performance enhancement of PCM based latent heat storage system within the context of materials, thermal stability and compatibility *Energy Convers. Manage.* **115** 132–58

[25] Nonnen T, Preißler H, Kött S, Beckert S and Gläser R 2020 Salt inclusion and deliquescence in salt/zeolite X composites for thermochemical heat storage *Microporous Mesoporous Mater.* **303** 110239

[26] Hauer A 2006 Thermochemical energy storage systems, CIMTEC5th Forum on New Materials.

[27] Hauer A 15 February 2011 Storage technology issues and opportunities, committee on energy research and technology (international energy agency), international low-carbon energy technology platform, strategic and cross-cutting workshop *Energy Storage Issues and Opportunities'. (Paris. France)*

# Chapter 6

## OTEC Rankine and Stirling engines

**B V Raghuvamshi Krishna**

## 6.1 Introduction

A naturally replenished energy—renewable energy is derived from endless sources such as wind, solar, hydroelectric action. On the other hand, fossil fuels, such as natural gas, oil are coal, are finite sources causing pollution and global warming, because of this most of the countries are depending on renewable energy resources. Since climate change is a threat to humankind across the world, it should stop reducing its dependency on fossil fuels. Based on the report given by the International Energy Agency, 67% of the electric power is produced from the fossil fuels and due to its depletion and fluctuating prices leads to use of renewable energy. Power generated from the renewable resources such as wind and solar are highest and around 227 GW of electric power is generated from solar in the year 2015, and the wind power boomed from 59 GW in 2005 to 433 GW in 2015 [1].

The Government of India is projected to generate electric power of 500 GW by 2030, as of now it stood at generating the 90.39 power, out of which 36.91 GW power comprised solar, 38.43 GW power from the wind and the rest by biomass (10.14 GW) and hydropower (4.74 GW) respectively.

Mankind is fenced in rotary machines such as steam and gas turbines, motors, pumps, gearboxes, aircraft engines, drive trains, and machine tools, many of the machines depends on fossil fuels and electricity and due to huge population, consumption of energy is very high. In a particular region if the population is approximately growing at a rate of 1.05% per year then the rate of consumption of energy increases by 0.52%, this leads to an increase in usage of fossil fuels and due to this greenhouse gases are released into the atmosphere, so it is mandatory to implement renewable energies such as wind, solar and geothermal power plants [2].

### 6.1.1 Wind energy

It is confirmed that wind energy is clean and finest renewable resource for producing the electricity because wind is abundantly available. The generation of electricity from the wind power increased enormously in the span of 10 years from 602 MW to 14 401 MW. Solar radiation and the Earth's rotation are the main factors that influences the air circulation in the atmosphere. Due to this direction and speed of wind is well defined, and electricity is generated as the kinetic energy of the wind hits the blades in turbine rotor. The wind power can be calculated based on the equation (6.1).

$$P = \frac{1}{2}\rho A_r v^3 C_p \tag{6.1}$$

where $P$ is the wind power in watts—W.
$\rho$ is the density of air in Kg m$^{-3}$.
$A_r$ is the area encompassed by the rotor—m$^2$.
$v$ is the wind speed—m s$^{-1}$.
$C_P$ is the aerodynamic coefficient of rotor power, and it is unitless.

The electricity from the wind farm turbines can be obtained as a product of the wind power from equation (6.1) and operating hours of the farm turbines. Considering the wind farm turbines plant operated continuously throughout the year and by using the equation (6.2) is used to compute losses due to some technological issues in the transmission system. Approximately this amounts to 3%–4% of the annual electric power.

$$E_{pw} = \frac{8760 \times 0.93}{2}\rho A_r v^3 C_p \eta \tag{6.2}$$

where $E_{pw}$ is the wind electricity in MWh [3].

### 6.1.2 Solar energy

Populated countries are suffering from energy problems and heavily depend on fossil fuels which causes air pollution and burning the fuel & diesel oil generates noxious gases such as $NO_x$ and $CO_2$. Due to electric disruption public must depend on private diesel generators, and this leads to air pollution because of the emissions from generators. As the citizens breathe these pollutants, that lead to physical as well as ill health problems such as depression, increased premature death, lung cancer, cardiovascular, and other types of diseases. It is proved that solar energy is outstanding renewable energy resource, and the country enjoys 3000 h of sunshine in a year and by using this solar energy fuel burning can be avoided thereby decreasing the carbon footprints and air pollution. As the huge amount of solar energy is available, replacing the diesel generators by microgrids is predicted to have lot of advantages. An analytical study was carried out to verify this at total cost that takes in consideration on comparing the cost paid for both diesel generators and microgrid energy production. A mathematical equation was developed to calculate the energy produced from diesel generator, which is given by:

$$TC_{Di} = IC_D + EC_{Di} + CC_{Di} + MC_{Di} \tag{6.3}$$

where,

IC$_D$ = financial cost of diesel generator per hour.

EC$_{Di}$ = environmental cost incurred when using diesel at the hour $i$ according to the energy Wi.

CC$_{Di}$ = consumption cost of diesel generator at the hour $i$ according to the energy Wi.

MC$_{Di}$ = maintenance cost of diesel generator at the hour $i$ considering maintenance costs, reliability data of diesel generator and the real energy Wi needed.

Investment cost is funds paid by the individual for installing the complete set-up and to make it operational. The diesel generator consist of many parts and it needs regular maintenance and always fuel oil is needed to run the generator driven by diesel engine. By installing the microgrid systems an average of 220 USD can be saved, once the microgrid is installed, it is driven by solar energy which is freely available and maintenance cost is less as compared to diesel engine [4].

### 6.1.3 Ocean thermal energy conversion

#### 6.1.3.1 Closed-cycle OTEC system

The majority of the Earth surface is covered by the water and 80% of solar energy received from the Sun is stored in oceans around the world. There is a method to utilize the maximum this enormous energy accumulated in ocean. A remarkable difference in temperature exists between the surface and depth of the ocean separated by a thermocline, the surface is warmer, and depth of ocean is cooler that forms a closed thermodynamic cycle. The warm source can be used as working fluid for driving a turbine. Figure 6.1(a) shows the solar—OTEC system and there are three main parts in this system which are discussed below:

- Solar collector: By transferring the warm water from ocean surface through solar collector it will get heated up and the same will flow through the evaporator. The main advantage of using this equipment is that additional heat will be added so that water will be warmer and by doing this it will be very effective for heating the working fluid.
- Organic Rankine cycle: Fluids with low boiling points such as methanol, toluene, ammonia, ethylbenzene, and ethanol may be used as working agents for this organic Rankine cycle. Working fluid that goes through the evaporator absorbs heat for phase change and then it will drive the turbine and the working fluid will be cooled by the condenser, cooled by the oceans cold side and it will be returned to the evaporator to form the closed cycle.
- Hydrogen producing unit: This unit will produce hydrogen by separating oxygen and hydrogen flows through the polymer exchange membrane (PEM) and it is transferred to tank as a clean fuel. PEM needs electric power and the electric power from turbine is transferred to PEM to perform chemical procedure for producing hydrogen.

Figures 6.1(a) and 6.1(b) both are similar S-OTEC with a difference of the thermoelectric generator [TEG] module and this module is added to increase the electrical output.

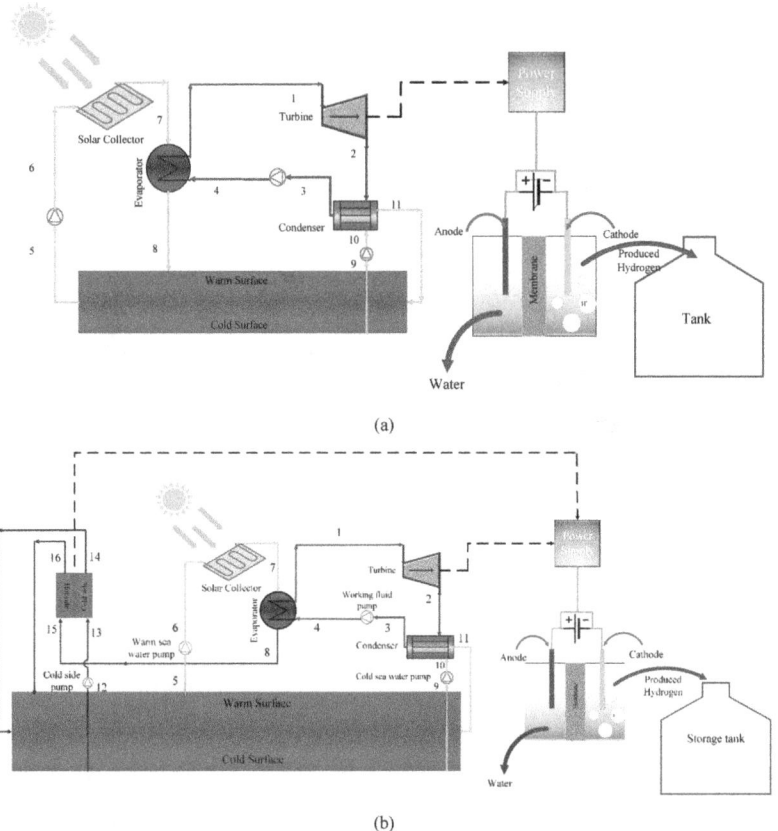

**Figure 6.1.** (a) OTEC system. (b) OTEC system with thermoelectric generator. Reprinted from [2], Copyright (2020), with permission from Elsevier.

The main purpose of this module is to increase the electric power output, and the electricity produced from this module will be transferred to PEM and the electricity produced from the turbine. Therefore S-OTEC with TEG system has high power output compared to normal S-OTEC system and energy and exergy efficiency can be improved.

In TEG module temperature gradients exist due to hot side at the outlet of evaporator and cold side from the ocean flow, so the electric power will be generated in TEG modules. The power generated is directly related to efficiency of TEG and value of transmitted heat from the cold side.

Low boiling point fluids/organic fluids such as ammonia, ethanol, methanol and ethylbenzene are used in cycle for hot source temperature and this working fluid absorbs heat of warm water in the evaporator of the OTEC unit and it will expand and drive the turbine. In the S-OTEC system working fluids methanol, toluene, ammonia, ethylbenzene, and ethanol are used and methanol has given the best performance for hydrogen production. Figure 6.2 shows the comparison of working fluids, and performance of methanol is best with and without the TEG module [2, 5]

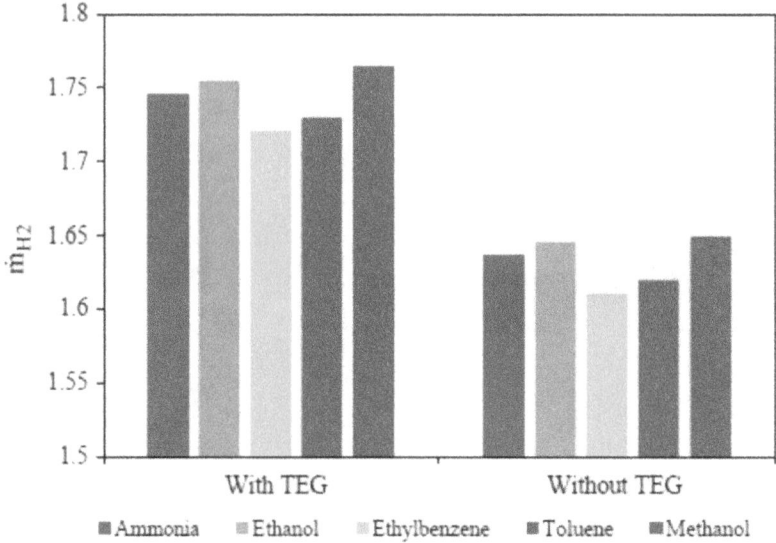

**Figure 6.2.** Rate of production of hydrogen in S-OTEC/TEG and S-OTEC for different working fluids. Reprinted from [2], Copyright (2020), with permission from Elsevier.

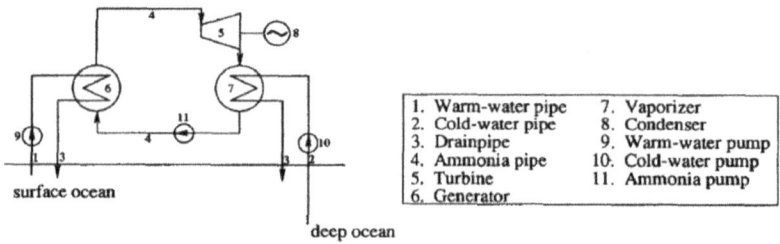

**Figure 6.3.** Closed-cycle OTEC system. Reprinted from [6], Copyright (1995), with permission from Elsevier.

Tahara *et al* [6] used ammonia ($NH_3$) as the working fluid in the closed-cycle OTEC system as shown in figure 6.3. Coal power plant produces large quantity of $CO_2$ emissions and this can be reduced by establishing the OTEC system at 140 000 t-C year$^{-1}$ [6].

The net energy output from system is $5.66 * 10^6$ GJ and 20% of the energy is consumed for operating the system, especially for deep water from the ocean. By replacing the coal-fired plant with the OTEC system, $CO_2$ emission can be reduced by 140 000 t-C year$^{-1}$. Concentration of inorganic $CO_2$ is very high in deep ocean water compared to surface ocean water and it is produced due to decomposition of phytoplankton. Therefore, the emitted $CO_2$ must be anchored immediately by the phytoplankton at the surface of ocean using the nutrients available in deep ocean water [6].

*6.1.3.2 Open-cycle OTEC*
Due to the use of expensive heat exchangers in the closed cycle OTEC system, it is proposed to use the steam generated from the warm sea water as the working fluid of

the OTEC system, Open cycle OTEC system is represented in figure 6.4, and shows the complete system from evaporator to condenser which operates at a pressure of 1 to 3 atmosphere. For this low-pressure OTEC system boiling system is needed to evaporate the warm water and flash evaporation is done by exposing the warm water to below saturation pressure to its corresponding temperature. The gases which are dissolved in sea water should be removed before prior to evaporation. Pure steam is produced in the flash evaporator and it is nothing but vapor. Pressure drop is established by cold water across the turbine, and steam condenses to 813 Pa at 4 °C. So, the turbine exit pressure should not fall below this value, maximum pressure drops in the turbine will be about 3000 Pa which corresponds to 3:1 pressure ratio. The pressure drop continues along the steam path where the differences in steam temperature and sea water temperature facilitates the heat transfer in evaporator and condenser. So the vapor gets condensed with cold sea water. In this open-cycle OTEC system, expensive heat exchangers are replaced by low system pressures. The main disadvantage of this open cycle OTEC system is low system operating pressure which requires sizable components for accommodating high volumetric flow rate of steam [7].

### 6.1.4 Stirling engine performance (SE)

Another alternative for producing power is the use of renewable energy resources and combined heat and power which is clean and efficient for energy production. Stirling engines depends on external combustion or external heat source and the working fluid will be helium, air, nitrogen or hydrogen and it operates on the closed cycle with cyclic expansion and compression process. Movement of the piston is due to pressure variation of working fluid that results in repeated cooling and heating phenomena, and it is converted to mechanical work. Researchers in the last two decades have designed, developed, and tested Stirling engines to evaluate their performance. The very first design in this field is that of Stirling engine with four cylinders with a maximum power output of 40 KW by combusting the wooden chips. Table 6.1 shows the use of SE to convert biomass energy to power energy.

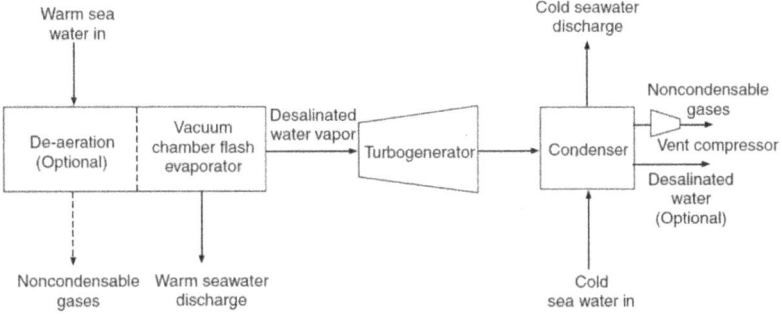

**Figure 6.4.** Open-cycle OTEC system. Reprinted from [8], Copyright (2020), with permission from Elsevier.

**Table 6.1.** Stirling engines are used to convert biomass energy to power energy. Reprinted from [8], Copyright (2020), with permission from Elsevier.

| Author | Year | Type of work | Fuel | Micro-generation Unit | Thermal output (kW) | Power output (kW) | Working gas | Pressure (bar) | Speed (rpm) | Hot Temperature (K) | Overall efficiency (%) | Electrical efficiency (%) |
|---|---|---|---|---|---|---|---|---|---|---|---|---|
| Cardozo et al. [62] | 2019 | Experimental | Wood and sugar cane bagasse pellets | Experimental test rig | 20 | 1 | Nitrogen | 15 | NA | ≈773 | >83 | >9 |
| Sowale et al. [75] | 2018 | Numerical | Human wastes | Bench-scale downdraft combustor test rig | 25 | ≈27.2 | Helium | 10 | 1431 | 663 | NA | 17.81 |
| Schneider et al. [76] | 2018 | Description of a new Pilot CHP plant | Wood pellets, chips and biogenic products | Pilot plant with a SE from FTM | 45–50 | 5 | Helium | 33 | NA | NA | NA | 25 |
| Damirchi et al. [68] | 2016 | Experimental and Numerical | Agriculture wastes | Experimental test rig | NA | 0.046 (max.) | Helium | 3–10 | 400–700 | 643–683 | NA | 16 (max.) |
| Schnetzinger et al. [77] | 2016 | Experimental | Wood pellets | Biomass pellet burner prototype with a SE from FTM | NA | 4–5 | NA | NA | NA | ≈923 | NA | 14.4–15.5 |
| Cardozo et al. [61] | 2014 | Experimental | Wood pellets | Experimental test rig | ≈15 | 0.46–0.49 | Nitrogen | 12–25 | NA | ≈773 | >72 | ≈3 |
| Kölling et al. [78] | 2014 | Experimental (proof of concept) | Wood chips | KÖB boiler with SE SOLO 2V | 12.7 | 4.5 | Helium | 120 | NA | 973 | 80 | 20 |
| Senkel et al. [66,79] | 2012 | Experimental | Wood chips | KÖB boiler with SE SOLO 2V | 12.7 | 4.5 | Helium | 120 | NA | 973 | 80 | 20 |
| Renzi et al. [80] | 2014 | Numerical | Biogas | – | 4.44 | 1 | NA | NA | NA | 773 | 90 | 22.5 |
| Müller et al. [81] | 2013 | Experimental | Wood pellets | Fluidized bed facility with Sunmachine SE | 15.5 and 16 | 2.05 and 2.45 | Nitrogen | 31 and 39 | NA | NA | 26.7 and 25.6 | 13.2 and 15.3 |
| Marinitsch et al. [82] | 2011 | Experimental | Wood chips | Updraft gasification unit | 141 | 22 | Helium | 46 | 1010 | ≈873 | 80.3 | 10.8 |
| Thiers et al. [59] | 2010 | Experimental and Numerical | Wood pellets | Sunmachine Pellet micro-CHP unit | ≤5.4 | ≤1.38 | Nitrogen | 33–36 | 500–1000 | 873–1073 | 72.1 | 14.3 |
| Biedermann et al. [56] | 2004 | Experimental | Wood chips | small-scale CHP pilot plant | 475 | 75 | Helium | 45 | 1000 | NA | 86 | 12 |
| Biedermann et al. [55] | 2003 | Experimental | Wood chips | small-scale CHP pilot plant | 220 | 35 | Helium | 45 | 1010 | 953–1053 | 90 | 9.2 |
| Podesser et al. [53] | 1999 | Experimental | Wood chips | Test biomass Stirling engine | 12.5 | 3.2 | Nitrogen | 33 | 600 | NA | | 25 |
| Lane et al. [54] | 1999 | Description of a micro CHP system | Chunk wood and pellets | Prototype pellet burner (sunpower) | ≈4 | ≈1 | Helium | 30 | 3000 | 823 | >85 | 23 |

Two renewable resources solar energy and biomass is considered for small and micro scale systems. Stirling engine performance is evaluated by comparing solar energy and biomass as the external source.

### 6.1.4.1 Modelling the Stirling engine: alpha configuration

The engine consists of two pistons housed by two cylinders with compression and expansion spaces. The engine parts are shown in figure 6.5. The five components of the machine are connected in series: the compression space (c), cooler (k), regenerator (r), heater (h), and expansion space (e).

In the alpha configuration, the mechanical losses are less compared to beta and gamma configuration, and the loss is due to friction (surface roughness) of the cylinder, the difference in diameter of cylinder and piston rings, variation in the mass of gas that flows between the inflow and outflow and the gap losses. As shown in figure 6.1, the displacer is not used in this configuration. Since there are two pistons in two separate cylinders having a uniform displacement in the same direction so that it provides a constant volume heating-cooling process of the working fluid. When the working fluid is transferred to the cylinder, one piston is fixed, and the other piston compresses or expands the working gas. As the working gas flows through the heater, heat will be added to it, the inner surface of the heater is at high temperature and high pressure and flow is turbulent whereas the outer surface is at high temperature, low pressure, and steady flow. When the working gas flows through the cooler heat is absorbed and rejects it to the coolant. Stirling engines are water cooled and the coolant should be at low temperature because if temperature increases its thermal efficiency drops.

As discussed above, the heater and cooler are constructed as a bundle of thin pipes because it provides a large area of contact for transferring the heat and flow of working gas with low pressure. The regenerator also has the same tubular

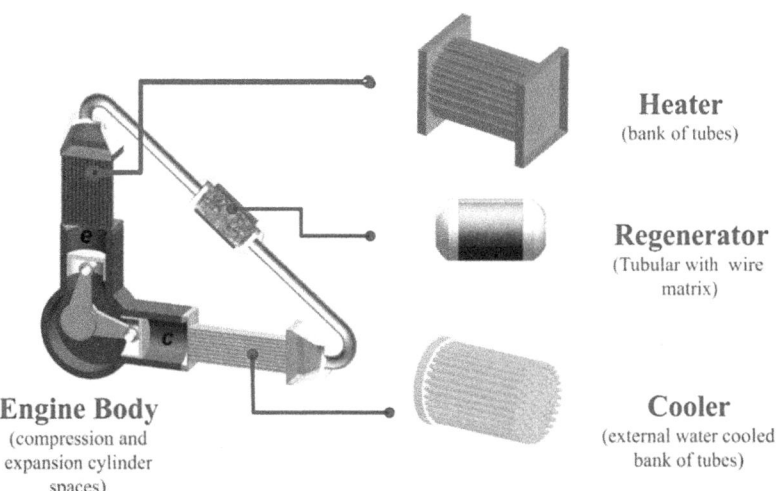

**Heater**
(bank of tubes)

**Regenerator**
(Tubular with wire matrix)

**Engine Body**
(compression and expansion cylinder spaces)

**Cooler**
(external water cooled bank of tubes)

**Figure 6.5.** Stirling engine configuration. Reprinted from [8], Copyright (2020), with permission from Elsevier.

configuration with a wire mesh, and it absorbs and releases heat from and back to the working gas. In this alpha configuration of the Stirling engine, helium is used as a working fluid due to its thermal properties such as high thermal heat transfer, lower thermal conductivity, and dynamic viscosity rather than hydrogen. Table 6.2 shows the geometric characteristics.

### 6.1.4.2 External energy source modeling for Stirling engine

Two external energy sources (renewable sources), solar energy and biomass, are used to assess the Stirling engine performance.

#### 6.1.4.2.1 Modelling of solar collector dish

Solar energy is modelled as concentric solar dish with a diameter of 8 m, an aperture of 12 cm and solar radiation vary between 500 and 900 W m$^{-2}$. The solar dish consists of a collector and cavity receiver. The receiver is located at the focal point to absorb solar radiation and transfer the heat energy to the working fluid of the Stirling engine. The design is aimed to reduce thermal losses. The available power for the Stirling engine ($Q_{engine}$) is calculated as a disparity between the heat in the receiver cavity ($Q_{receiver}$) and the emissivity losses ($Q_{losses,\ emissivity}$) and the convection losses ($Q_{losses,\ emissivity}$). The amount of heat absorbed by the receiver cavity is calculated based on the solar irradiation average value ($I_{receiver}$) and the area of the receiver ($A_{receiver}$), as shown in equation (6.4).

$$Q_{receiver} = I_{receiver} A_{receiver}. \tag{6.4}$$

There are two heat losses. One is surface emissivity ($\delta$), and it can be calculated as a function of Stefan–Boltzmann constant ($\sigma$), receiver area ($A_{receiver}$), the average temperature at receiver ($T_{receiver}$) and ambient temperature ($T_{amb}$) as shown in equation (6.5).

$$Q_{losses,\ emissivity} = \delta \sigma A_{receiver} \left( \overline{T}_{receiver}^4 - T_{amb}^4 \right). \tag{6.5}$$

The second loss is due to convection heat loss and this loss requires calculating the convection heat transfer coefficient in the cavity receiver ($h_{receiver}$) as a function of the Nusselt number ($Nu_D$), thermal conductivity ($k$), and receiver diameter ($D_{receiver}$). $Nu_D$ is calculated by using equation (6.6).

$$Nu_D = 0.3 + \frac{0.62 Re^{1/2} Pr^{1/3}}{[1 + (0.42 Pr^{\frac{2}{3}})]^{1/4}} \left[ 1 + \left( \frac{Re}{282000} \right)^{\frac{5}{8}} \right]^{4/5} \tag{6.6}$$

Re and Pr numbers are Reynolds and Prandtl numbers. Assuming the wind velocity 3 m s$^{-1}$ and Prandtl number 0.7 the convection heat loss can be calculated using equation (6.7).

$$Q_{losses,\ conv} = h_{receiver} A_{receiver} (\overline{T}_{receiver} - T_{amb}). \tag{6.7}$$

**Table 6.2.** Geometric characteristics. Reprinted from [8], Copyright (2020), with permission from Elsevier.

| Parameter of each engine component | Value |
|---|---|
| **Engine cylinders** (pistons, compression and expansion spaces) | |

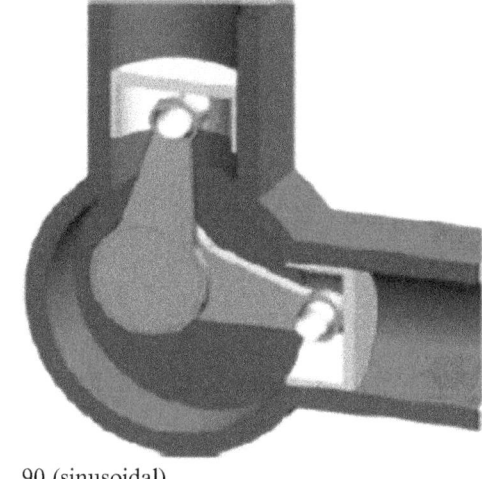

| | |
|---|---|
| Phase angle (configuration) [°] | 90 (sinusoidal) |
| Engine swept volume [cm$^3$] | 130.0 |
| Engine clearance volume [cm$^3$] | 25.0 |
| Engine rotational speed [rpm] | 1500 |
| Mean operating pressure [bar] | 5–80 |
| **Regenerator** (tubular regenerator with wire matrix) | |

| | |
|---|---|
| Regenerator length ($L_r$) [mm] | 60.0 |
| Matrix porosity ($\varphi_{r,matrix}$) [–] | 0.7 |
| Wire matrix diameter ($d_{r,wire}$) [mm] | 0.3 |
| Regenerator volume ($V_r$) [cm$^3$] | 69.8 |
| **Heater** (arrangement of smooth pipes) | |

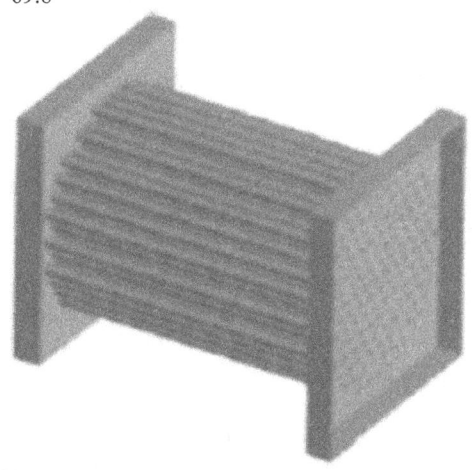

| | |
|---|---|
| Heater length ($L_h$) [mm] | 150.0 |
| Number of tubes ($n_{th}$) [–] | 80 |
| Heater volume ($V_h$) [cm$^3$] | 84.8 |

**Cooler** (bundle of parallel smooth tubes)

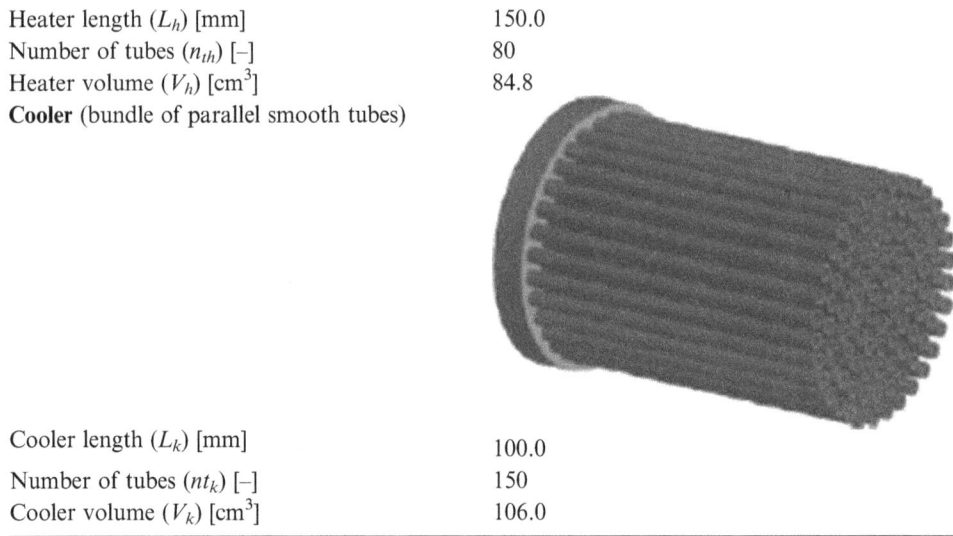

| | |
|---|---|
| Cooler length ($L_k$) [mm] | 100.0 |
| Number of tubes ($nt_k$) [–] | 150 |
| Cooler volume ($V_k$) [cm$^3$] | 106.0 |

**Table 6.3.** Parameters of solar dish collector. Reprinted from [8], Copyright (2020), with permission from Elsevier.

| Design parameter | Value |
|---|---|
| Dish diameter [m] | 8.0 |
| Receiver aperture diameter [m] | 0.12 |
| Reflectance [–] | 0.94 |
| Rim angle [°] | 45.0 |
| Incident angle [°] | 23.0 |
| Optical total error [mrad] | 8.0 |

Table 6.3 shows the parameters of solar dish collector to evaluate the thermal performance of the cavity receiver.

### 6.1.4.2.2 Biomass boiler modeling

Figure 6.6 shows the experimental setup composed of a pellet boiler and gas analysis unit. The boiler works under a forced draught system, and its operation is programmed in LABVIEW software, and the airflow rate is regulated automatically by adjusting the fan speed. The flue gases are forced to come out from the boiler through the fan. By doing so, fresh air enters the combustion chamber due to differences in pressure from primary and secondary air channels. The purpose of the primary and secondary air channels is to introduce the air through orifices. Combustion is initiated with the aid of electrical resistance placed on the grate to burn the pellets.

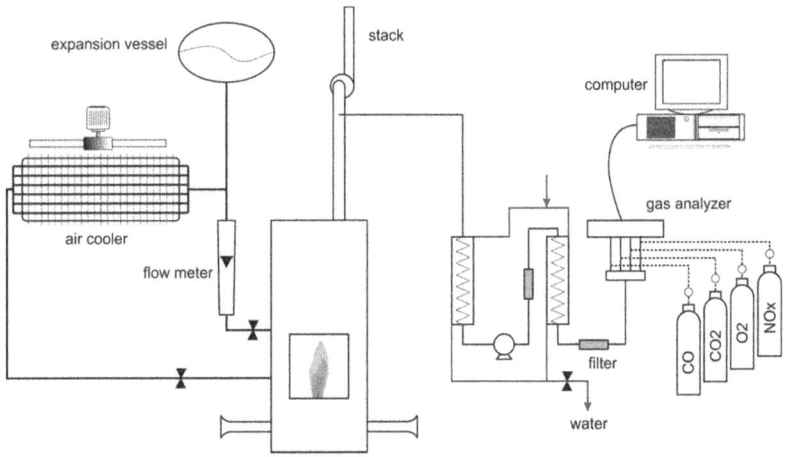

**Figure 6.6.** Combustion test facility.

Compared to a solar energy source, anbiomass-fuelled Stirling engine provides 87.5% more power output. A biomass-fuelled Stirling engine provides a power output of 4.3 KW with a total efficiency of 46.67% and the solar system results in 2.3 KW power with an efficiency of 31.33%. Biomass is considered a good renewable source for generating electric power [8].

## 6.2 Conclusion

High-power consumption leads to massive usage of fossil fuels, and it pollutes the environment. To protect the environment, it is necessary to generate energy from the renewable energy sources. Many renewable sources are available to generate power; OTEC is one of the best renewable resource systems to generate energy. By implementing it, $CO_2$ emissions can be heavily reduced. The Stirling engine is used to generate energy by using biomass and solar energy. 4.3 KW of power is generated using biomass as fuel and from solar 2.3 KW of power is generated. To reduce $CO_2$ emissions and to protect environment, energy should be generated from renewable resources.

## References

[1] Aboagye B, Gyamfi S, Ofosu E A and Djordjevic S 2021 Status of renewable energy resources for electricity supply in Ghana *Sci. African.* **11** e00660
[2] Malik M Z, Musharavati F, Khanmohammadi S, Baseri M M, Ahmadi P and Nguyen D D 2020 Ocean thermal energy conversion (OTEC) system boosted with solar energy and TEG based on exergy and exergo-environment analysis and multi-objective optimization *Sol. Energy.* **208** 559–72
[3] Aquila G, de Queiroz A R, Balestrassi P P, Rotella Junior P, Rocha L C S, Pamplona E O and Nakamura W T 2020 Wind energy investments facing uncertainties in the Brazilian electricity spot market: a real options approach *Sustain. Energy Technol. Assess.* **42** 100876

[4] Julian M, Bassil N and Dellagi S 2020 Lebanon's electricity from fuel to solar energy production *Energy Reports* **6** 420–9

[5] Wang M, Jing R, Zhang H, Meng C, Li N and Zhao Y 2018 An innovative Organic Rankine Cycle (ORC) based Ocean Thermal Energy Conversion (OTEC) system with performance simulation and multi-objective optimization *Appl. Therm. Eng.* **145** 743–54

[6] Tahara K, Horiuchi K and Kojima T 1995 OTEC as a countermeasure for CO2 problem *Energy Convers. Manage.* **36** 857–60

[7] Masutani S M and Takahashi P K 2001 Ocean thermal energy conversion (OTEC) *Encycl. Ocean Sci.* 167–73

[8] Ferreira A C, Silva J, Teixeira S, Teixeira J C and Nebra S A 2020 Assessment of the Stirling engine performance comparing two renewable energy sources: solar energy and biomass *Renew. Energy* **154** 581–97

# Chapter 7

## The Goswami cycle and its applications

**Gokmen Demirkaya, Martina Leveni, Ricardo Vasquez Padilla and D Yogi Goswami**

Thermodynamic cycle innovations have been a critical factor in incorporating renewable energy into the energy sector. This chapter overviews the Goswami cycle, which combines the Rankine and absorption refrigeration cycles, producing power and cooling simultaneously. The Goswami cycle uses an ammonia–water mixture as the working fluid, although other combinations of miscible fluids with different boiling points can also be used. Parametric studies cover low temperature solar and industrial waste heat sources. Recent studies for geothermal applications and coupling the Goswami cycle with a desalination system are also presented in this chapter.[1]

## 7.1 Introduction

The Rankine thermodynamic cycle is the most commonly utilized power cycle in thermal power plants and provides approximately 85% of worldwide electricity production [1]. A Rankine cycle may be combined with a bottoming refrigeration cycle or modified to include refrigeration within the same cycle. Binary mixtures provide an efficient heat transfer mechanism from the heating sources as they have a variable boiling temperature [2–4]. Moreover, binary mixtures have the ability to work at lower operating temperatures compared to steam. The innovative cycle proposed by Goswami [3, 5] utilizes a binary working fluid mixture to produce both power and refrigeration simultaneously. Although the Goswami cycle can use a number of binary fluid mixtures, the ammonia–water mixture is used in the most basic form of Goswami cycle. This cycle consists of an ammonia-based Rankine cycle and an absorption refrigeration cycle. It can be utilized as a bottoming cycle with waste heat from a coal- or natural gas-based thermal power cycle or as a standalone cycle with low-to-medium temperature solar or geothermal energy. In

---

[1] Parts of this chapter have been adapted from [1, 2] with permission from Elsevier and Wiley.

most applications, refrigeration is a more expensive product than power since it requires refrigeration equipment as well as power to produce conventional refrigeration. Therefore, the additional output of refrigeration by the combined cycle provides a greater benefit than conventional power systems.

The Goswami cycle (GC) initially used a mixture of ammonia and water; however, various other fluid mixtures such as $CO_2$ and various amines, and mixtures of certain hydrocarbons have been studied recently [6, 7].

Vasquez *et al* [8] investigated the Goswami cycle performance at low temperatures up to 170 °C. The maximum theoretical effective first law and exergy efficiencies for the heat source and absorber temperatures of 130 °C and 10 °C were found as 21% and 92%, respectively. The cycle was optimized for different thermodynamics outputs i.e. cooling and power output. When the cycle is optimized for maximum cooling output, the theoretical limit of net work and cooling output from the cycle are approximately 100 kJ $kg^{-1}$-solution and 75 kJ $kg^{-1}$-solution, respectively [9]. For the maximum power output case, the theoretical limits of net work and cooling outputs from the cycle are 160 kJ $kg^{-1}$-solution and 15 kJ $kg^{-1}$-solution, respectively. The maximum effective first law efficiency was 20% for the heat source and absorber temperatures of 170 °C and 30 °C, respectively. In addition, they showed [8] that a lower absorber temperature produces a higher vapor flow rate. The turbine exit quality was another parameter that was investigated and showed that only high-pressure ratios such as eight and very high expander efficiencies could cause 10% or more wetness at the turbine exit. The cooling output is always at a maximum when rectification is provided and no superheater is used; therefore, installation of a rectifier column is necessary to maximize the refrigeration output. Irrespective of the turbine efficiency, higher rectification is required to produce higher cooling, which reduces the work output and overall efficiency. Cooling output is limited at low pressures by higher turbine exhaust temperatures and bounded at higher pressures by the low production of vapor.

Demirkaya [9] showed that when ammonia vapor is superheated after the rectification process, the cycle efficiencies increase but cooling output decreases. Superheating enhances the power output but the difference is not significant at high-pressure ratios. If the cycle is used for maximum power, it is not necessary to use a rectifier or a superheater, which reduces the initial costs. The maximum net work and cooling outputs are decreased approximately by 35% and 53%, respectively, when the absorber temperature increases from 10 °C to 30 °C. Demirkaya *et al* [10] conducted an optimization study of the Goswami cycle used as a bottoming cycle (heat source temperature <150 °C) and topping cycle (heat source temperature between 150 °C and 250 °C). When the Goswami cycle is used as a bottoming cycle, optimization for work as well as effective exergy efficiency leads to a single optimal point, which means that improving one of these objective functions leads to an improvement in the output. However, when the objective functions are effective first law and exergy efficiencies and cooling output, it was seen that there is a conflict between these functions. As soon as the cycle produces some amount of cooling, efficiency terms are reduced considerably. For the other scenario, if the Goswami cycle is used as a topping cycle, the cycle produces no cooling. The optimal point of the cycle occurs when the temperature is at the maximum and ammonia concentration is at the minimum of their ranges to provide the highest possible work, first law and exergy efficiencies.

Demirkaya *et al* [9, 11] used a scroll expander in an experimental study to measure its effectiveness in the Goswami cycle. The scroll expander efficiency was measured as 30% when the boiler and the rectifier temperatures were 85 °C and 55 °C, respectively. It was shown that superheated inlet conditions improved the efficiency of the modified expander; superheating the rectifier exit by 3 °C and 6 °C resulted in an expander efficiency of 40%. The experimental system's fairly small scale had an impact on the testing parameters and results. Therefore, they concluded that a customized smaller scroll device designed for the test conditions would likely perform better and provide higher expander efficiency. Leveni *et al* [12] carried out the custom scroll turbine design work for the Goswami cycle; they developed a scroll turbine model with a volume ratio of 2.53 and a diameter of 20 cm. The mathematical model predicted that the expander efficiency could reach up to 69.7% at a pressure ratio of 4. This theoretical study showed that a proper design can achieve a high turbine efficiency in future experimental studies.

Kumar *et al* [13] scaled up the experimental cycle to minimize the losses associated with small-scale setup. They used an orifice plate, which was suitable for a typical pressure ratio of 9.0, to simulate the expansion process in the experimental setup. Due to the simulated turbine device, their focus was on the cooling performance of the Goswami cycle; they investigated the cooling alone (CA) and combined cooling and power (CCP) modes for a heat source temperature of 150 °C. The maximum cooling performance was found as 34.26 kW for the CA case. In the CCP mode, the cooling capacity and net power output were about 15.6 kW and 2.21 kW respectively with the power to cooling ratio of 0.14. The effective first law and effective exergy efficiency for the CCP case were around 13% and 48%, respectively.

Hosseinpour *et al* [14] presented a cogeneration system that included wood-fed biomass gasification, a solid oxide fuel cell (SOFC) and a Goswami cycle. The Goswami cycle was used to produce cooling and electrical power by recovering waste heat of the SOFC. The syngas produced by the top gasification cycle provides the SOFC system fuel and the waste heat obtained from the complete oxidization of the unreacted gases provides the required energy to drive the bottom Goswami cycle. The first law and exergy efficiencies of the proposed cogeneration system were 58.5% and 33.7%, respectively, compared to the values of 40% and 26.9% for the system without the bottoming Goswami cycle. By coupling the Goswami cycle with SOFC, the system net electrical power output rose from 382 kW to 478 kW. It is worth mentioning that additional cooling output of 81.6 kW was generated by the Goswami cycle.

Rivera *et al* [15] proposed two different derivatives of the Goswami cycle. The proposed cycle derivatives use an additional component to condense a fraction of the working fluid produced in the generator; the first derivative is the cycle with flow extraction after the rectifier, while the second uses flow extraction in the turbine. In the first case, a condenser was added to condense a part of the ammonia produced in the boiler and rectifier column. The ammonia leaving the turbine and the condenser enters the evaporator, producing the cooling output. The cooling effect is mainly produced by the latent heat of ammonia. In the second case, a fraction of the ammonia in the vapor phase entering the turbine is extracted at an intermediate pressure to be condensed in the condenser as in the first case. The proposed derivative cycles have the capability of producing a considerably higher amount of cooling power compared to that produced with the basic Goswami cycle. From the comparison among the two cases and the original Goswami cycle, it was proven that the Goswami cycle produced the highest amount of electric power; however, the cooling power was considerably lower than that produced with the two derivative cases. In exchange for 23% to 50% less work output, the cooling power was 6 times higher than that produced with the basic Goswami cycle.

Sayyaadi *et al* [16] compared four bottoming cycles, namely the Kalina, organic Rankine, the Goswami and trilateral flash cycles, which recover heat from a combined 30 MW gas-turbine and absorption chiller system. Bottoming cycles were compared in a configuration where the exhaust gases at the outlet of the absorption chiller's generator with low-grade thermal energy are directed to a bottoming cycle for further exploitation of waste energy. In order to compare the cycle performances fairly, each bottoming cycle was optimized in a multi-objective optimization process with thermal objectives (efficiency and generated power). Their results showed that the Goswami cycle is the foremost bottoming cycle as it generates the highest values of the mechanical and cooling powers, highest thermal and exergetic efficiency, the lowest production cost, and the lowest payback period within a reasonable time (3.8 years for their case study). However, they found that the highest required capital investment was associated with the Goswami cycle due to a higher number of components required. Their case study showed that the Goswami cycle increased the plant's thermal and exergy efficiencies by 8.4% and 13.5%, respectively. Moreover, Goswami cycle cooling capability reduces the inlet air system's absorption chiller capacity by 3.2%.

Guillen *et al* [17] studied the performance of the Goswami cycle working with a topping supercritical Brayton cycle, which uses $CO_2$ as a working fluid. The topping Brayton cycle converts the concentrated solar thermal power into electricity and the Goswami cycle produces electricity and cools the compressor inlet. Guillen *et al* [17] considered a gas turbine inlet temperature and pressure of 700 °C and 250 bar, respectively, and their results showed that the highest overall thermal and exergy efficiencies were 45.3% and 21.8%, respectively. The Goswami cycle improved the overall thermal and exergy efficiencies by 0.95% and 0.70%, respectively.

The previous research has shown that Goswami cycle is effective for recovering both low and medium waste heat sources. This chapter presents Goswami cycle's performance using low temperature solar and industrial waste heat sources. Several researchers studied the performance of the cycle for geothermal energy and desalination applications [18]. Recent studies for geothermal applications and coupling the Goswami cycle with a desalination system are also presented.

## 7.2 Process description

The Goswami cycle uses a unique binary mixture working fluid to produce power and refrigeration simultaneously in the same loop. The cycle requires less equipment compared to top and bottoming combined cycles; namely, an absorber, separator, boiler, heat recovery and refrigeration heat exchangers and a turbine as illustrated in figure 7.1.

The cycle is shown in figure 7.1, in which a binary ammonia–water mixture leaves the absorber (state 1) at the cycle low pressure in a saturated solution form with a comparatively high ammonia concentration. The high-pressure (state 2) solution recovers heat from the returning weak ammonia liquid solution in the recovery heat exchanger (state 2 → state 3). The boiler produces two mixture streams (state 4):

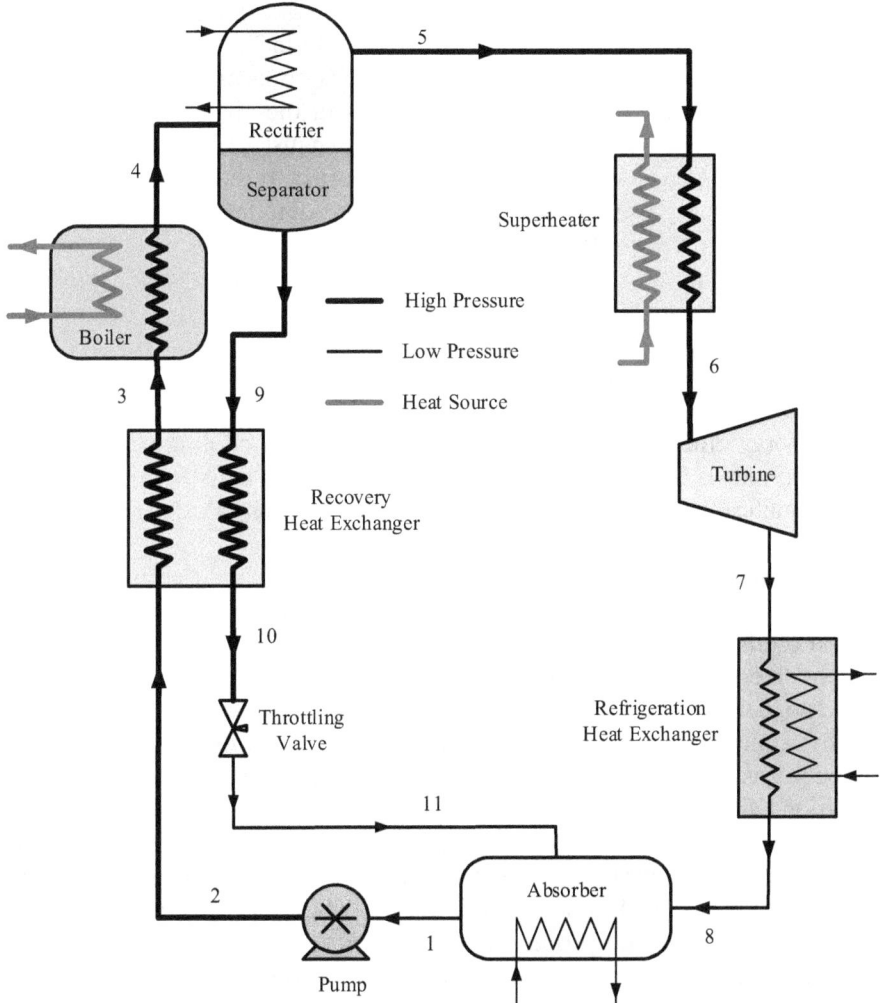

**Figure 7.1.** Schematic description of the Goswami cycle.

a relatively low concentration ammonia liquid, and a high concentration ammonia vapor. These two-phase mixtures are separated into a liquid mixture (state 9) and a vapor mixture (state 5). The liquid stream, which is weak in ammonia concentration transfers heat (state 9 → state 10) to the strong solution (state 2 → state 3). The liquid stream (state 10) is then throttled (state 11) and sprayed into the absorber. A cold stream may be used to chill the saturated rich ammonia vapor (state 4) at the rectifier in order to condense out any remaining water; however, this application reduces work output. On the other hand, stream 5 could be superheated before entering the expander, although the cooling output would suffer as a result of this application.

Under specific operating conditions, the temperature of the fluid leaving the expander at state 7 can be much lower than the ambient temperature, providing cooling from the sensible heating of the expander exhaust (state 7 → state 8). The key to this process is the employment of a binary working fluid mixture. The condensing temperature of an ammonia rich vapor (state 7) at constant pressure can be lower than the saturation temperature of a lower concentration liquid (state 1). The stream (state 8) then rejoins the weak liquid in the absorber where the basic solution is restored. This permits the expander exhaust temperature to be much lower than the temperature at which absorption occurs. It differentiates this process from the pure working fluid operation, in which the expander exhaust temperature is same as the vapor condensation temperature. As a result, it is achievable to expand the vapor to sub-ambient temperatures resulting in a low temperature condition that can be used for cooling output.

The basic solution stream from the absorber can provide rectification cooling or it can be supplied externally from any other renewable source. Demirkaya [9] investigated the cycle performance parameters and found that the effective first law and exergy efficiencies were in favor of an internal cooling source when the turbine efficiency is high (>50%). However, for ammonia mass fractions more than 0.35, no significant difference was noted between the external and internal cooling sources.

The modeling requires thermodynamic property data for the ammonia–water working fluid. The thermophysical property correlations used in this investigation are based on those provided by Xu and Goswami [19], which are a combination of the Gibbs free energy method for mixture properties, and empirical equations of bubble and dew point temperatures for phase equilibrium. The accuracy of the thermodynamic properties is critical to the study's results, thus Xu and Goswami [19] model was compared to a theoretical model based on Helmholtz free energy formulation [20] used in the National Institute of Standards and Technology (NIST) Reference Fluid Thermodynamic and Transport Properties Database (REFPROP) [21]. As there is no significant difference between the results of Xu and Goswami [19] and the Tillner-Roth and Friend [20] models, it can be concluded that the Gibbs free energy method for mixture properties and empirical equations of bubble and dew point temperatures for phase equilibrium [19] is appropriate for the calculation of the ammonia water mixture properties.

### 7.2.1 Efficiency evaluation of the combined power and cooling cycles

Combined power and cooling cycle efficiency is a critical challenge that needs to be solved. Adding two simultaneous outputs of different energy qualities, namely power and refrigeration, would be akin to adding 'oranges to apples'. Vijayaraghavan and Goswami [22] addressed this issue in detail and suggested that exergy change of the working fluid fits better instead of refrigeration output, therefore the first law efficiency (equation (7.1)) is given as:

$$\eta_I = (W_{net} + E_c)/Q_h \tag{7.1}$$

where the exergy change of the refrigeration output is given as:

$$E_c = m_{cf}[h_{cf,\,in} - h_{cf,\,out} - T_o(s_{cf,\,in} - s_{cf,\,out})]. \tag{7.2}$$

Other parameters in the above equation, $m_{cf}$, $h_{cf}$ and $s_{cf}$ are mass flow rate, enthalpy and entropy of the chilled fluid and the subscripts in and out refer to inlet and outlet values. Rosen and Le [23] further elaborated the efficiency expression (7.1) and they suggested the use of an exergy efficiency as below:

$$\eta_{exergy} = (W_{net} + E_c)/(E_{hs,\,in} - E_{hs,\,out}) \tag{7.3}$$

where $E_{hs}$ refers to the exergy of the heat source. Moreover, Vijayaraghavan and Goswami [22] suggested that the refrigeration part should be correctly weighted for efficiency expression in order to allow fair comparison of cycles. The refrigeration output is replaced by the electric power equivalent to generate the same cooling effect by a conventional refrigeration system. They called the efficiencies 'effective' efficiencies, which are given below:

$$\eta_{I,\,eff} = \left(W_{net} + E_c/\eta_{II,\,ref}\right)/Q_h \tag{7.4}$$

$$\eta_{exergy,\,eff} = \left(W_{net} + E_c/\eta_{II,\,ref}\right)/(E_{hs,\,in} - E_{hs,\,out}). \tag{7.5}$$

The term $\eta_{II,\,ref}$ in the above equations refers to the second-law efficiency of refrigeration, which could be replaced by $COP_{Practical}$ for realistic values.

## 7.3 Low-temperature implementations

Waste heat sources can be characterized as low-grade (<100 °C), medium-grade (100 °C–400 °C) or high-grade (>400 °C). Low-grade waste heat can be found in many fields of industry and buildings; but it is the most challenging to recover cost-effectively. Ventilation, hot water applications, low-temperature flat plate solar panels, and solar ponds are typical examples of recovering low-grade waste heat. Mid-grade waste heat is most commonly found in the chemicals, food and beverage industries, oil refineries, and other process industries, as well as building utilities. High-grade heat can be found in manufacturing industries and power plants, especially gas turbines that produce a significant amount of waste heat. For any grade of waste heat, heat recovery is a significant opportunity to improve the overall energy conversion efficiency.

The Goswami cycle can be sized and scaled from a kilowatt to megawatt system, and it can use waste heat from a conventional power cycle or low-temperature heat sources such as solar or geothermal energy. In this section, two different implementations of the cycle will be discussed. Firstly, an independent cycle that utilizes low-grade heat source will be discussed and secondly the Goswami cycle as a bottoming cycle for medium grade heat sources will be examined.

### 7.3.1 Simulation details

The performance of the cycle is studied using simulations for various factors such as boiler temperature and pressure, strong solution concentration and rectifier exit conditions. The main parameters that can be varied to influence the cycle are the heat source temperature, system high pressure, basic solution mass fraction, and absorber pressure and temperature. Saturation in the absorber reduces the number of independent parameters that govern the cycle to four. Rectifier and superheater temperatures can also be modified, as well as the conditions of heat transfer from the source to the ammonia–water mixture.

Assumptions for the simulations of the Goswami cycle are given in table 7.1. Commercial programming software, Matlab [24], was used to simulate the combined power and cooling cycle. The following assumptions were used in the analysis:

- Pressure drops were neglected.
- The system low pressure was determined by the basic solution concentration ($x_{strong}$) and the absorption temperature of 20 °C.
- The boiling conditions were completely described, with inputs including boiling temperature, pressure, and basic solution concentration.
- The heat recovery heat exchanger employed an effectiveness value, while the pinch point for the boiler, superheater, and refrigeration heat exchangers was limited to 5 °C.
- The rectification process was limited by either the rectifier exit temperature or an ammonia mass fraction of 0.999, whichever was encountered first.

**Table 7.1.** Goswami cycle assumptions for the simulation.

| Parameter | Value |
|---|---|
| Ambient temperature, $T_{ambient}$ | 15 °C |
| $\eta_{II, ref}$ | 30% |
| Isentropic turbine efficiency | 85% |
| Pump efficiency | 85% |
| Recovery heat exchanger effectiveness, $\varepsilon$ | 85% |
| Pinch point | 5 °C |
| Minimum turbine exit vapor quality | 90% |

- The turbine exhaust temperature was specified to be equal to or below 10 °C to generate refrigeration; the quantity of cooling produced (if any) was calculated as the energy needed to heat the turbine exhaust from the exhaust temperature to 10 °C ($T_{\text{cooling-limit}}$). For any exhaust temperature above $T_{\text{cooling-limit}}$, it was assumed that no refrigeration was produced.
- The cooled fluid in the refrigeration heat exchanger was water.
- Superheating was not considered in this simulation.

### 7.3.2 Low-grade solar implementation

Performance values of an independent Goswami cycle, which utilizes a solar heat source equal or below 100 °C, are shown in this section. Flat plate solar collectors can be easily adapted as a boiler to provide heat to the Goswami cycle. To show the combined power and cooling characteristics of the cycle, a low-temperature heat source is selected, which is 85 °C. The strong solution of the cycle is 0.4 kg $NH_3$ $kg^{-1}$-solution, and no rectification cooling is considered for the first simulation. Pressure ratio ($P_r$) is calculated as the ratio of system high pressure and low pressure, respectively. The absorber pressure determines the system low pressure; the bubble point pressure of the mixture calculates the maximum pressure that Goswami cycle can operate. The system high pressure is incrementally increased up to the bubble point pressure of the mixture.

Figure 7.2 presents computed results of cycle outputs for varying pressure ratio. The relative position of the maxima for work production, effective first- and second-law efficiencies, and cooling production are shown. Similar to work output, cooling also has a maximum, which is limited at low pressures by higher turbine exhaust

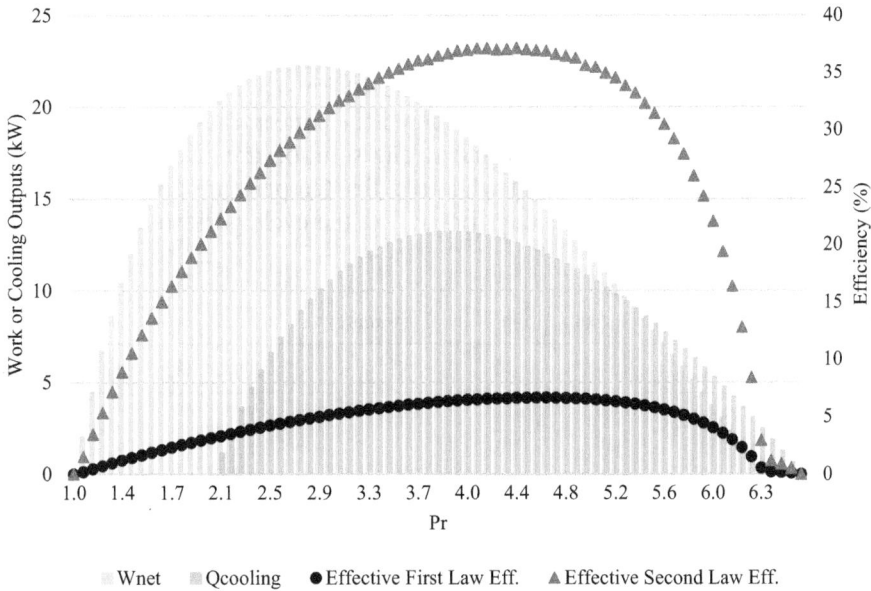

**Figure 7.2.** Goswami cycle performance parameters.

temperatures and bounded at higher pressures by the low production of vapor. For a unit mass flow rate (kg s$^{-1}$) of strong solution flow, the cycle produces 22.2 kW of power and 13.2 kW$_{th}$ of cooling at the respective maximum points. The maximum effective first- and second-law efficiencies are 6.7% and 37.2%, respectively.

The cycle's effective first law efficiency is often compared to ideal cycle efficiencies. The respective Carnot cycle efficiency is 19.6% (by using heat source and sink temperatures of 85 °C and 15 °C), which shows that the cycle first law efficiency is significantly lower than the ideal cycle. Lee and Kim [25] evaluated the efficiency of an ideal Lorenz cycle, which uses finite heat sources to maximize power. The Lorenz cycle is the suitable ideal cycle since heat input and rejection cannot occur at constant temperature.

$$\eta_{\text{Lorenz, finite}} = 1 - \left( \frac{T_{\text{heat sink, inlet}} + \text{pinch}}{T_{\text{heat source, inlet}} - \text{pinch}} \right). \quad (7.6)$$

Later, DiPippo [26] addressed that the triangular cycle is more appropriate for geothermal binary power cycles than the Carnot cycle since the heat source is not isothermal; the hot fluid (e.g., brine) cools as it transfers heat to the cycle working fluid. The ideal triangular cycle's thermal efficiency is defined as below:

$$\eta_{\text{triangular}} = \frac{T_{\text{heat source}} - T_{\text{heat sink}}}{T_{\text{heat source}} + T_{\text{heat sink}}}. \quad (7.7)$$

The triangular and Lorenz cycle efficiencies are calculated as 10.8% and 7.3%, respectively. In practice, the triangular and Lorenz cycles with finite heat sources represent the top theoretical limit and are more suitable for comparison. When we compare the maximum effective first law efficiency of 6.7% (Goswami cycle) with 10.8% (triangular) and 7.3% (Lorenz with finite heat sources), the outcomes appear to be encouraging for practical future applications.

The ammonia concentration has a significant impact on the maximum value, therefore the next parametric study considered the concentration as a variable. The simulation details of boiler, rectifier and strong solution are given in table 7.2. The boiler temperature is kept constant while rectifier temperature and strong solution parameters were changed to see their impact on the cycle performance values.

**Table 7.2.** Inputs parameter for the simulation of the Goswami cycle.

|  | $T_{\text{boiler}}$ (°C) | $T_{\text{rectifier}}$ (°C) | $X_{\text{strong}}$ (kg NH$_3$ kg$^{-1}$ solution) |
|---|---|---|---|
| Case 1 | 95 | 95 | 0.3 |
| Case 2 | 95 | 50 | 0.3 |
| Case 3 | 95 | 95 | 0.4 |
| Case 4 | 95 | 50 | 0.4 |
| Case 5 | 95 | 95 | 0.5 |
| Case 6 | 95 | 50 | 0.5 |

The variation of net work, cooling output, effective first law efficiency and effective exergy efficiency with the $P_r$ ($P_{boiler}/P_{absorber}$) is shown in figures 7.3–7.6. The boiling pressure in the power cycle is controlled by the rate of vapor production and the rate at which vapor exits the turbine. For the lower pressure ratios, the vapor rate amount is higher than at higher-pressure ratio values, which causes higher work output at lower pressure ratios. For the specified boiler condition, the vapor mass flow rate decreases when the concentration rate is increased; all the strong solution is vaporized at 95 °C for Case 1. The vaporization rate is 33.8% and 32.9% for Cases 3 and 5, respectively. On the other hand, the situation is opposite for the rectified cases, Case 6 has the highest mass flow rate which leads to highest work output among the rectified cases. The rectifier condenses significant amounts of vapor for Cases 1 and 3, 83.3% and 36.0%, respectively.

The cooling outputs for the cases under consideration are given in figure 7.4. The temperature limit of the turbine exhaust is specified as 10 °C in the simulation details, therefore the exhaust temperature has to be below this limit to produce refrigeration. As seen in figure 7.4, the temperature never goes below 10 °C for the non-rectification cases, namely Cases 1 and 3, and Case 5, produces only a minor cooling which is around 9.5 $kW_{th}$. A small amount of refrigeration can also be obtained for these cases when the rectification temperature is 85 °C. As explained above for the work output results, Case 6 has the highest vapor rate for the rectified cases and with a turbine exhaust temperature of −10.9 °C, the cycle produces highest cooling output among the rectification cases, even though Cases 2 and 4 have lower exhaust turbine temperatures, −26 °C and −21.1 °C, respectively. As shown in

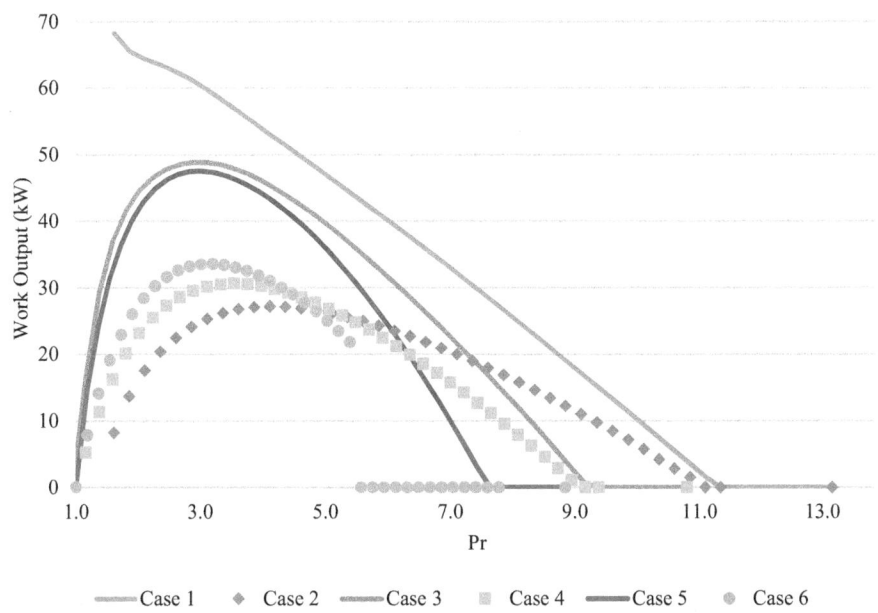

**Figure 7.3.** Effect of ammonia mass fraction, pressure ratio and rectification on work output.

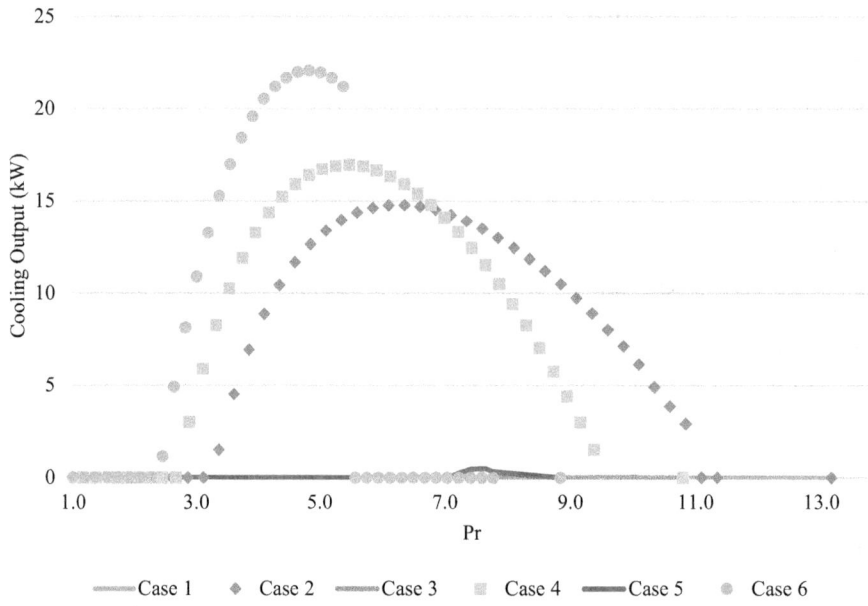

**Figure 7.4.** Effects of ammonia mass fraction, pressure ratio and rectification on cooling output.

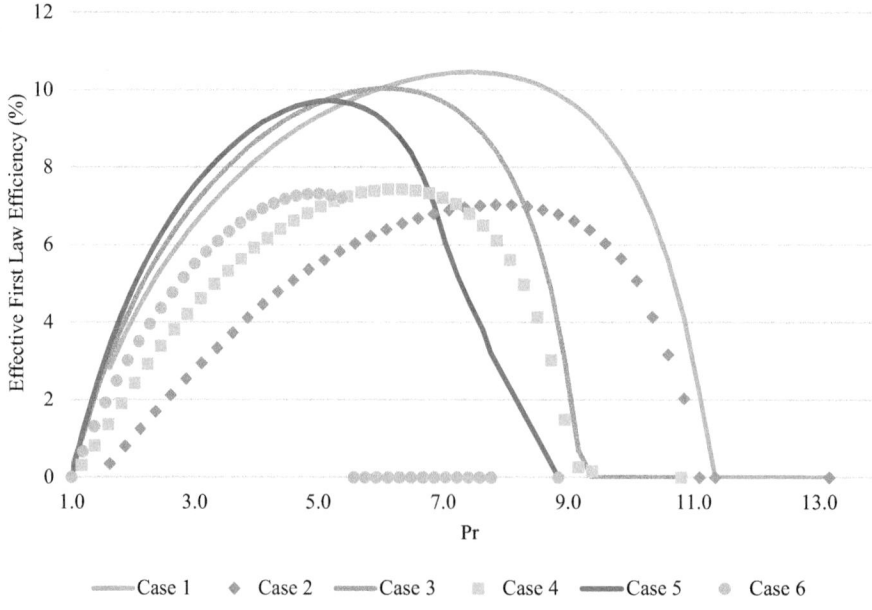

**Figure 7.5.** Effects of ammonia mass fraction, pressure ratio and rectification on effective first-law efficiency.

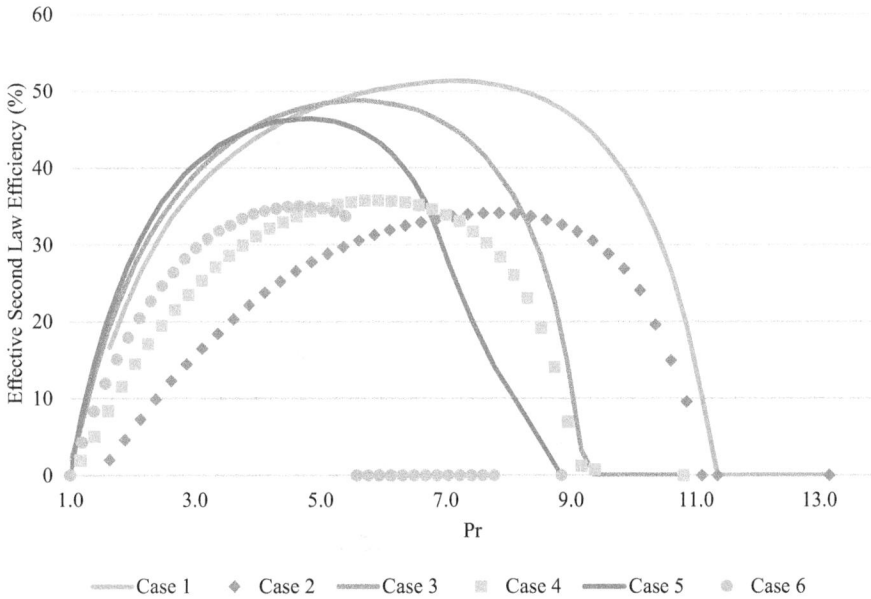

**Figure 7.6.** Effects of ammonia mass fraction, pressure ratio and rectification on effective second-law efficiency.

figure 7.4, when the pressure ratio increases, cooling output decreases due to low production of vapor.

The cycle efficiencies for the cases under consideration are given in figures 7.5 and 7.6. The effective first-law and second-law efficiencies can be above 10.5% and 51.4%, respectively. The Carnot, ideal triangular and Lorenz cycle efficiencies are calculated as 22.8%, 12.9% and 9.4%, respectively. The Goswami cycle effective first law efficiency is between the triangular and Lorenz cycle. It is worth noting that Goswami cycle results shown here may also be assumed as ideal cycle results, as some parameters are idealized in the simulations such as no pressure and thermal losses are considered. Other than these two assumptions, the operation parameters are realistic when the cycle is scaled for conventional generation (>1–5 MW), some parameters such as turbine efficiency, pinch points etc. will be offset from the assumed values when the cycle is aimed for small-scale applications (such as 1–100 kW work output).

### 7.3.3 Case study: industrial waste heat implementation

Conventional power plants mostly operate with Brayton or Rankine cycle. Combined-cycle (CC) plants have been widely used since 2000s with the usage of aero derivative gas turbines in electricity power generation. A schematic of a combined cycle is given in figure 7.7. The waste heat from the gas turbine is directed into the heat recovery steam generator (HRSG) and steam energy is converted into electricity in the bottom Rankine cycle. Since the 2000s, gas turbine technology has been improving in both output and efficiency due to the developments in advanced

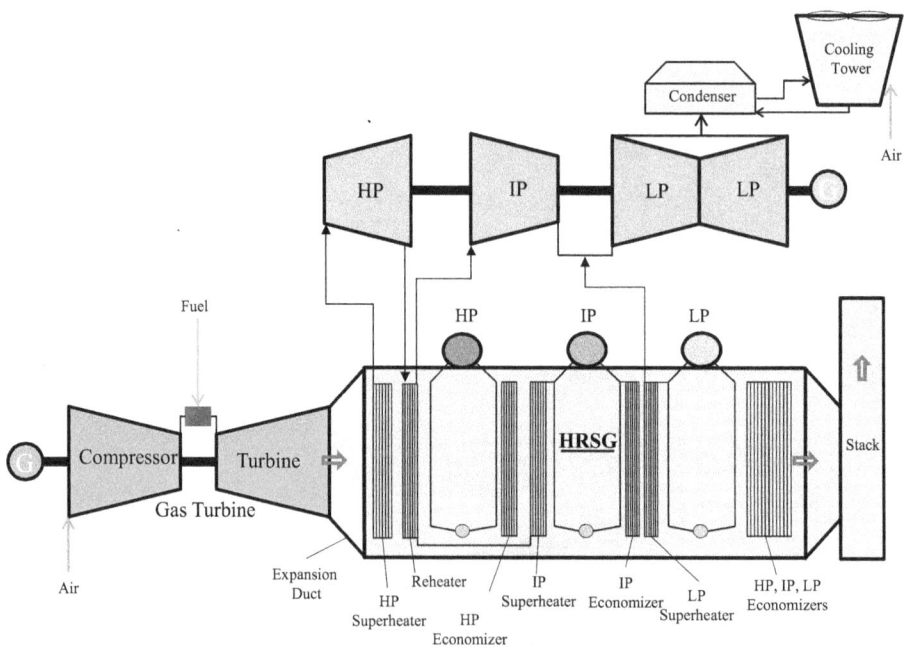

**Figure 7.7.** Combined cycle schematic configuration.

**Table 7.3.** Gas turbine performance values (50 Hz machines).

| Class | Power output (MW) | Efficiency (%) | References |
|-------|-------------------|----------------|------------|
| H | 448–571 | 42.9–44.0 | (General Electric [27]) |
| | 450–593 | 41.0–42.8 | (Siemens Energy [28]) |
| F | 288–314 | 38.6–38.7 | (General Electric [29]) |
| | 329 | 41.0 | (Siemens Energy [28]) |
| E | 132–210 | 34.6–38.0 | (General Electric [30]) |
| | 187 | 36.5 | (Siemens Energy [28]) |

material coating and manufacturing. Due to these advancements, several gas turbine classes are developed based on their power output capabilities, namely H class (>400 MW), F class (>250 MW) and E class (>120 MW).

Combined cycle efficiency depends mainly on the gas turbine efficiency, typical heavy-duty gas turbine efficiencies taken from two major gas turbine OEMs are given below in table 7.3. For modern combined cycles, which use H class gas turbines, efficiencies are above 60% and continue to improve with the advancements in H class gas turbines.

In this section, the Goswami cycle is integrated with a combined cycle power plant operating on natural gas, to seek and determine the potential overall power and efficiency improvement. A typical combined cycle power plant (1 GE 9HA gas turbine, 1 HRSG, 1 steam turbine, 1 cooling system) is developed with a commercial GTPRO software (Thermoflow Inc. [31]). The results and heat and mass balance of this plant are taken as benchmark points, and integrated combined and Goswami cycle performance values are compared with this benchmark point.

Schematic drawing of the first case of the implementation is given in figure 7.8; the waste heat from the HRSG will be utilized in the Goswami cycle. The simulation details are the same as given in 3.1, the flue gas temperature at the end of the economizers is 87.8 °C (taken from the GTPRO [31]), which will be used as heat input to the Goswami cycle. The Goswami cycle using this waste heat source is studied and results are tabulated in table 7.4. $W_{net}$ given in the table is Goswami cycle's net work output for a unit mass flow rate, therefore the mass flow rate of the integrated Goswami cycle is calculated by dividing the recovered energy at the Goswami module by the required boiler input energy, and then the calculated mass flow rate is multiplied by $W_{net}$ to calculate the total electricity generation from the Goswami cycle. The highest output that can be obtained from the waste heat is when the strong solution of the Goswami cycle is 0.9, and it is found that additional 3272 kW can be produced. For each strong solution row given in the table 7.4, the optimized results are given based on the best effective second-law efficiency.

Although the first case seems to be optimistic, more complex and high capex implementation of the Goswami cycle is conducted to seek for a higher energy recovery from the gas turbine waste heat. There are several modules within the HRSG, which produce high-pressure (HP), intermediate-pressure (IP) and low-

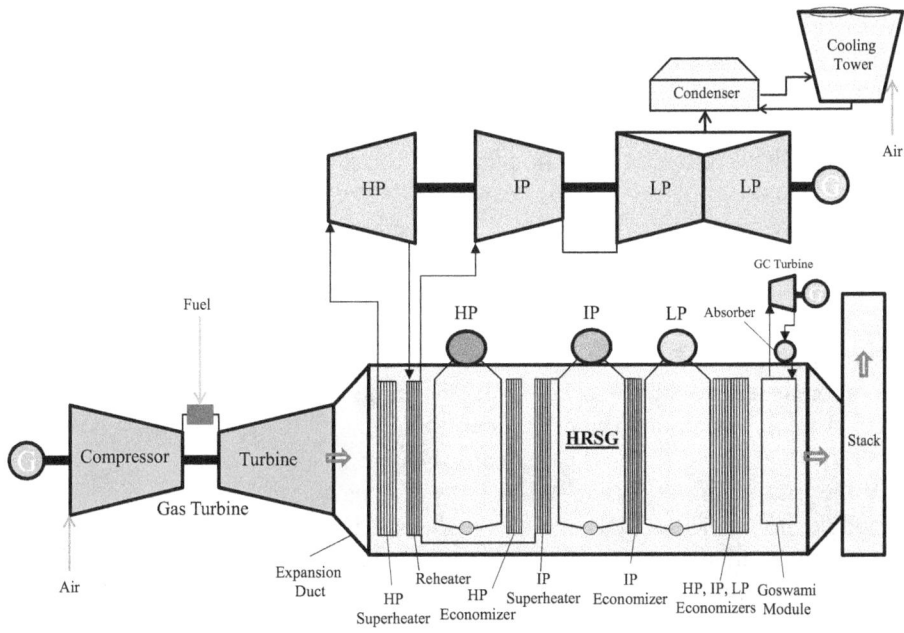

**Figure 7.8.** Combined cycle integration with the Goswami cycle–Case I waste heat implementation.

**Table 7.4.** Goswami cycle using a low-grade heat source.

| $X_{strong}$ kg $NH_3$ kg$^{-1}$ solution | Effective 1st law eff. % | Effective 2nd law eff. % | $W_{net}$ for a unit solution flow rate (kW) | $W_{el}$-Case I (generated from the HRSG) (kW) |
|---|---|---|---|---|
| 0.1 | 10.0 | 55.0 | 27.3 | 3195 |
| 0.2 | 9.4  | 51.8 | 23.8 | 3003 |
| 0.3 | 9.1  | 50.1 | 22.6 | 2898 |
| 0.4 | 8.6  | 47.3 | 24.5 | 2752 |
| 0.5 | 8.3  | 44.7 | 27.5 | 2656 |
| 0.6 | 8.4  | 45.1 | 33.9 | 2682 |
| 0.7 | 8.9  | 48.5 | 44.1 | 2839 |
| 0.8 | 9.5  | 54.2 | 61.7 | 3046 |
| 0.9 | 10.2 | 61.2 | 87.6 | 3272 |

pressure (LP) steam. The HP steam goes to the turbine and return back (hot reheat heater—HRH) to the HRSG, it is reheated and mixed with IP steam and directed to the turbine. The HRH expands at the IP section of the turbine and directed to the LP section of the turbine and it is mixed with the LP steam before entering the turbine. The low-pressure (LP) steam constitutes only 8.3% of the total steam that goes to the low-pressure section of the steam turbine. It is aimed that the Goswami cycle could replace the low flue gas section of the HRSG, such as the LP section, to recover more energy due to its binary mixture working fluid. The schematic drawing of the second case is shown in figure 7.9; the LP boiler and superheater section of the HRSG and steam cycle is replaced by the two modules, which will feed the Goswami cycles (a top GC and a two bottoming GCs), to get more benefit from the binary mixture working fluid. The top and bottoming Goswami cycle concepts are shown in figure 7.10, while more information can be found in the reference [9]. The two HRSG modules will feed the top Goswami cycle, and bottoming cycles will be heated by the top Goswami cycle. The flue gas temperature entering the LP superheater is 336.6 °C, the LP superheated steam is 290 °C, although the binary mixture has more advantages in the heat transfer pinch point, the boiler temperature of the top Goswami cycle is set to 290 °C to be on the safe side, and the boiler pressure is taken as 95 bar.

In order to see the difference and improvements of the two cases, the LP section power output is calculated using the turbine efficiency given in the GTPRO, which is overall 91.9% (mechanical and electrical losses are included). This efficiency is valid for all the total steam flow (which is 591.1 t h$^{-1}$) passing through the steam turbine LP module, however the fraction of the replaced LP boiler and superheat in the total flow is 8.3% as mentioned above. If there is a dedicated LP turbine for this small amount of flow, the efficiency would be lower than 91.9%; it is worth noting that the Goswami cycle expander efficiency is assumed to be 85%.

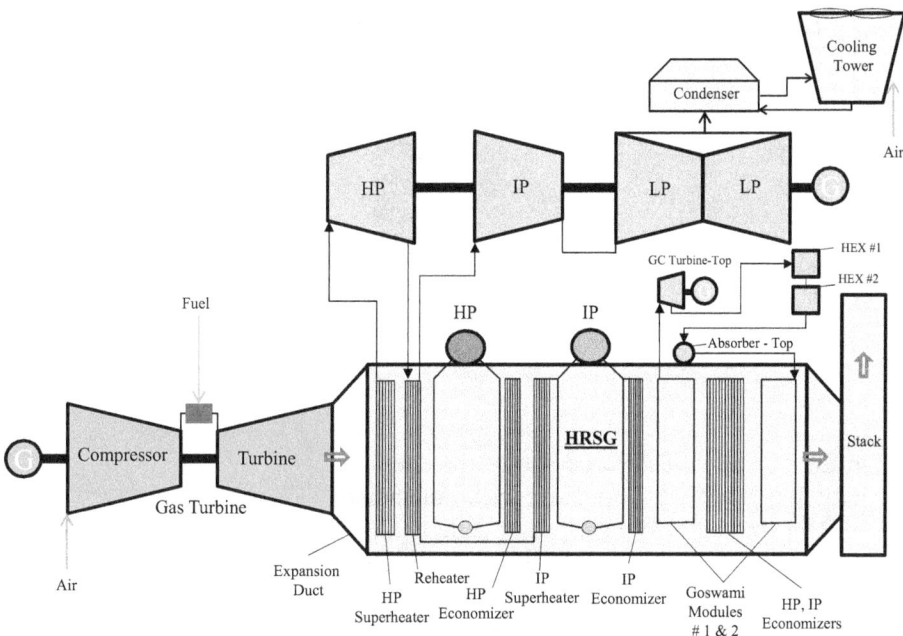

**Figure 7.9.** Combined-cycle integration with the Goswami Cycle — Case II LP replacement.

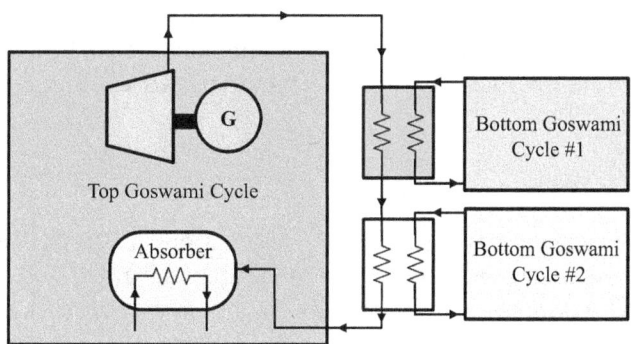

**Figure 7.10.** Schematic conceptual description of the top and bottoming Goswami cycles.

The results of Cases I and II are given in table 7.5. As seen, utilizing a waste heat Goswami module, Case I improves the overall plant efficiency by 0.31%. The efficiency of the plant further improves by replacing the LP boiler and superheater by Goswami modules. However, it should be mentioned that Case II capex would be much higher than Case I, as there would be a top and two bottoming Goswami cycles. The flue gas temperature after Case II modules is still high (>90 °C) to implement Case I, so both cases can be implemented at the same time if desired. The last column in table 7.5 shows that the cycle efficiency improved by 0.78% if both Case I and Case II are implemented (see figure 7.11). The heat rate term is mostly

**Table 7.5.** Goswami cycle implementation with a combined cycle.

|  | LP-Rankine cycle | Case I (figure 7.8) | Case II (figure 7.9) | Case III (figure 7.11) |
|---|---|---|---|---|
| Net power output | 8486 | 3272 | 14,464 | 17,736 |
| Plant efficiency (%) | 62.38 | 62.69 | 62.90 | 63.16 |
| Net heat rate (kJ kWh$^{-1}$) | 5771.3 | 5742.9 | 5723.5 | 5700.1 |
| Difference with the benchmark point |  |  |  |  |
| Plant efficiency (%) | 0.00 | 0.31 | 0.52 | 0.78 |
| Net heat rate (kJ kWh$^{-1}$) | 0.0 | −28.4 | −47.8 | −71.2 |

used in the energy industry and shows the energy consumption to produce 1 kWh electricity. It is seen that Goswami cycle implementation reduces the energy usage by 28.4–71.2 kJ for each kWh electricity generated.

Another advantage of the Goswami cycle is the ability to generate refrigeration as well as power simultaneously. The gas turbine performance typically degrades during the summer time as the ambient air temperature increases. To eliminate the hot summer period drawback, OEMs typically propose to use power augmentation systems such as evaporative cooling system at the air intake to cool the air. However, this system requires additional refrigeration equipment, air or water-cooled refrigerators; the disadvantage is that these equipment are expensive and high power consumers. The Goswami cycle refrigeration output can be used at the intake, which can improve the gas turbine power output and efficiency during the hot summer period.

## 7.4 Geothermal implementation

The use of renewable energy resources represents a possible response to satisfy the growing energy demand and reduce the environmental impact of fossil fuels and therefore global warming [32]. Geothermal energy is a promising solution due to worldwide availability among all renewable energies [33], weather-independent resource reliability, and efficient energy conversion and management [34]. In 2021, the installed capacity of geothermal plants around the world was 15.6 TW [35]. This increasing trend shows how geothermal energy exploitation has accelerated and it is expected to keep growing in the future.

The most investigated geothermal systems are complex structures based on the integration of different subsystems with a single output, characterized by a large number of devices that can cause greater heat losses, low energy conversion efficiency, higher costs, and a more complex management of the plant. In addition, the use of some organic substances with a high global warming potential (GWP), high costs, high flammability, and toxicity [36], represent additional drawbacks of these systems. With the aim of reducing these drawbacks, a geothermal system with a simple structure based on a modified Goswami cycle configuration will be

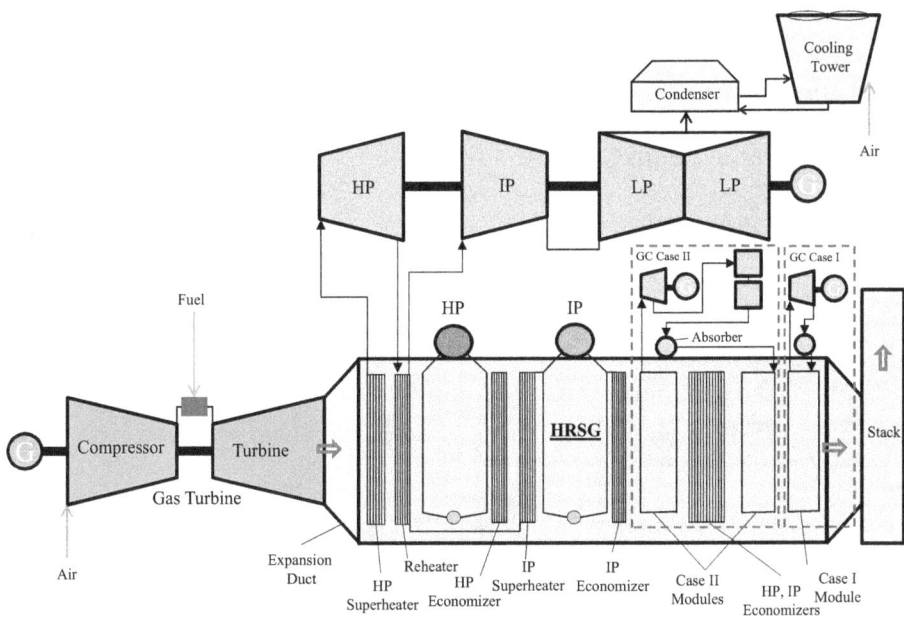

**Figure 7.11.** Combined-cycle integration with Goswami Cycle — Case III (Case I + Case II).

presented in this section. In this system, the heat rejected to the environment by the rectifier is internally recovered to preheat the strong solution, which also reduces the system energy losses. The energy, exergy, costs and performance analysis, considering the geothermal reservoir of Torre Alfina (TA) in central Italy, were investigated. The exergy destruction and losses, the overall performance and the total absolute component costs of the systems are presented. Finally, a comparison with a complex system based on an ORC coupled with a water/LiBr absorption chiller is outlined.

### 7.4.1 Goswami cycle costs analysis

The first step in order to evaluate the component costs is the calculation of the heat exchanger areas of the systems. To this aim, the logarithmic mean temperature difference method (LMTD) is used. The LMTD is defined by the logarithmic mean as follows:

$$\text{LMTD} = \frac{\Delta T_A - \Delta T_B}{\ln\frac{\Delta T_A}{\Delta T_B}} \tag{7.8}$$

where $\Delta T_A$ is the temperature difference between the two streams at the hot side and $\Delta T_B$ is the temperature difference between the two streams at the cold side of the heat exchanger. Finally, the LMTD is used to find the product $UA$ of the heat exchangers.

$$UA = \frac{\dot{Q}}{LMTD} \tag{7.9}$$

where $U$ is the overall heat transfer coefficient, and $Q$ is the thermal power exchanged.

Cost correlations have been selected for each system component and reported in table 7.6. All the heat exchangers are assumed as shell and tube configuration. The effect of the pressure in the heat exchangers is taken into account by means of the correction factor ($a$) defined in equation (10), as proposed in [37]. The pressure correction factor is calculated as follows:

$$a = 10^{(a_1 + a_2 \log(p) + a_3 \log^2(p))} \tag{7.10}$$

where $p$ (bar) is the cycle pressure, $a_1$, $a_2$, $a_3$ are constants for different equipment and are given in table 7.7.

For the first equation in table 7.6, the number of the turbine stages ($n$) was selected considering a limitation related to the maximum enthalpy drop exploitable for each stage as equal to 65 kJ kg$^{-1}$. This limit is set in order to avoid high mechanical stresses [38].

The last stage size parameter ($SP$) of a typical turbine used in geothermal application has been calculated as follows:

$$SP = \frac{\sqrt{\dot{V}}}{\Delta h_{is}^{1/4}} \tag{7.11}$$

**Table 7.6.** Component cost equations [38].

| Component | Cost (€) | Assumptions |
|---|---|---|
| Turbine | $C_0 \left(\frac{n}{n_0}\right)^{0.5} \left(\frac{SP}{SP_0}\right)^{1.1}$ | $C_0 = 1230$ k€ <br> $n_0 = 2$ <br> $SP_0 = 0.18$ m |
| Pump | $C_0 \left(\frac{W_{el}}{W_{el,\,0}}\right)^{0.67}$ | $C_0 = 14$ k€ <br> $W_{el,\,0} = 200$ kW |
| Boiler | $C_0 \left(\frac{UA}{UA_0}\right)^{0.9} a$ | $C_0 = 1500$ k€ <br> $UA_0 = 4000$ kW K$^{-1}$ |
| Generator | $C_0 \left(\frac{W_{el}}{W_{el,\,0}}\right)^{0.67}$ | $C_0 = 200$ k€ <br> $W_{el,\,0} = 5000$ kW |
| Heat exchangers | $C_0 \left(\frac{UA}{UA_0}\right)^{0.9} a$ | $C_0 = 260$ k€ <br> $UA_0 = 650$ kW K$^{-1}$ |

**Table 7.7.** Constants values of pressure factors for the heat exchangers [37].

| Type | Pressure (bar) | $a_1$ | $a_2$ | $a_3$ |
|---|---|---|---|---|
| Boiler | <6 | 0 | 0 | 0 |
| | 6–141 | 0.038 81 | −0.112 72 | 0.081 83 |
| Heat exchangers | <6 | 0 | 0 | 0 |
| | 6–141 | −0.001 64 | −0.006 27 | 0.0123 |

where $V$ is the volume flow rate and $\Delta h_{is}$ is the enthalpy drop of the turbine last stage.

Furthermore, accounting for inflation, all components' costs are updated to the year 2019 using the CEPCI (Chemical Engineering Plant Cost Index). The purchased cost of each component in the year 2019 can be calculated by:

$$C_{i,\,2019} = C_{i,\,\text{ref}} \frac{\text{CEPCI}_{2019}}{\text{CEPCI}_{\text{ref}}} \tag{7.12}$$

where the average value of $\text{CEPCI}_{2019}$ is 607.5 and $\text{CEPCI}_{\text{ref}}$ is 576.1 for 2014 and 394.1 for 2000, respectively taken from the website of Chemical Engineering Plant Cost Index [39] and [37].

### 7.4.2 Case study: the geothermal reservoir of Torre Alfina (Italy)

#### 7.4.2.1 Description of the Torre Alfina

In Italy, the Legislative Decree n.22, promulgated in February 2011, promotes research and development of new geothermal power plants with reduced environmental impact, and states that fluids with medium and high enthalpy are both of national interest. According to this decree, the releases for mining concession can reach as far as 50 MW$_e$ of installed power [40]. The new experimental plants must be compliant with geothermal fluids re-injection on national scale in the same original formations, zero emissions in the atmosphere and a nominal installed power not exceeding 5 MW$_e$ for each plant. Castel Giorgio-Torre Alfina area (CGTA) is one of the mining concessions released by the Italian Ministry of the Economic Development according to the above-mentioned legislative decree. The CGTA, whose study area covers about 480 km$^2$, is located in Central Italy between Lazio and Umbria regions, close to Bolsena Lake, as shown in figure 7.12.

The reservoir fluid is water with a salinity of about 6 g l$^{-1}$, gas saturated (1%–2% by weight, mostly $CO_2$ and traces of $H_2S$), and water level is at 200 m below ground level. The single well flow rate is generally of about 400 t h$^{-1}$. At the reservoir top, at a depth of about 550 m, a $CO_2$ gas cap has been found in the central part of the field, with a thickness of about 100 m, a gas pressure of about 44 bar. The average temperature of the geo-fluid is 140 °C [41–46]. A summary of the characteristics of the brine and the project above-mentioned are shown in table 7.3, while in figure 7.1 the exact location of CGTA and the wells is pointed out.

Profitable use of electricity and cooling production by the Goswami cycle in this context could be in the tourism sector (i.e. hotels) and food-processing industries— in particular for food preservation—widely present in the rural areas.

The operating conditions considered in this analysis for the Goswami cycle are summarized in table 7.9.

#### 7.4.2.2 Results

A parametric analysis is used to define the optimal ammonia mass fraction of the strong solution and the optimal high pressure of the cycle. The performance variation – net electrical power ($\dot{W}_{\text{NET}}$), net cooling power ($\dot{Q}_C$), first law efficiency

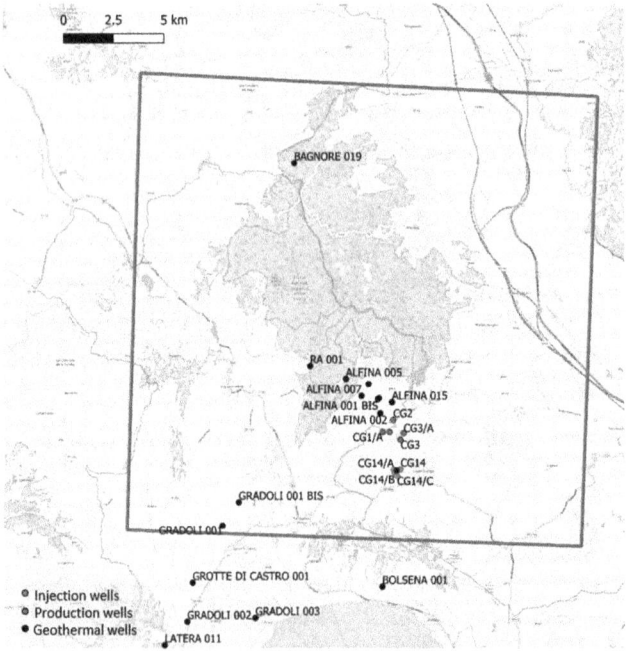

**Figure 7.12.** Torre Alfina mining concession and wells location.

**Table 7.8.** Torre Alfina reservoir and CGTA pilot plant data [41–46].

| | |
|---|---|
| Pressure (bar) | 44 |
| Average reservoir temperature (°C) | 140 |
| Salinity (g L$^{-1}$) | 6 |
| $CO_2$ mass fraction (%) | 1–2 |
| Number of production wells | 5 |
| Number of reinjection wells | 4 |
| Total brine mass flow rate (t h$^{-1}$) | 1050 |

**Table 7.9.** Operating conditions for Goswami cycle [18].

| Parameter | Value | Parameter | Value |
|---|---|---|---|
| $p_0$ (bar) | 1.01 | $\eta_{turbine}$ (−) | 0.85 |
| $T_0$ (°C) | 25 | $\eta_{pump}$ (−) | 0.85 |
| $T_{GB}$ (°C) | 140 | $T_{boiler}$ (°C) | 80 |
| $p_{GB*}$ (bar) | 44 | $T_{absorber}$ (°C) | 30 |
| $\dot{m}_{GB*}$ (kg s$^{-1}$) | 299 | Pinch (°C) | 10 |

*Note: GB stands for geothermal brine.

and exergy efficiency for variations in ammonia concentration ($x$), and high pressure of the cycle ($P_{high}$), are shown in figure 7.13.

In the considered ranges of $x$ and $P_{high}$, the maximum $\dot{W}_{NET}$ (3.15 MW) is obtained for a maximum pressure of 15.45 bar, an ammonia mass fraction of 0.4632, with a first-law efficiency of 13.12% and an exergy efficiency of 35.53%. These operating conditions are referred to as GWC-MW case in the following description. Conversely, the maximum $\dot{Q}_C$ (2.15 MW) is reached for a maximum pressure of 23.45 bar, an ammonia mass fraction of 0.4992, with a first-law efficiency of 16.6% and an exergy efficiency of 34.03%. These operating conditions are referred to as GWC-MC case in the following section.

The exergy efficiencies of the two cases differ by 1.5% in favor of the GWC-MW. This means that the resource can be well exploited in both cases, and expand the range of applicability based on the costumers' needs. From the parametric analysis, it is possible to observe that the maximum cooling and the maximum net power are reached at higher ammonia concentrations, which is also discussed in detail in section 7.3.2. This allows a higher mass flow rate for the refrigerant stream (from state point 5 to 7, figure 7.15), thanks to the separation and rectification processes, which is favorable for the production of cooling and power. In addition, GWC-MC

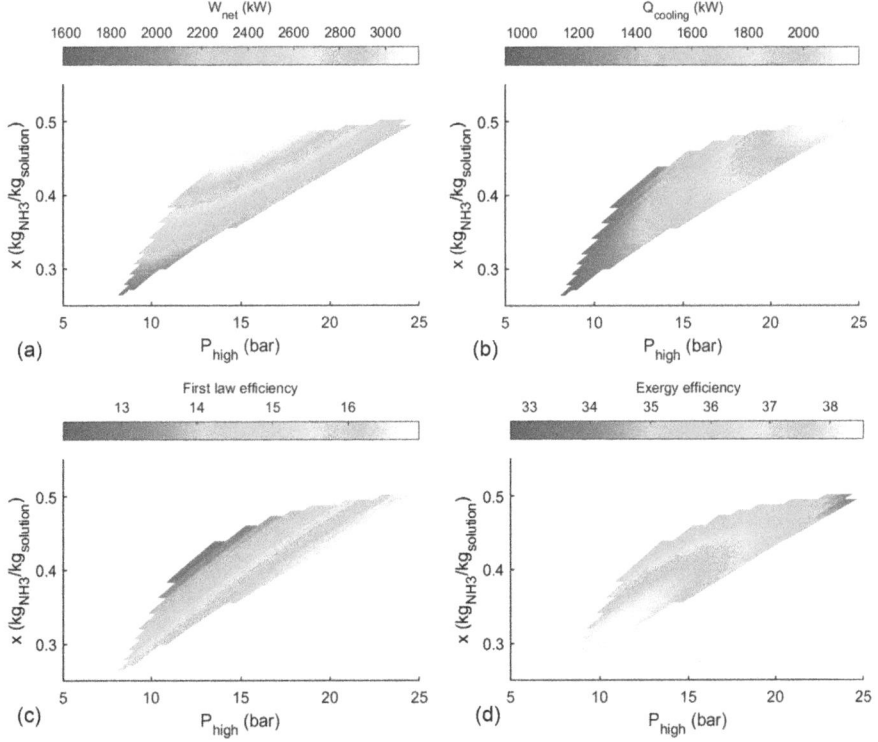

**Figure 7.13.** GWC performance evaluation varying $x$ and $P_{high}$ for CGTA application: (a) $W_{net}$, (b) $Q_{cooling}$, (c) first-law efficiency, (d) exergy efficiency.

case compared to GWC-MW requires higher pressure at the turbine inlet to reach a cooler exhaust.

It is also worth noting that the maximum $\dot{W}_{\text{NET}}$ and $\dot{Q}_C$ do not correspond to the maximum overall first law and exergy efficiency of the system. The exergy efficiency tends to maximize at lower ammonia mass fractions (0.35–0.25) and lower maximum pressures (8.00–14.00 bar). The first law efficiency tends to maximize at higher maximum pressures (20.05–24.73 bar) and ammonia mass fractions (0.45–0.51).

The GWC-MC case shows a higher-pressure ratio (6.27) and maximum pressure. A higher-pressure ratio enables the expansion of the fluid to a slightly lower temperature, which, consequently, allows a consistent increment of $\dot{Q}_C$ (an increase of more than 0.5 MW). On the other hand, the GWC-MW case shows a lower maximum pressure and lower pressure ratio (5) compared to GWC-MC.

The exergy analysis shows where the cycle's main irreversibilities are, and helps in the identification of potential improvement areas. The relative exergy destructions and losses are shown in figure 7.14. The main sources of exergy destruction are absorber, separator-rectifier column, turbine and heat exchangers. For both cases, the dominant exergy destruction source is the absorber; the absorber has the highest relative exergy destructions of 23.3% and 19.2%, for GWC-MW and GWC-MC, respectively. These irreversibilities are due to the processes involved in this component: the condensation and the mixing of ammonia in water [47]. Other large exergy destructions occur at the rectifier-separator (Sep-Rec), equal to 11.4% and 14.1% for GWC-MW and GWC-MC respectively.

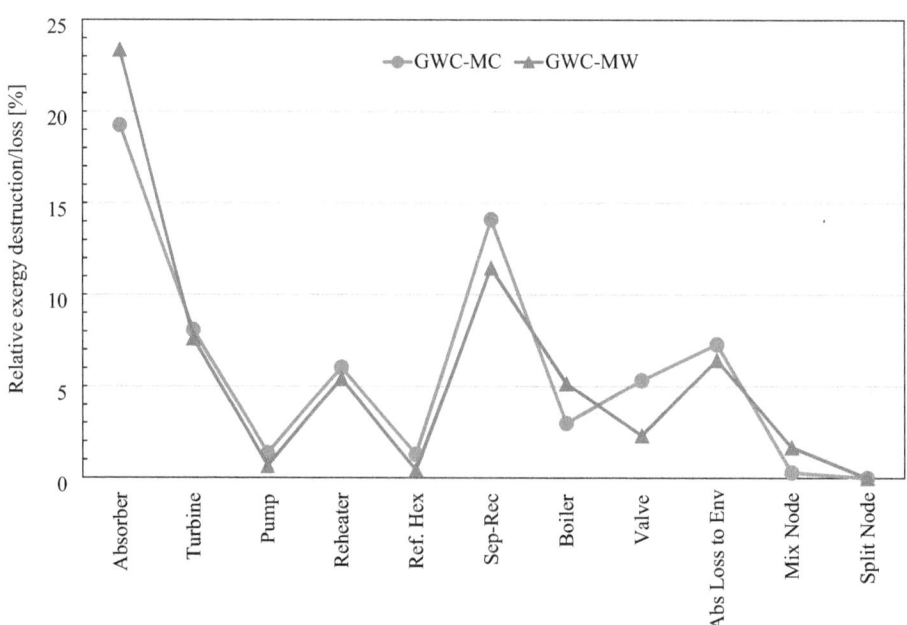

**Figure 7.14.** Relative component exergy destructions/losses of Goswami cycles.

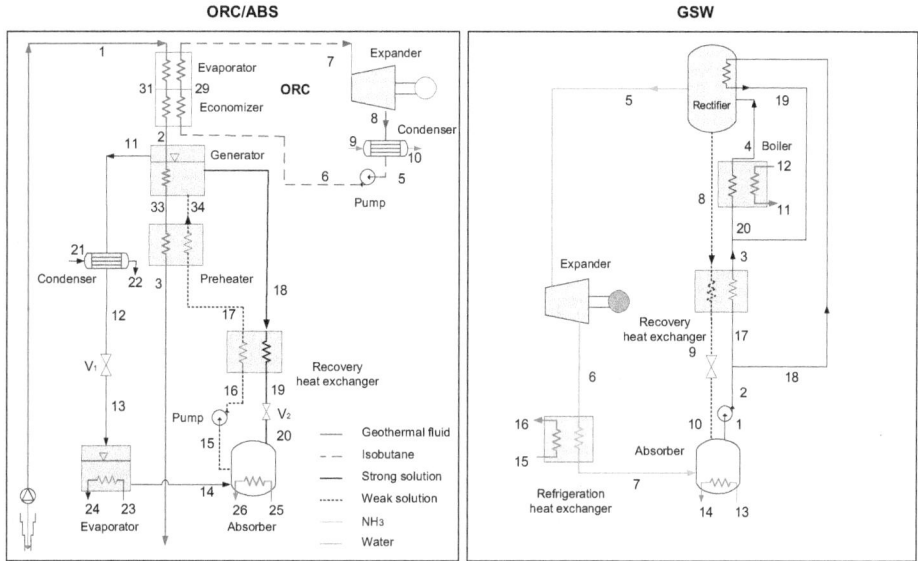

**Figure 7.15.** Plant schematics: cascade ORC/ABS (left) and Goswami cycle (right).

The temperature at the outlet of the rectifier plays an important role in the Goswami cycle output products. In general, a lower temperature allows higher cooling output but at the same time, more heat is rejected to the strong solution and recovered. Considering the GWC-MC scenario, the irreversibilities are higher at the rectifier-separator, but they are lower at the boiler (Boiler) compared to the GWC-MW scenario because in the first case the heat rejected to the strong solution is relatively higher and it is recovered by the strong solution reducing the boiler heat input. Consequently, the irreversibilities at the boiler are lower in the GWC-MC as the heat exchanged is lower compared to the heat input.

Figure 7.14 also shows that the turbine is a critical component as well as the recovery heat exchanger (Reheater), which gives a relatively high contribution to the overall exergy destruction. Evidently, the turbine behavior can be improved by increasing its efficiency.

The relative absorber losses to the environment (Abs Loss to Env) are equal to 6.4% and 7.2%, respectively. Since the absorber also works as the condenser, it rejects to the environment a huge quantity of heat at lower temperature than the heat source. In order to improve the overall efficiency, this amount of heat could be exploited for other applications, such as preheating water in desalination processes.

Still, in figure 7.14, it is shown that the recovery heat exchanger gives a contribution of the overall exergy destruction between 5% and 6%, which it is due to imperfect matching of the composite curves of strong and weak solution.

The heat exchanger areas are shown in table 7.10. The values of the overall heat transfer coefficient for each component of the cycle have been assumed from literature [48–50].

**Table 7.10.** Heat exchangers' areas of Goswami cycle for GWC-MC and GWC-MW scenario [18].

| Component | $\dot{Q}$ (kW) | $U$ (kW m$^{-2}$ K) | $A$ (m$^2$) |
|---|---|---|---|
| GWC-MC | | | |
| Absorber | 27 282.0 | 0.91 | 4696.9 |
| RHE | 29 010.9 | 0.99 | 1832.1 |
| Boiler | 27 434.9 | 1.10 | 2494.1 |
| RefHE | 2199.7 | 1.34 | 164.2 |
| Rectifier | 14 417.1 | 1.00 | 566.7 |
| GWC-MW | | | |
| Absorber | 33 902.6 | 0.91 | 5836.8 |
| RHE | 25 971.4 | 0.99 | 1299.1 |
| Boiler | 35 544.5 | 1.10 | 2659.3 |
| RefHE | 1512.8 | 1.34 | 112.9 |
| Rectifier | 19 529.2 | 1.00 | 922.2 |

In both cases, the absorber area has the highest value that ranges between 4696.9 and 5836.8 m$^2$, while the lowest values correspond to the refrigerant heat exchanger areas. It can be concluded that the GWC-MW requires higher areas compared to the GWC-MC.

### 7.4.2.3 Comparison between the Goswami cycle and a cascade organic Rankine cycle/absorption chiller system

Figure 7.15 shows the plant schematics of both systems. The cascade ORC/ABS system has already been investigated and additional details can be found in [51]. In order to compare the cascade ORC/ABS system and the Goswami cycle, two scenarios have been considered for the cascade ORC/ABS system, one (called OAC1) with a maximum power production of 5 MW and the other (called OAC2) with the same maximum power production as the Goswami cycle. Therefore, the OAC2 scenario aims to evaluate the performance under the same power output conditions of the Goswami cycle. As previously described, the maximum power output of the GWC is 3.154 MW (GWC-MW) for the Torre Alfina heat source.

Therefore, this value has been set as the power output for the OAC2 case with the objective of evaluating the influence on the second-law efficiency in both cases. It is worth noting that the chilled water reaches 8 °C in all the cases, which ensures a fair comparison between the systems and an actual application for the Torre Alfina area. Table 7.11 shows the summarized results, and the exergy efficiency for all cases, while figure 7.16 compares the component exergy destructions/losses between GWC and ORC/ABS systems.

As shown in figure 7.16, the absorber (ABS) of the GWC is the component with the highest relative exergy destruction (23.3% and 19.2%) among all the components of both systems. In light of these findings, the system can be improved by exploiting the rejected heat from the absorber through a number of applications. In the context

**Table 7.11.** Performance and outputs comparison [18].

| Case | $\dot{W}_{NET}$ MW | $\dot{Q}_C$ MW | $\eta_{x,d}$ % |
|---|---|---|---|
| OAC1 | 5.00 | 27.10 | 40.98 |
| OAC2 | 3.15 | 38.92 | 32.77 |
| GWC-MW | 3.15 | 1.51 | 35.53 |
| GWC-MC | 2.35 | 2.20 | 34.03 |

**Table 7.12.** Assumptions of the system Goswami-desalination study.

| $T_4$ (°C) | 100 | $m_1$ (kg s$^{-1}$) | 1 |
|---|---|---|---|
| $T_1$ (°C) | 35 | Pinch (°C) | 8 |
| $T_{13}$ (°C) | 25 | $q_{10}$ (–) | 0 |

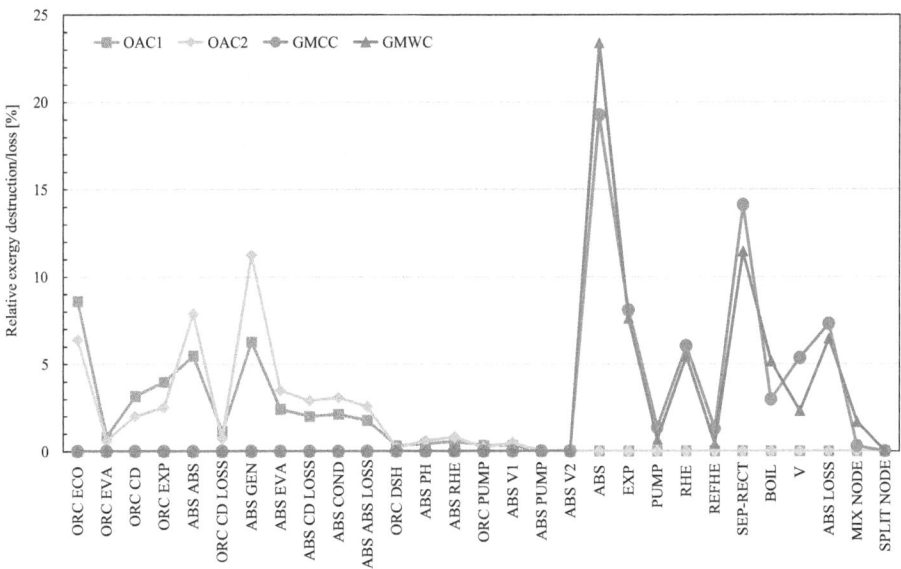

**Figure 7.16.** Relative component exergy destructions/losses comparison between GWC and ORC/ABS cycles.

of the Castel Giorgio-Torre Alfina area, this heat can be beneficial for food processing industries, and district heating.

Considering the ORC/ABS cases (OAC1 and 2), the generator of the absorption cycle (ABS GEN) is the component with the highest relative exergy destruction for both with a value that ranges between 11.3% and 6.3%. Because the GWC has less components compared to the cascade ORC/ABS, the average exergy destruction per component is higher. Comparing OAC1 and OAC2, it is possible that the decrease of cooling production corresponds to an increase of the exergy destructions at the generator (ABS GEN) and absorber (ABS ABS) of OAC2. On the other hand, when the power production increases, i.e. in OAC1, an increase in the exergy destructions at the expander (ORC EXP), economizer (ORC ECO) and condenser (ORC CD) are observed.

From the exergetic point of view (see table 7.11), the best scenario is the OAC1 ($\eta_{x,d} = 40.98\%$), which is the case where the resource is better exploited. However, if the same power output is considered—OAC2 vs. GWC-MW—the best scenario becomes the GWC-MW. Even though the cooling output of OAC2 is much higher compared to GWC-MW, the exergy efficiency is lower because of the higher heat source input, thus the higher exergy input to the system.

Another important consideration to highlight is from the point of view of the geothermal resource. The outlet temperature of the geothermal resource is 356 K for the cascade ORC/ABS cases, while for GWC cases it varies between 391 K and 385 K. This means that in the latter case the resource can be exploited for other purposes before reinjection (reinjection temperature should not be lower than 353.15 K in order to preserve the TA reservoir). For example, the remaining sensible heat in GWC cases could be employed for district heating. Thus, the choice between the systems depends on the needs of the customers. Regarding the cost analysis, figure 7.17 gives an overview of the heat exchangers areas for each case.

The biggest contribution in terms of area is given by the absorber (ABS) for GWC cases and this is mainly due to the necessity of the system to reject a huge amount of heat to the environment (33.9 MW for GWC-MW, 27.3 MW for GWC-MC) in order to allow the process of absorption. Regarding ORC/ABS systems, the evaporator (ABSC, EVA) in the absorption cycle had the highest area among all heat exchangers due to the high production of cooling. It is worth noting that a great

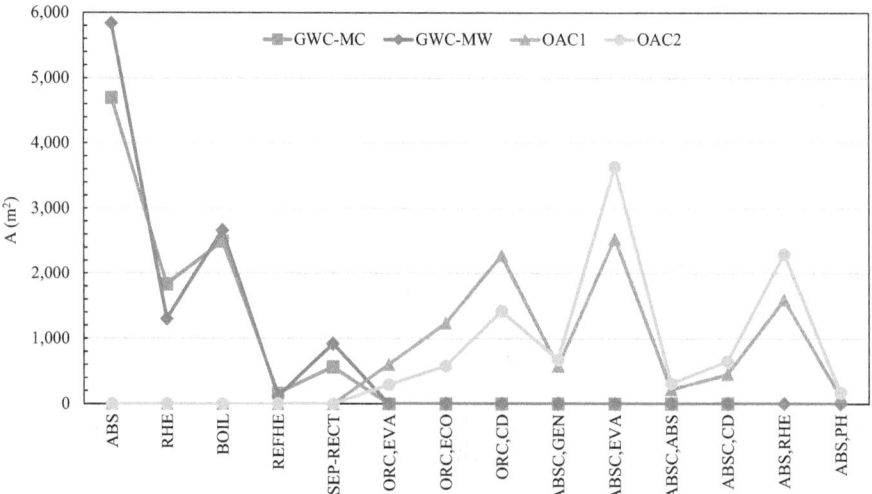

**Figure 7.17.** Comparison of the heat exchangers' areas of both systems.

contribution in terms of area is also given by the heat exchangers that transfer heat from the brine to the organic fluid (evaporator ORC, EVA and economizer ORC, ECO) and the ORC condenser (ORC, CD). This is also reflected in terms of exergy destructions in both systems. Therefore, a better thermal matching could be found in order to reduce the irreversibilities, improve the heat transfer and reduce the areas. To do so, further studies are necessary in terms of working fluid selection regarding the ORC and in terms of optimization of the GWC.

The component cost analysis is shown in figure 7.18. The cost of the wells (perforation, installation and pumps) is taken from [52] and updated to 2019. The total cost for the planned wells is 16 296 378 € (5 production wells with a depth of 1200 m, 4 reinjection wells with a depth of 2300 m). Typically, the highest cost of a geothermal power plant is for the wells and thus the depth of the reservoir affects the total costs.

Figure 7.18 shows that the highest total absolute component cost (TACC) is found for the OAC1 with a total of 16.8 M€, while the GWC-MC shows the lowest TACC of 10.8 M€. The cost of the wells (16.3 M€) exceeds the TACC in all cases, except for OAC1.

Regarding the turbines, when the working fluid is isobutene a two-stage turbine is considered with a $\Delta h_{is}$ per stage equal to 29.7 kJ kg$^{-1}$ (both cases) while in the case of ammonia the enthalpy drops per stage are 64.8 (GWC-MC) and 61 kJ kg$^{-1}$ (GWC-MW), requiring a three-stage turbine. In these cases, the number of stages slightly affects the costs of the turbine, while the most affecting parameter is the last stage size parameter (SP) which is also related to the working fluid through the volume flow rate. Basically, the higher the volume flow rate, the higher the cost of the turbine.

In all four cases, the cost of the turbine has the highest contribution, ranging from 87% to 64% of the TACC.

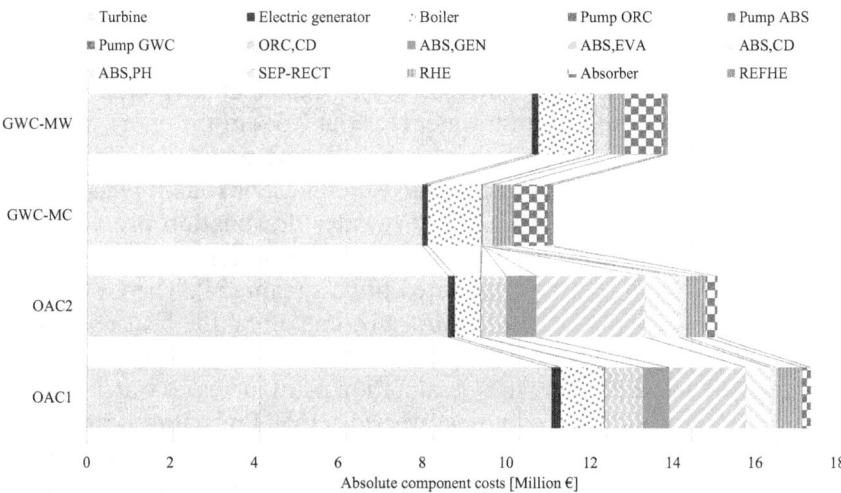

**Figure 7.18.** Absolute component cost repartition for the four cases.

The second-most expensive component is the evaporator for the cascade ORC/ABS system (ABS, EVA), ranging from 11% to 17% of the TACC. Considering all cases, the boiler gives a contribution to the costs that ranges from 4% up to 13% of TACC for GWC-MW.

Finally, comparing the systems with the same power production, total absolute component cost for OAC2 (14.6 M€) is higher than GWC-MW (13.4 M€).

Main findings are summarized below:

- Goswami cycle: Two cases were selected, the maximum net electrical power case ($\dot{W}_{NET}$ = 3.15 MW) and the maximum cooling production case ($\dot{Q}_C$ = 2.2 MW). The highest exergy destructions: the absorber (GWC-MC: 19.2%, GWC-MW: 23.4%), separator-rectifier (GWC-MC: 14.1%, GWC-MW: 11.4%) and expander (GWC-MC: 8.07%, GWC-MW: 7.60%). The exergy efficiency of the GWC-MC is 34.03%, while it is 35.53% for the GWC-MW. This means that the resource can be well exploited in both cases, and expand the range of applicability based on the costumers' needs.
- From the geothermal resource point of view, the outlet temperature of the geothermal resource is 356 K for the cascade ORC/ABS, while for GWC it varies between 391 K and 385 K. This means that in the second case the resource can be exploited for other purposes before reinjection (reinjection temperature should not be lower than 353.15 K in order to preserve the TA reservoir).
- The component cost analysis showed that the highest total absolute component cost (TACC) for the OAC1 is 16.8 M€, while the TACC for GWC-MC it is 10.8 M€, which is the lowest.
- For the case of same power output in both the systems OAC2 and GWC-MW, and comparing the exergy efficiency and the TACC, the best scenario is the GWC-MW with 35.53% and 10.8 M€, against 32.77% and 14.6 M€ for OAC2.

## 7.5 Desalination implementation

The worldwide demand for 'fresh' water is growing exponentially, but its availability on the planet is limited: natural fresh water is about 0.5% of the entire water supply on Earth. In order to meet future needs of the growing population and increasing industrialization, more fresh (and/or potable) water must be found. Desalination is a successful technique and is growing [53]. Seawater desalination process separates saline seawater into two streams: a fresh water stream containing a low concentration of dissolved salts and a concentrated brine stream [54]. The Goswami cycle can be extended for desalination applications by exploiting the heat rejected to the environment thanks to the heat recovery from the absorber to preheat the seawater.

Higher feed water temperature in a desalination plant increases water production, and improves performance and economic efficiency [55]. Preheating seawater before it enters a desalination plant has been proven to increase fresh water production rate for all membrane-based desalination processes [56, 57] since the viscosity of seawater decreases as the temperature goes up. As the absorber also works as the condenser, it rejects a huge quantity of heat to the environment at lower temperature than the heat source, as shown in the previous section. In order to improve the overall efficiency, this amount of heat could be exploited for other applications, such as preheating water in the desalination process.

Figure 7.19 shows the plant schematic considered in this section and table 7.12 shows the assumptions. The seawater properties used for the following analysis are based on [58–60].

Figure 7.20 shows the mass flow rate of the cold stream at the absorber with varying ammonia mass fraction ($x$) and the high pressure of the cycle ($P_{high}$). When the cold stream is water, the mass flow rate ranges between 2 and 9.8 kg s$^{-1}$. However, when the cold stream is seawater, the mass flow rate necessary to cool down the absorber is slightly higher (3 and 10.2 kg s$^{-1}$). This is beneficial when the Goswami cycle is integrated with a desalination plant since more seawater can be preheated compared to fresh water. Lower mass flow rates are observed mostly where the ammonia mass fraction is lower (0.2–0.4).

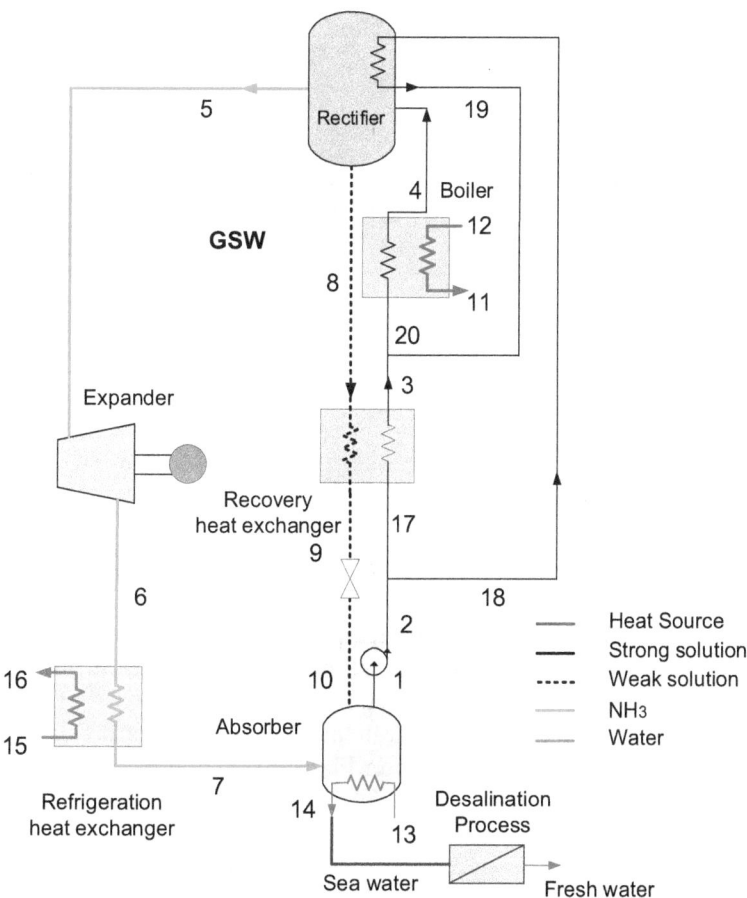

**Figure 7.19.** Goswami cycle integrated with the desalination process.

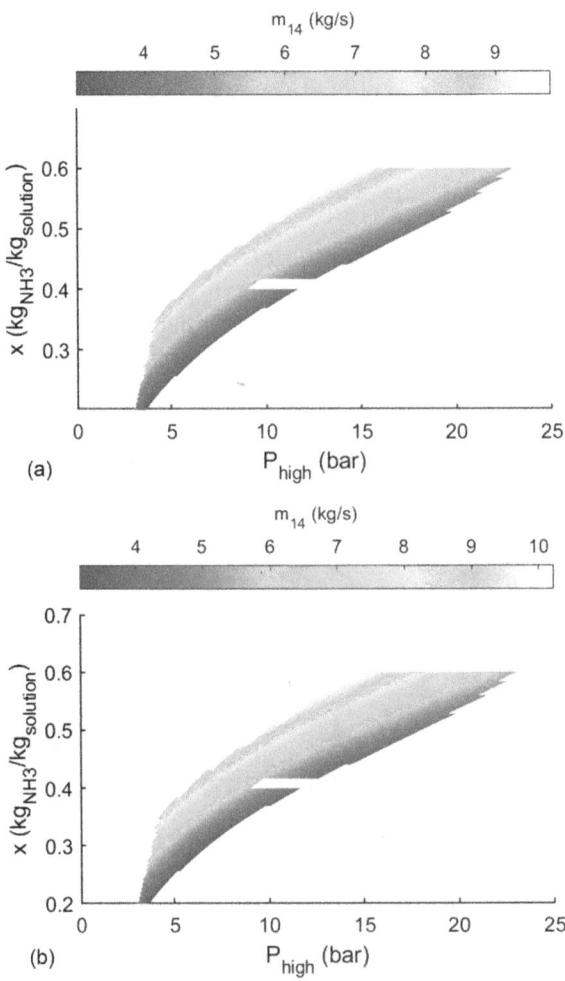

**Figure 7.20.** Mass flow rate at the absorber: (a) water properties and (b) sea water properties.

Considering a cold stream inlet temperature of 25 °C ($T_{13}$), seawater can be heated up to 38.5 °C, as shown in figure 7.3. Higher concentration of ammonia and high pressure of the cycle ensure higher temperatures at the outlet of the absorber, while lower concentration of ammonia allows lower outlet temperatures at the absorber. When the ammonia mass fraction is lower, less heat is rejected to the environment and thus lower mass flow rate and lower temperatures are reached.

The Tampa Bay Seawater Desalination Plant [61] is one of the biggest seawater desalination plants in the United States and it is a membrane-based reverse osmosis (RO) system. It can capture up to 44 million gallons daily and provide up to 25 million gallons of desalinated water daily ($1.1 \text{ m}^3 \text{ s}^{-1}$) to the area. Currently, this RO system has a 57% permeate recovery rate [62].

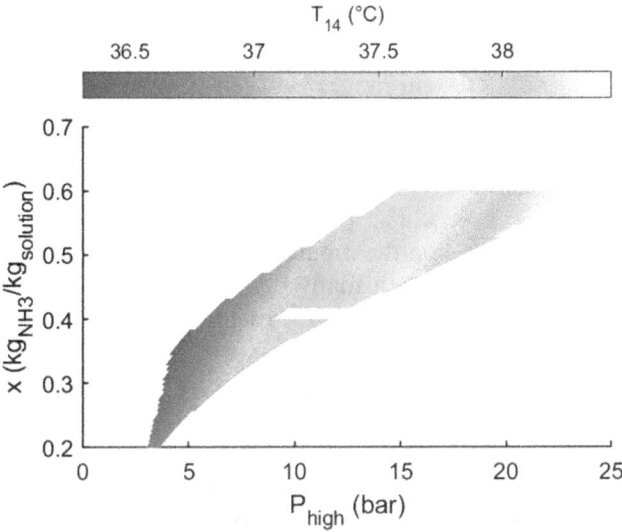

**Figure 7.21.** Temperature at the outlet of the cold stream at the absorber.

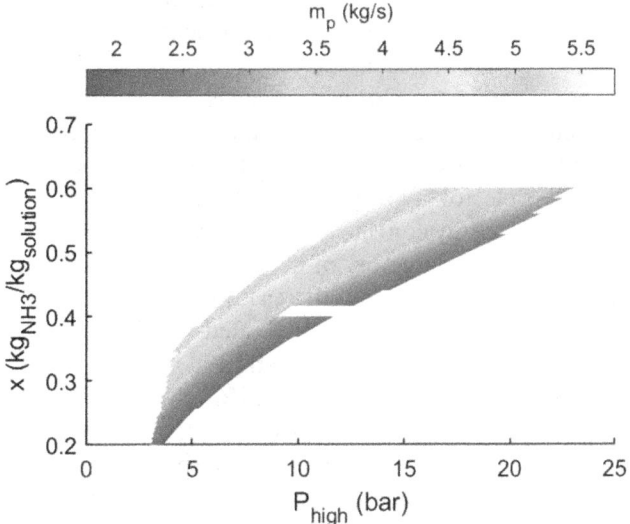

**Figure 7.22.** Potential permeate production from the cold stream at the absorber.

Considering such a permeate recovery rate, the permeate that can be potentially produced ($m_p$) by the cold stream at the absorber as shown in figure 7.22.

The permeate production ranges between 1.77 kg s$^{-1}$ and 5.76 kg s$^{-1}$ for a strong solution mass flow rate ($m_1$) of 1 kg s$^{-1}$. Electric energy intensity of a reverse osmosis water treatment for ocean water is estimated to be 12 000 kWh day$^{-1}$ per MGD of fresh water produced by the plant [63]. Thus, the total electric consumption for the RO plant integrated in the system ranges between 19 kW and 63 kW. Based on the input parameters, power produced by the Goswami cycle can cover between 0.5% and 49.3% of the RO electric load. It is worth noting that the percentage of RO electric load covered by the Goswami cycle does not increase linearly as the permeate production decreases. The power production and the permeate are not proportional, but they depend on the inputs selected. In conclusion, Goswami cycle has a potential for integration with many desalination processes, enhancing their performances and efficiencies, while also providing electricity and cooling to the desalination plants.

## 7.6 Remarks and conclusions

In this chapter, the theoretical performance of a combined power and cooling cycle was investigated for a variety of boiler pressures and ammonia concentrations to determine their effects on the net work, cooling, and effective efficiencies. First, a low-temperature heat source case is considered. The maximum theoretical effective first-law and exergy efficiencies were 10.5% and 51.4%, respectively, for a heat source temperature of 100 °C, and an outlet temperature at the absorber of 20 °C. The Carnot, ideal triangular and Lorenz cycle efficiencies for the same heat source and sink conditions are calculated as 22.8%, 12.9% and 9.4%, respectively. The effective first-law efficiency of the Goswami cycle is comparable with the efficiencies of ideal trilateral and Lorenz cycles while using a low-grade heat source. Secondly, an industrial waste heat recovery is presented. For this case, the Goswami cycle is used as a bottoming cycle for a Brayton–Rankine natural gas power plant. The results show that overall energy conversion efficiency of the power plant is improved significantly. Depending on the number of Goswami cycles used as bottoming cycles in the plant, the overall plant efficiency increases between 0.31% to 0.78%.

The Goswami cycle performance was evaluated for the conditions of an actual geothermal reservoir of Torre Alfina (Italy), and the performance values compared with a cascade ORC coupled with a water/LiBr absorption chiller system. The results showed that the Goswami cycle performs better than an ORC/Absorption system for the exergy efficiency, however ORC/Absorption can produce higher cooling output. A cost analysis was also performed and it was found that the total absolute component cost for the ORC/Absorption system was highest with a total of 16.8 M€, while the GWC-MC showed the lowest cost of 10.8 M€.

Goswami cycle has the potential for integration with many desalination processes, enhancing their efficiencies, while also providing electricity and cooling to the desalination plants.

The Goswami cycle studies are summarized in table 7.13. For the cases where only work output is given in the table, the cooling output can be generated with the rectification process, though the effective second-law efficiency will be degraded due to higher exergy losses at vapor production. It is seen that the cycle can operate at different conditions and produce work, cooling and water as useful products.

**Table 7.13.** Summary of Goswami cycle performances.

| Cycle parameter | Units | Section 7.3.2 Low-grade solar implementation | | Section 7.3.3 Industrial waste heat implementation | | Section 7.4 Geothermal implementation | Section 7.5 Desalination implementation |
|---|---|---|---|---|---|---|---|
| Boiler temperature | °C | 80.0 | 95.0 | 95.0* | 82.8 | 290.0** | 130.0 | 100.0 |
| Absorber temperature | °C | 20.0 | 20.0 | 20.0 | 20.0 | 20.0 | 30.0 | 35 |
| $x$ | kg $NH_3$ kg$^{-1}$-solution | 0.4 | 0.3 | 0.5 | 0.9 | 0.1 | 0.46/0.50 | 0.2–0.6 |
| Mass flow rate | kg s$^{-1}$ | 1.0 | 1.0 | 1.0 | 1.0 | 1.0 | 82.0 | 81.5 |
| Phigh/plow | | 4.4 | 7.2 | 4.7 | 3.9 | 781.9 | 5.0/6.3 | 2.2–5.8 |
| | | | Useful products | | | | | |
| Work | kW | 15.9 | 32.2 | 27.8 | 87.6 | 388.5 | 3151/2352 | 0.4–16 |
| Cooling | kW | 12.6 | | 22.0 | | | 1514/2201 | 0–11 |
| Water | kg s$^{-1}$ | | | | | | | 1.77–5.76 |
| Note | | | | $*T_{rectifier}$ is 50 °C | | $**P_{high}$ = 95 bar, $P_{low}$ = 0.12 bar | | |

7-35

# References

[1] Leyzerovich A S 1997 *Large Power Steam Turbines: Design and Operation* vol 1 (Tulsa, OK: Pennwell Books)

[2] Kalina A I 1984 Combined cycle system with novel bottoming cycle *J. Eng. Gas Turbines Power* **106** 737–42

[3] Goswami D Y 1995 Solar thermal power: status of technologies and opportunities for research *Proc. of the 2nd ISHMT-ASME Heat and Mass Trans. Conf.*, Tata McGraw Hill, New Delhi

[4] Ibrahim O M and Klein E S A 1996 Absorption power cycles *Energy* **21** 21–7

[5] Goswami D Y 1998 Solar thermal power technology: present status and ideas *Energy Sources* 20 137–45

[6] Vijayaraghavan S and Goswami D Y 2005 Organic working fluids for a combined power and cooling cycle *J. Energy Res. Technol.* **127** 125–30

[7] Chen H, Goswami D Y, Rahman M M and Stefanakos E K 2011 A supercritical Rankine cycle using zeotropic mixture working fluids for the conversion of low-grade heat into power *Energy* **36** 549–55

[8] Padilla R V, Demirkaya G, Goswami D Y, Stefanakos E and Rahman M M 2010 Analysis of power and cooling cogeneration using ammonia-water mixture *Energy* **35** 4649–57

[9] Demirkaya G 2011 Theoretical and experimental analysis of power and cooling cogeneration utilizing low temperature heat sources

[10] Demirkaya G, Besarati S, Padilla R V, Archibold A R, Goswami D Y, Rahman M M and Stefanakos E L 2012 Multi-objective optimization of a combined power and cooling cycle for low-grade and midgrade heat sources *J. Energy Res. Technol.* **134** 5

[11] Demirkaya G, Padilla R V, Fontalvo A, Bula A and Goswami D Y 2018 Experimental and theoretical analysis of the Goswami cycle operating at low temperature heat sources *J. Energy Res. Technol.* **140** 3

[12] Leveni M, Narasimhan A K, Almatrafi E and Goswami D Y 2019 Performance improvement of a combined power and cooling cycle for low temperature heat sources using internal heat recovery and scroll expander *ASME 2019 Gas Turbine India Conference* (*Chennai, India*)

[13] Kumar G P, Saravanan R and Coronas A 2017 Experimental studies on combined cooling and power system driven by low-grade heat sources *Energy* **128** 801–12 6

[14] Hosseinpour J, Chitsaz A, Eisavi B and Yari M 2018 Investigation on performance of an integrated SOFC-Goswami system using wood gasification *Energy* **148** 614–28 4

[15] Rivera W, Sánchez-Sánchez K, Hernández-Magallanes J A, Jiménez-García J C E A and Pacheco E A 2020 Modeling of novel thermodynamic cycles to produce power and cooling simultaneously *Processes* **8** 320

[16] Sayyaadi H, Khosravanifard Y and Sohani A 2020 Solutions for thermal energy exploitation from the exhaust of an industrial gas turbine using optimized bottoming cycles *Energy Convers. Manage.* **207** 112523

[17] Guillen D, Leveni M, Manfrida G and Sanjuan M 2019 Integration and optimization of supercritical carbon dioxide brayton cycle and Goswami cycle *Volume 6: Energy*

[18] Leveni M and Cozzolino R 2021 Energy, exergy, and cost comparison of Goswami cycle and cascade organic Rankine cycle/absorption chiller system for geothermal application *Energy Convers. Manage.* **227** 113598

[19] Xu F and Goswami D Y 1999 Thermodynamic properties of ammonia-water mixtures for power-cycle applications *Energy* **24** 525–36

[20] Tillner-Roth R and Friend D G 1998 A Helmholtz free energy formulation of the thermodynamic properties of the mixture {Water+Ammonia} *J. Phys. Chem. Ref. Data* **27** 63–96

[21] McLinden M O, Klein S A and Lemmon E E W 2006 NIST Reference Fluid Thermodynamic and Transport Properties REFPROP V7. 0

[22] Vijayaraghavan S and Goswami D Y 2003 On evaluating efficiency of a combined power and cooling cycle *J. Energy Res. Technol.* **125** 221–7

[23] Rosen M A and Le M N 1995 Efficiency measures for processes integrating combined heat and power and district cooling, In *Thermodynamics and the Design, Analysis and Improvement of Energy Systems* **35** 423–34

[24] The MathWorks Inc. 2020 MATLAB, version 9.8.0.1323502 (R2020a) https://mathworks. com/products/new_products/release2020a.html

[25] Lee W Y and Kim S S 1992 The maximum power from a finite reservoir for a Lorentz cycle *Energy* **17** 275–81

[26] DiPippo R 2007 Ideal thermal efficiency for geothermal binary plants *Geothermics* **36** 276–85

[27] G. E. Company 2020 9HA Power Plants https://ge.com/content/dam/gepower/global/en_US/ documents/gas/gas-turbines/prod-specs/9ha-power-plants.pdf

[28] S. Energy 2019 We power the world with innovative gas turbines-Siemens gas turbine portfolio https://assets.siemens-energy.com/siemens/assets/api/uuid:82efd37c-db81-45c0– 972b-c5dc25365775/gas-turbines-siemens-interactive.pdf

[29] G. E. Company 2018 9F Power Plants https://ge.com/content/dam/gepower-pgdp/global/ en_US/documents/product/gas turbines/Fact Sheet/2018-prod-specs/9F_Power_Plants_R3. pdf

[30] G. E. Company 2018 9E&GT13E2 Power Plants https://ge.com/content/dam/gepower-pgdp/ global/en_US/documents/product/gas turbines/Fact Sheet/2017-prod-specs/GEA32931A 9E-GT13E2_Power_Plants_R2.pdf

[31] Thermoflow Inc., GT PRO 2020 https://thermoflow.com/

[32] Adams S, Klobodu E K M and Apio A 2018 Renewable and non-renewable energy, regime type and economic growth *Renewable Energy* **125** 755–67

[33] Kana J D, Djongyang N, Raïdandi D, Nouck P N and Dadjé A 2015 A review of geophysical methods for geothermal exploration *Renew. Sustain. Energy Rev.* **44** 87–95

[34] Moya D, Aldás C and Kaparaju P 2018 Geothermal energy: power plant technology and direct heat applications *Renew. Sustain. Energy Rev.* **94** 889–901

[35] Richter A 2021 Think Geoenergy 7 January https://thinkgeoenergy.com/thinkgeoenergys-top-10-geothermal-countries-2020-installed-power-generation-capacity-mwe/ [Consultatoilgiorno 4 March 2021]

[36] Dinçer İ and Kanoğlu M 2010 *Refrigeration Systems and Applications* (New York: Wiley)

[37] Turton R, Bailie R C, Whiting W B and Shaeiwitz J A 2008 *Analysis, Synthesis and Design of Chemical Processes* (New York: Pearson)

[38] Astolfi M, Romano M C, Bombarda P and Macchi E 2014 Binary ORC (Organic Rankine Cycles) power plants for the exploitation of medium–low temperature geothermal sources— Part B: Techno-economic optimization *Energy* **66** 435–6

[39] C. Engineering 2019 Chemical Engineering Plant Cost Index Annual Average

[40] D. G. S.-U. M. of economic development 2019 Research of resources geothermalized to the testing of Pilot Plants 2019

[41] Dell'Ambiente M 2020 Geothermal Pilot Plant named Castel Giorgio in the Province of Terni, in the municipality of Castel Giorgio (TR)

[42] Buonasorte G, Cataldi R, Ceccarelli A, Costantini A, D'Offizi S and Lazzarotto E A 1988 Ricerca ed esplorazione nell'area geotermica di Torre Alfina (Lazio-Umbria) *Boll. Soc. Geol. It.* **107** 265–337

[43] Nardi L D, Pieretti G and Rendina E M 1977 *Stratigrafia dei terreni perforatidaison daggienelnell'areageotermica di Torre Alfina Boll. Soc. Geol. Ital.* pp 403–22

[44] Studio dellepotenzialitàgeotermichedelterritorioregionaleumbro Report finale 2013

[45] Buonasorte G, Cataldi R, Pandeli E and Fiordalisi A The Alfina 15 well: deep geological data from Northern Latium (Torre Alfina Geothermal area)

[46] Costantini A, Ghezzo C and Lazzarotto A 1984 *Carta Geologicadell'areageotermica di Torre Alfina (prov. Di Siena-Viterbo-Terni) scala* **1** 25.000

[47] Hasan A A and Goswami D Y 2003 Exergy analysis of a combined power and refrigeration thermodynamic cycle driven by a solar heat source *J. Sol. Energy Eng.* **125** 55–60

[48] White S D and O'Neill B K 1995 Analysis of an improved aqua-ammonia absorption refrigeration cycle employing evaporator blowdown to provide rectifier reflux *Appl. Energy* **50** 323–37

[49] Incropera D and Bergman E L 2006 *Fundamentals of Heat and Mass Transfer* (New York: Wiley)

[50] Gebreslassie B H, Medrano M, Mendes F and Boer E D 2010 Optimum heat exchanger area estimation using coefficients of structural bonds: application to an absorption chiller *Int. J. Refrig.* **33** 529–37

[51] Leveni M, Manfrida G, Cozzolino R and Mendecka E B 2019 Energy and exergy analysis of cold and power production from the geothermal reservoir of Torre Alfina *Energy* **180** 807–18

[52] Dell'Ambiente M 2015 Geothermal pilot plant named 'Torre Alfina' in the municipality of Acquapendente (VT)

[53] Kucera J 2019 *Desalination: Water from Water* 2nd edn (New York: Wiley)

[54] Khawaji A D, Kutubkhanah I K and Wie E J-M 2008 Advances in seawater desalination technologies *Desalination* **221** 47–69

[55] Eshoul N, Almutairi A, Lamidi R, Alhajeri H and Alenezi E A 2018 Energetic, exergetic, and economic analysis of MED-TVC water desalination plant with and without preheating *Water* **10** 305

[56] Li C, Goswami D Y, Shapiro A, Stefanakos E K and Demirkaya E G 2012 A new combined power and desalination system driven by low grade heat for concentrated brine *Energy* **46** 582–95

[57] Li C, Besarati S, Goswami Y, Stefanakos E and Chen E H 2013 Reverse osmosis desalination driven by low temperature supercritical organic Rankine cycle *Appl. Energy* **102** 1071–80

[58] Sharqawy M H, Lienhard J H and Zubair S M 2010 Thermophysical properties of seawater: a review of existing correlations and data *Desalination and Water Treatment* **16** 354–80

[59] Nayar K G, Panchanathan D, McKinley G H and Lienhard J H 2014 Surface tension of seawater *J. Phys. Chem. Ref. Data* **43** 043103

[60] Nayar K G, Sharqawy M H, Banchik L D, John E V and Lienhard H 2016 Thermophysical properties of seawater: a review and new correlations that include pressure dependence *Desalination* **390** 1–24

[61] Water T B 2021 Tampa bay seawater desalination, Tampa Bay Water, 2021 https://tampabaywater.org/tampa-bay-seawater-desalination [Consultatoilgiorno 6 March 2021]

[62] Benjamin J, Arias M E and Zhang E Q 2020 A techno-economic process model for pressure retarded osmosis based energy recovery in desalination plants *Desalination* **476** 114218

[63] Pabi S, Reekie L, Amarnath A and Goldstein R 2013 Electricity use and management in the municipal water supply and wastewater industries (Palo Alto, CA: Electric Power Research Institute)